ROUTLEDGE LIBRARY EDITIONS: CULTURAL STUDIES

T0227564

Volume 4

MISUNDERSTANDING MEDIA

MISUNDERSTANDING MEDIA

BRIAN WINSTON

LONDON AND NEW YORK

First published in 1986 by Routledge & Kegan Paul

This edition first published in 2017
by Routledge
2 Park Square, Milton Park, Abingdon, Oxon OX14 4RN

and by Routledge
711 Third Avenue, New York, NY 10017

Routledge is an imprint of the Taylor & Francis Group, an informa business

British Library Cataloguing in Publication Data
A catalogue record for this book is available from the British Library

ISBN: 978-1-138-69145-2 (Set)
ISBN: 978-1-315-45997-4 (Set) (ebk)
ISBN: 978-1-138-69998-4 (Volume 4) (hbk)
ISBN: 978-1-138-69999-1 (Volume 4) (pbk)
ISBN: 978-1-315-51221-1 (Volume 4) (ebk)

Publisher's Note
The publisher has gone to great lengths to ensure the quality of this reprint but
points out that some imperfections in the original copies may be apparent.

Disclaimer
The publisher has made every effort to trace copyright holders and would welcome
correspondence from those they have been unable to trace.

Misunderstanding
MEDIA

Brian Winston

Routledge & Kegan Paul
London and New York

For all my beloved necessities!

Adele, 'cool' Jessica and this book's coeval, Matthew

First published in 1986 by
Routledge & Kegan Paul plc
11 New Fetter Lane, London EC4P 4EE

Set in Times 10/12pt
by Columns, Reading
and printed in Great Britain
by T J Press (Padstow) Ltd
Padstow, Cornwall

British Library Cataloguing in Publication Data

Winston, Brian

Misunderstanding media.

1. communication   2. Technological innovations
I. Title
302.2   P91

ISBN 0-7102-0002-1

CONTENTS

ACKNOWLEDGEMENTS

I cannot think when I first became disenchanted with gadgets – in my head that is, for my heart still leaps at every new electronic toy. The disenchantment must have arisen from my general failure to make things go as a child, a sense that automobiles ceased to function at my gaze and, most of all, my conviction – nurtured by bitter experience in live television production – that cameras and all other machines were just as likely to break as to work. Anyway, for whatever reason, some 15 years ago – before the 'information revolution' had really got underway – I was already vaguely upset with the concept. It was not, however, my ineptness with machinery alone that set me on this negative path. The structure of what follows is in embryonic form in a book I published at that time,* during the research for which I first began to glimpse a pattern to the history of media technologies. It was this combination of an historical overview coupled with a certain wariness about modern electrical wonders that constitutes the genesis of this essay. The unformed insight, which I got at that time, into the nature of technologies – how they are *invented* and how diffused – came to be more and more important to me.

During my life I have heard the threat or promise of technological upheaval constantly touted but, despite the fact that I work in a high-tech industry supposedly at the cutting edge of many of these developments, I have not really experienced any technologically based disruption of my professional work. Instead, all I see is evolution and the ways in which various forces operate to contain change so that one can live with it. For instance, 23 years ago when I joined the broadcasting industry, videotape was already nearly a decade old, yet here am I in the mid-1980s still happily working with film and teaching others to do the same. This is not to say that film might not give way to tape – even, that it will inevitably do so; rather it is to point out that this changeover is taking all my working life to occur. Between talk of revolution, increasingly insistent in the last decade, and the everyday

* *Dangling Conversations*, vol. II – *Hardware, Software* (Davis-Poynter, 1974).

reality of the slowly evolving world of work was a gap which vividly impressed itself on me.

It seemed to me that a crucial factor contributing to the false picture of rapid change was a certain failure of vision, of knowledge of those events in technological history which I thought might indicate contrary non-'revolutionary' interpretations of the present situation. Even within my trade, never mind beyond its closely guarded mysterium, people know little of the history of the machines with which they daily labour and most times could not care less.

By the early 1980s, I wanted to recover this technological history and, more, to put historical accounts of technologies, normally kept apart, together – not least because that is what we are told is happening: the technologies are coming together. And as I did this work I became aware that the pattern I had earlier detected indeed held for the development of each of the central technologies of the so-called 'information revolution' and that the historical record supported my work experience (that change was slow) rather than the rhetoric of the day (that there was a revolution under way).

This book is then a challenge to that dominant rhetoric, the rhetoric of technological revolution, especially in the field of information processing.

In the nature of the case, I have had liberally to garner the flowers of others. Although the history of technology is an under-tilled corner, which I would argue is significant given the centrality of technology in our lives, yet some toil there and I am – as is quite obvious from what follows – most mightily in their debt.

I have learned much about American broadcasting from my colleagues in the Cinema Studies Department of New York University, most notably William Boddy, Michelle Hilmes and Aaron Nmugwen whose doctoral researches have influenced my understanding in ways that cannot be adequately footnoted. Bill Boddy was kind enough to read the manuscript for me and I am therefore doubly in his debt.

To the members of my seminar group on technology and ideology, especially Svein Bergum and Jimmy Weaver, I am equally grateful. To my colleagues and friends Martin Elton and Mitchell Moss I am most indebted for careful advice, early sight of papers, supply of references and corrections. Indeed the Interactive Telecommunications Programme at NYU, whence come all these last, has given me repeated opportunities to test ideas in the forum of open debate.

More friends were of crucial assistance. Daniel Zwanziger kindly cast an eye over my science and if it is still wanting it is no fault of his. Bernard Abramson gave me the benefit of a computer professional's insight and a most thoughtful critique of the central ideas of the book. Bertell Olhman and Patrick Watson took on the deficiencies of the book's end as did Ted Conant those of its opening.

And I am indebted to others for various references, hints and leads, among them Robert Horwitz and Herb Schiller of the University of California, San Diego; Janet Staeger of my department, Michael Wreen of Marquette University, Steve Scheuer and Nick Hart-Williams. Wayne Baden guided me into the strange thickets of American law. He and Frederick Houston also suffered long hours as AMTRAK commuters doomed to listen to the results of my research and my arguments. In fact so much of this book was researched, corrected and discussed between Rhinecliff and Grand Central on AMTRAK's Hudson Valley line that I also need to thank the conductors and staff of those trains, especially 'Mo' Fink.

In the spot once reserved in acknowledgements for the professional typist, I would like to thank the Computer Shop, Kingston, New York, the suppliers and maintainers of my word processor – Tim Klepeis, who sold it to me, Lillie Rothe who kept me in supplies, and Clem Haneke who made sure the machine always worked. (Lest the disappearance of the manuscript typist seem to contradict my above remarks, I refer the reader to chapters 3 and 4 below in which are detailed the events of all those decades when we could have had word processors but didn't.) (And I would also note that the Computer Shop went out of business while this manuscript was in press.)

The book could not have been written without the help of Lynne Jackson. Her furnishing of material, references and ideas affected the whole underlying structure of the work. Her ability to grapple with the NYU Library computer, which far exceeds my own, was a most necessary skill and her overview of the text contributed enormously to whatever readability it might have.

To Peter Hopkins, as ever, goes the responsibility of onlie begetter and lastly, but of course not least, to my wife go my usual thanks for pedantic wretchery. Any infelicities of style are not for her want of care.

Brian Winston
Marbletown, New York
January 1986

INTRODUCTION

The once and future of telecommunications technologies

This book will attempt to demonstrate why there is little or no reason to join the epithet 'revolution' to the epithet 'information'. This purpose is perhaps best symbolised by the fact that although every word on this page has been in the clutches of various computers, from my personal word processor to the publisher's more elaborate devices, what you hold in your hands is a user-friendly, portable, randomly accessible retrieval device half a millennium old and of a design elegance unmatched by any of the vaunted machines of the 'information revolution' – a book.

The persistence of books, and the ironies of books about the 'information revolution', are too glibly ignored. This glibness can be attributed to a general lack of historical sense, for the 'information revolution' exists only as a consequence of far-reaching misunderstandings about electronic media, their development, diffusion and present forms. This book will be exactly concerned with these matters; and it will be a central thesis that the history of the technologies of information reveals a gradual, uncataclysmic progress. No telecommunications technology of itself or in aggregate suggests revolutionary development. On the contrary, each of them can be seen as a technological response to certain social relations which, at least in the West, have remained basically unchanged during the entire industrial period; the technology, far from being a disruptive force, actually reflects the comparative stasis of these relations.

The devices in question range from the telegraph, and its immediate non-electronic predecessors, to holography. Included are computers – macro, micro, personal and pocket – cameras, xerography, telephones and videophones, satellites and videotape, radio and other electromagnetic/photokinetic distribution systems including fibre optics, Polaroid, photographic printing processes (and the developments in the press that preceded them) and, above all, television – in short, the range of machines which, it is now claimed, has reached

1

such a critical mass that not only are deep, radical alterations in our society being effected but even our very sensoria are changed under the impact, to make us new women and men.

Received opinion is that a combination of developments has brought us to this revolutionary point. The by-now-ancient telephone wire has had its capacity increased to accommodate the most complex of electrical signals, those encoding visual information, in great numbers. The communications satellite has created an elegant worldwide method of communication, equally capable of carrying 'broadband' audiovisual signals. The computer has rendered vast amounts of information accessible in wholly new ways. And finally, various other methods of duplication and storage have been developed to allow us to copy instantly visual and audiovisual messages into a number of media. Joining the computer to the telephone and both to satellites creates a McLuhanesque patchwork of electronic nerves stretching down into every last household, linking them together into the global village. And our window on to the village square is the television cathode ray tube.

Such noise and hubbub have the proponents of this 'revolution' made that alternative readings are all but drowned out. However, it is my contention that far from a revolution we have business, and I mean business, as usual. All of the following could be just as viable a set of predictions as those promising revolutionary 'world boxes', 'wired cities', 'electronic cottages' and all sorts of 'future shocks'.

- Entertainment-led cable television, that is cable systems relying on traditional television broadcasting forms, will have discernible effects only in situations where free-air broadcasting widely delivers poor signals (as in the USA) or where the population at large prefers signals originating in a neighbouring nation to its own (as in Canada).
- The free transnational propagation of audiovisual signals elsewhere will be contained by both governmental action and public (non) response.
- Entertainment channels, whether delivered by cable or by other means, will probably never exceed one dozen – as the slow process of shakeout in the United States is revealing.
- Videocassette recorders are *the* crucial device to expand entertainment television. They will have the most significant effect on all current and proposed systems for the mass distribution of audiovisual signals, including cable.
- All new means of distribution and expansion of service will not produce new content for television. (Expanding television to include replications of bourgeois 'high culture' in societies where such coverage has hitherto not been much seen does not, self-evidently,

create new content.) A more varied range of distribution systems (including the 'bicycling' of videocassettes) will exist without national audiovisual establishments being essentially changed.

- Subteens may, for a time, give up dancing together and substitute the watching of videos for audiorecords. This, in effect a dance (or rather non-dance) style, will pass as all such do. Music videos will develop via an increasing reliance on narrative forms.
- Flat screen, component television will replace current receivers. Big screen projection television will not – unless accompanied by the wholesale remodelling of the housing stock to make bigger rooms. People will not walk about watching personal TV screens either, for fear of bumping into each other or getting run over.
- Videotext devices, whether two-way interactive or not, will never replace print except in very limited situations.
- Interactive services will be provided, in so far as they are required, by updating the basic telephone to broadband capacity using waveguides and fibre-optic technologies, rather than by the laying down of alternative national systems. Containing the power of the telephone company over these expanded services will pose a major problem.
- Narrowband intreractive services (burglar and fire alarms, metering, etc.) will be provided, in so far as they are needed, by this updating rather than by entertainment-led cable television systems. The telephone will gain in intelligence and scope. It will never become a 'videophone'.
- Interactive television uses will not include shopping (except perhaps of a limited mail-order type), schooling, political decision-making by the entire electorate or by any substantial proportion thereof or any other 'global village' use. In fact, interactive television will meld into the intelligent telephone.
- Videodisk, except for laser-based systems which will be used only in training or archival situations (and might therefore be uneconomic of themselves), are doomed to oblivion – and are already well on their way there.
- Videodisks masquerading as audiorecordings will offer the recording industry its only hope of destroying the public's ability inexpensively to copy their product.
- Light-sensitive polymers will render nitrate-based film stocks obsolete but, except for still photography processes, will fail to compete with electronics. Thus videocameras will replace 8mm film cameras as the primary tool of personalised audiovisual image creation. Home videos will be as important (and unimportant) as home movies are.
- Professional film use will inexorably be replaced by electronic systems except at the final stage of theatrical print preparation.

Cinemas will, like theatres, opera houses and concert halls, survive.

- The personal computer will be an essential of the academy and businesses however small. It will also find a variety of professional uses. But the basic structure and functions of the academy, business and the professions will remain unaltered.
- Home computing will prove to be a fad – albeit a widespread and probably quite persistent one – like railroad modelling or philately. The central thrust of home computing will continue to be games and this play will constitute the dominant fad within the fad, although in more affluent homes the computer might replace the typewriter.
- Computer literacy will function like driving instruction, only be less complex. It will be limited to understanding the operation of extremely user-friendly computers with cheap, prepared software for word processing, book-keeping, graphics and data-base management – by definition, no major feat.
- The marketing of previously free information via databases will result in people learning to live without that information. Effective databases will be limited to professional situations with the result that the home computer will remain a largely isolated device.
- Holography, true stereoscopicy, is, given this culture's addiction to realism, a necessity. It will therefore be marketed. And it will have as much and no more effect than any other advance (including film, radio and television) has done.

(In parenthesis it can be added that medicine will contribute evidence as overwhelming as that connecting smoking to cancer as to the (physically) deleterious effect of the television screen and with parallel limited effects on behaviour. The evidence is already being gathered. The US National Institute for Occupational Safety and Health found that in the United Airlines office in San Francisco, an environment with a high density of VDTs (Visual Display Terminals, known in UK as VDU – Visual Display Units, i.e. televisions), half of 48 pregnancies between 1979 and 1984 had ended in miscarriages, birth defects or other abnormalities. Working with VDTs can also increase risk of seizure in epileptics, according to the British Health and Safety Executive. The HSE also found facial dermatitis occurred in VDT work environments with low humidity.[1] The Newspaper Guild commissioned a study from the Mount Sinai School of Medicine in New York in which 1100 VDT workers in six locals were monitored for six months. Increased eye and radiation problems were found. The clincher is that the American Electronics Association (who make the things) testified before the Congress in 1984 that there was no evidence as to the deleterious effects of television. Their spokesman said: 'Regulation of VDTs on any health and safety basis is unwarranted.'[2])

Business, media, alienation, nuclear families, right-wing governments, technologically induced health hazards, traffic jams, deep-fried foods, dating – all as usual. No revolution – just 'the constant revolutionising of production, uninterrupted disturbance of all social conditions, ever-lasting uncertainty and agitation', as the Communist Manifesto puts it – in effect, a continuation of the developments of the last two centuries.

The justification for the above provocations is to be found in what follows. The 'information revolution' is said to depend on the chip, the bird, the wire and the screen. In order to make the case against the 'revolution' a pattern of change must be demonstrated which would hold good for the histories of these dominant devices – computer, satellite, telephone and television – and sustain the predictions listed above. Television is crucial in all of this – a paradigm of how telecommunications technologies develop, of their supposed radical impact as they are diffused through society, of their revolutionary potential. We shall begin with television.

A kind of a glow

A quarter of a century ago, when the television set first became firmly established in the living rooms of the West, considered opinion was that our civilisation had taken a further turn for the worse. The age of television ('chewing-gum for the eyes' was a 1950s phrase) was yet another tread in the staircase of that civilisation's decline.[3] Those McLuhan was to call 'the elders of the tribe' saw the warp and weft of social life under threat. 'There is now a vast crowd that is a permanent audience waiting to be amused, cash customers screaming for their money's worth, all fixed in a consumer attitude. They look on more and more and join in less and less.'[4] As we slumped before the electronic Cyclops it was clear that the end was drawing ever nigh.

In the quarter of a century that has followed these initial reactions, considered opinion – informed, intelligent, well-educated, middle-class opinion – has changed little. The elders of the tribe and their cohorts, the guardians of public morality and the taste-making lackeys of the haute bourgeoisie, always respond thus to every new, popular mode of communication. As Raymond Williams has pointed out, they are always wrong: wrong about Shakespeare in the 1590s, about Austen in the 1810s; wrong, it might be added, about circuses and quilts, vaudeville and narrowboat decoration, Marvel Comics and Hollywood, jazz and brass bands.

One central reason why all this huffing and puffing has been utterly misplaced is that television, particularly, is no subversive demon. High-minded critics have worried endlessly about the circumstances and social consequences of viewing, never, beyond an obsession with sex and violence, taking cognisance of what was viewed; how traditional, hidebound and safe, how tied to the past were the new medium's programme forms and contents. Television's function has been to reinforce the central value system of the society it serves and the dominant mode it has used has been primarily dramatic.

Take mode. Television adopted, from film, a system of representing reality which is extremely naturalistic. Actors needed to learn almost nothing for work in the electronic studio that they had not already learned for the big screen; in turn, the rules governing the presentation of personality used there were imported with little change from the stage. Television, film and stage in the West are a seamless web, unlike the various different traditions of performance in the East, each with its own elaborate presentation code – and what is true of acting is equally true of setting, lighting, directing. The very fact that studios have multiple cameras, four being the norm, is a tribute to the need to emulate, live and therefore instantaneously, the rules of special and temporal continuity created by the classic discontinuous and non-instantaneous shooting of the film camera.

The charge that television has rotted society by the imposition of ersatz and debased entertainment forms is not sustainable. Every television form has an honourable and lengthy history. At best the new medium has privileged some forms, giving them considerable vibrancy – the short dramatic piece, the one-act play or extended sketch, now situation comedies, would be a good example. But in many other areas – variety and the whole range of 'classical performance' arts for instance – the television does little but preserve previous theatrical modes in aspic. Even in the programming types which thrive on the small screen tradition looms very large. Almost every single-set situation comedy obeys realistic *theatrical* conventions, including the downstage sofa and the canted side walls of the set. What takes place on that sofa owes much to either Aristophanes or Menander, performances utterly obedient to the conventions of Greek comedy either old or new. And in general all of television is as time-honoured. It has created no new story-forms; its narrative structures, fictional and non-fictional alike, are dominated, like all others in this culture whatever the medium, by puzzles and the slow revelation of solutions, the liquidation of the audience's initial lack of knowledge about the story being told.

The drama intrudes into everything television does. It makes all of life dramatic, obeying an injunction in our culture which says that this is the price of our attention. Television, just as the radio and the movies before it, has failed to escape from the culturally imposed imperatives of conflict in its coverage of sports and news, religion and dance, children and the weather. Conflict and deviancy are the mainstays of its documentaries.

Its presentational codes in these non-fiction areas, too, reflect a close understanding of the past. The setting for current (or public) affairs television, often a cross between the contemporary sitting room and an airport departure lounge, also reference the schoolroom, with a magic movie screen behind the presenter in place of the blackboard. Television knows that in our culture, from mediaeval magistrates' benches and clerical pulpits on, authority hides its knees. Newscasters do the same. The rank and file are ranked and filed on unenclosed chairs. And it is a mark of the common touch for talk show hosts and hostesses to be similarly exposed.

And what is the dominant image of non-fiction television? A person addressing the camera – newscaster to correspondent to reporter to expert. The vaunted visual imperative to which television newspeople claim to be in thrall does not amount, on close examination, to much. When the head is replaced (although its voice never stops) by other visual material, the film or tape, the still photograph or dynamic computer graphic symbolise as much as illustrate the subject. Shots of the exteriors of the sites of power, courts, parliaments and white houses, stand for politics. Industry is workers at assembly line or factory gate. Trade is shots of the docks. Inflation is bank clerks and supermarket shelves. Defence is old film of weapons being tested. War is distant puffs of smoke or horrific corpses. The weather is seen through the blizzard-swept screen of the camera-car. The pictures, unless of sporting events, press conferences, hostage situations or fires, are all of aftermaths. Deviate from these norms and a Pulitzer prize is the certain reward.

Instead of the newspaper, we have a photogenic journalist reading the newspaper. Instead of the stage, we have the studio, set and lit like the stage. Instead of the fairground booth, we have the game show. Instead of the opera or the sporting event we have the outside (or, as the American terminology more accurately and aptly has it, the remote) broadcast replicating the event. Instead of the Tories of Oxfordshire spending, in 1745, £40,000 to unseat the Whigs, we have the Democrats spending millions of dollars on commercials to dislodge the Republicans.[5]

The mode is traditional and the function essentially conservative although, since we live in complex and contradictory societies, that conservatism is often masked by conflicting inconsistencies. But in setting a social agenda for an otherwise atomised society – giving people something to talk to each other about, reinforcing the ways in which they see the world – an overall conservative tendency, as evidenced by its forms, is also the mark of television's central function.

Television's deeper purposes are concealed, even, in the main, from those who work within the industry it has spawned, behind the transparency of its mode of representation. Our taste for realistic presentation has resulted in almost all production being unperceived by the audience. The scene within is unaffected by any interferences on the part of the professionals creating the message. They are the glaziers who, having situated and meticulously cleaned the pane of glass, can now leave the viewer wondering if it is really in place. This transparency is crucial. It is what allows newspeople and their audiences to confuse a certain type of ideological production with objectivity. It removes any underlying notion of effectiveness from the idea of entertainment, allowing audiences to receive everything from law-and-order dramas to the daily wash of the soap operas as being without social meaning.

The world created parallels our own in detail but it is also a world, even at its most serious, curiously unbeset by many of our quotidian concerns – the need to work, the divisions of class and race, the strain of alienation.

The agenda set by television for society is a propaganda agenda of the very best, most convincing type. It does not preclude problems. It highlights them, concentrates on them even, only to displace them – in fictional and non-fictional modes alike – from real consideration into the realm of melodrama. The television world is largely a world of two dominant groups – the fictional rich and/or criminal and the non-fictional poor and their rulers. It is a world of minute surface detail underpinned by a hidden and simplistic account of the complex ideology that governs the way we live our lives. It is the world of entertainment, where a show that is really a show keeps you in (rather than sending you out) with a kind of a glow. So television has become what, in 1953, NBC-TV network chief 'Pat' Weaver hoped it would become – 'the shining center (sic) of the home'.[6]

Despite all attempts at creating 'moral panics' because of the medium's supposedly baleful influences (and without prejudice to the reality of those influences), television has fitted into our lives exactly because it is so much a product of our culture and poses so little threat to it. It also fitted in because, contrary to common belief but like almost all the other devices we shall be discussing, it was a very long time coming. Telecommunications in general, and television as the example *par excellence*, do not suddenly descend upon us. 1984 was not only the year of Orwell's fictional nightmare, in which interactive television – in its most repressive surveillance mode – plays such a prominent role; it was also the year, in reality, in which television celebrated its first century.

The first television century

Early in January of 1884, Paul Nipkow, a Berlin science student, filed patents for an 'electric telescope'. He had, over that previous Christmas, placed a small disk perforated with a spiral of holes between a lens and an element of selenium which was inserted into an electrical circuit. It had been known for a decade that selenium, when exposed to light, would vary any electrical current passed through it in response to the intensity of the light. In this selenium was not unlike other substances, such as carbon, which varied in their resistance to current when exposed to pressure. The effects of the pressures exerted by the human voice on an element of carbon in an electrical circuit had been incorporated into a device seven years previously. It was called a telephone. The principle of transforming sound pressure into a modulated electrical signal was, and still is, its heart. The same principle, but now of transforming light into a modulated electrical current, was at the heart of Nipkow's telescope.

When Nipkow spun his disk, he scanned the image before the lens, breaking it down into a series of varying light impulses. These, as they hit the selenium plate, created variable resistance in the circuit. At the other end of the circuit, the process could be reversed. The electric current could be reconstituted into a series of light waves which, when passed through an exactly synchronous spinning disc, would reconstruct the picture. This could then be viewed through an eyepiece.

After another year's work Nipkow filed a master patent for television. Although he had established a viable system of 'scanning' with the disk, he then did nothing more. The rest is silence.

It could not have been that he had no real concept of what he was about. Since the phenomenon of selenium had been announced in 1873 and the telephone patented in 1876, the search for 'seeing by electricity' had been going on. The science informing that search dated back to the isolation of selenium in 1818. Photoelectricity, the creation of electricity through the operation of light, had been observed in 1839. In 1877, exploiting selenium, a French lawyer Senlecq had described a **telectroscope**, adapting the telephone to create a facsimile apparatus. (These had existed in various forms for telegraphy since 1843 and in 1847 a **copying telegraph** using a scanning technique had been introduced.[7] Phototelegraphy begins with a device, developed by the Abbé Caselli and introduced in 1862, which could transmit **daguerreotypes** or a facsimile of the sender's handwriting. With the support of Napoleon **III**, Caselli established a number of commercial stations, but the slowness of the system prevented him from mounting a real challenge to Morse.[8])

Senlecq reported: 'The picture is, therefore, reproduced almost instantaneously; . . . . we can obtain a picture, of a fugitive nature, it is true, but yet so vivid that the impression on the retina does not fade.'[9] It is unlikely that Senlecq's electrically driven pencil would have created the half-tones necessary to duplicate a photographic effect. He did, however, suggest a scanning system, involving moving the selenium across the ground-glass screen of a **camera obscura**.

Senlecq's announcement stirred professional electricians all over the old and new world. For the next few decades schemes were put forward of one sort or another – even the great Alexander Graham Bell reportedly deposited plans for a television system at the Smithsonian in 1880. The usual arguments as to who suggested what, when, proliferate. A selenium camera is credited to an American, Carey, in 1875, although he did not publish until 1879. Probably most of these devices remained unbuilt, but an Englishman, Bidwell, adapted a common laboratory device to demonstrate the possibilities of picture transmission by selenium in 1881. He used it in that year to obtain a still image on chemically treated paper, and his machinery still exists in the London Science Museum.

Nipkow was in a not inconsiderable company. Even more were to be inspired by him and experiments with mechanical scanning persisted well into this century. John Logie Baird, a showman as much as a scientist who kept up British popular interest in television in the 1920s, was still using mechanical scanning a decade later. His pioneering demonstration of April 1925 – and the work of Charles F. Jenkins (another independent inventor), publicly revealed in America in June of the same year – both used mechanical scanning systems. CBS demonstrated a colour television in the 1940s which used spinning disks.

In the early period it was not the fallibility of the mechanical scanning principle that held up progress, although viable systems of synchronisation for camera and receiver were hard to develop, but rather the low sensitivity of selenium.

A separate strand, in the realm of pure science, comes into play at the turn of the twentieth century. In 1897 J.J. Thomson discovered the electron, thus explaining, among much else, that what was happening in photoelectric emission was the liberation of electrons from the atoms of the substance through the action of the light. In that same year an electron beam or cathode ray tube was developed. Between 1904 and 1906 Fleming and De Forest built the first electron vacuum tubes (or valves), devices which can be used to amplify signals. In 1907 a Russian scientist, B.L. Rozing, applied for a patent – 'A Method of Transmitting Images Over A Distance' – in which he proposed electrical instead of mechanical methods for scanning an image. He took the cathode ray tube, by then a common laboratory device in the form of an oscilloscope, and adapted it so that it could be used to scan a scene. On 9 May 1911 he had got so far as to transmit over a distance 'a distinct image . . . consisting of four luminous bands'.[10] Rozing's scanning electron beam at the transmission end was joined, in a suggestion by the distinguished British scientist Campbell Swinton in 1908, by a similar beam at the receiving end. Campbell Swinton also described a different theoretical basis for modulating the electrons in the camera tube. He built no devices.

By the First World War the photoelectric quality of substances with greater sensitivity than selenium was being used, notably potassium. After the war Vladimir Zworykin, a student of Rozing's, now in America and working first for Westinghouse and then for RCA, developed an effective camera tube along the lines suggested by Campbell Swinton. It used within it a plate sensitised with silver and cesium. As a stream of electrons was fired at the plate in a zigzag pattern, its direction controlled by electromagnets, a sequential variation in the current was created at the anode, the positive terminal of the tube, behind the plate. This information then modulated a carrier wave for transmission by either cable or wireless into the receiver. At the home end the process was reversed. The internal signal plate of the camera became the phosphor-treated front end of the tube, the screen. The electron beam, again generated by a cathode and controlled by electromagnets, was modulated by the incoming wave. These variations were translated by the scanning electron dot into variations in intensity which became, through the phosphors, perceptible to the human eye.

The scanning system produced by the zigzagging dot was one which created the image sequentially, line by line, just as Nipkow's spinning disk had done. In the British version of this electronic camera, built by EMI, the pattern or raster was to scan the picture area twice, every even line in one 25th of a second and every odd line in the next 25th of a second. This way a sufficiently fast rate of change (fifty a second in effect) was created from frame to frame for the physiological requirements of critical fusion fequency (CFF), the point at which the eye ceases to see discrete pictures, to operate. Achieving this sort of speed was necessary because a single scan in a 25th of a second gave a flickering impression. Film projectors with double or treble blades in their shutters effectively reveal each image twice or thrice at rate of anything from 32 to 72 times per second.[11] The interlaced raster achieved the same range of CFF for the electronic image. The American version, because of the different characteristics of American electrical supply, scanned at a 30th of a second.

Other electronic cameras were developed on slightly different principles which ultimately failed to compete with RCA's in the studio. Of these the most significant is one built by Philo T. Farnsworth and patented in 1927, which produced better images than the earliest Zworykin camera tubes and continued to out-perform them well into the 1930s, when transmitting films.

In 1929 Baird had begun a series of experiments for the BBC using low-definition mechanical systems. By 1932 the BBC took over the enterprise. Two years later the Postmaster General appointed a committee to consider the development of television and 'advise on the general merits of the several systems' then available.[12] The BBC was entrusted with the experiment. Baird's company had by now refined mechanical scanning to give 240 lines and, to consolidate its strengths, it produced (using a system created in Germany the previous year) a film camera with a rapid developing tank underneath it. In effect the camera could expose the scene and then develop and fix the film in about 60 seconds. This was then placed in the telecine device, and scanned mechanically at a 25th of a second. The EMI camera was placed in competition with this device. It effectively scanned at twice the Baird rate and also had the further advantage of producing 405 lines instantaneously. Yet the two technologies were closer than might be supposed from this.

Sir Archibald Gill, a member of the 1934 committee, recalls the case was not quite open-and-shut. The Baird system did indeed, even with the line and frame disadvantage, produce a slightly better picture than the EMI system when transmitting film.[13] And, as film transmission was held by all experts in every country to be vital as a major source of television images, this was no small advantage.

The world's second long-running public television service had been inaugurated from studios in Alexandra Palace, London on 2 November 1936 using both systems. The Germans were actually first with a public – as opposed to an experimental – system using technology comparable to both Baird's and EMI's for the Berlin Olympics that summer. The network reached to five German cities and the service, as in Britain, was continued until the war.

The committee supervising the British run-off gave it until February 1937 before opting for EMI. Not only was the stability of the all-electronic picture superior, but the complexities of the intermediate film stage in studio-based originations counted against Baird. Beyond that, mechanical scanning was reaching the end of the road, while electronic systems were obviously capable of much refinement. Clearly the future was with the electronic camera.

But, paradoxically, in America the progress of these same devices was subject to various delays. The technological development was subjected to intra-industry argument and government regulation. Low-definition experimentation had commenced with the granting of a licence, by the Federal Radio Commission (FRC), to Charles Jenkins, Baird's opposite number, in 1927. In 1929 twenty-two more stations were licensed and between twenty and forty a year were operating experimentally until 1944. The great corporations in the field and the Federal Communications Commission (the FCC, successor, in 1934, to the FRC), in various combinations, fought about everything during the 1930s: the system, the number of lines, the place on the radio frequency spectrum. The FCC declared that RCA should not be allowed to go ahead with a viable 441 line system in 1938 because it would freeze technical standards at this level; yet the BBC had been publicly running virtually the same system with fewer lines (405) for over a year. By 1939 Philco was using 605 lines. Despite this the US standard of 525 lines, one 30th of a second per frame, and the VHF part of the spectrum was finally agreed in 1941.

A further delay then occurred because of the war, no more receivers being built from exactly the year in which the go-ahead came. Following the war, the FCC again intervened to slow the granting of licences for stations, with a consequential brake on the selling of sets. Between 1946 and 1950 the number of stations licensed increased from 30 to 109 and the number of sets from 5,000 to just under 10 million. In the next five years, stations jumped to 573 and sets to nearly 33 million. And between 1955 and 1960 another 80 stations and 36½ million sets finally made America into the earth's first televisual nation.[14]

This then is how the beast television came amongst us. For those living through the explosive expansion of the television universe in America and northern Europe during the 1950s, forgetting the flickering demonstrations of the previous decades, it might well have seemed that a gadget on the frontiers of science had been turned loose with little thought as to its overall social effect. But, as this brief retelling of what ought to be a familiar story reveals, the pace of the development was quite leisurely, for the 'explosion' followed at least two decades of careful preparation. The pace of these events cannot be accounted for by the progress of technology alone. Zworykin, when he patented the first effective electronic camera in 1923, had no new tools and little more theory than did his teacher Rozing. During the 1930s, the most significant developments were all to do with subsidiary circuits improving the performance of the tubes. Everything else had been to hand since, at least, the last part of the previous decade.

To understand this history in terms of the technology alone, the stance of the technological determinist, is inadequate. Information revolutionists, in predicting the future, take exactly such a technologically determined view. If the technology makes it possible, they seem to be saying, then it will happen (normally, most of them seem to add, through the beneficence of the market). But as even this cursory glance reveals, this was not the pattern with television. The technology could have been made more widely available sooner, even in Germany and the UK; the factors delaying it were not limited to technology. In America in the late 1930s they were not technological at all. Such external factors persist, an inevitable concomitant of technology being produced by, with and for a society.

More than this, beneath the confusions of the chronological history of television a pattern, a sequence of phases which the technology went through on its journey from the realm of pure science to everybody's living room, can be discerned. The primary task before us is the explication of this pattern, repeated, as is shown below, in every one of the telecommunications technologies that we have accepted thus far.

BREAKAGES LIMITED

The past is prologue

The suggestion that we are not in the midst of monumental and increasingly frequent change in telecommunications runs so counter to our whole underlying philosophy of progress, as well as the particular rhetoric of the 'information revolution', that it must surely be doubted by right-thinking persons. But the position taken here, rather, is that Western civilisation over the past three centuries has displayed, despite enormous changes in detail, fundamental continuity. While it is impossible to predict that such continuity will be sustained over either the short or the long term, it is contended that any discontinuities, should they occur, will not *primarily* be attributable to telecommunications technologies. Other more traditionally disruptive social forces – disaffected proletarian youth, for example – rather than communications, will make greater contributions to such upheavals as might occur. This is to deny telecommunications the role of engine of change and also thereby to deny the possibility that a revolution in information technology (however that be defined) is any species of general revolution.

'Revolution' is used here in its commonly understood sense of alteration and change, rather than in its original technical sense of recurrence or turning. This is the meaning, with its modern connotation of rapid political change, intended by those who coined the phrase 'information revolution'.

> *Revolution* and *revolutionary* and *revolutionize* have of course also come to be used, outside of political contexts, to indicate fundamental changes, or fundamentally new developments, in a very wide range of activities. It can seem curious to read of 'a *revolution* in shopping habits' or of the '*revolution* in transport' and of course there are cases when this is simply the language of publicity to describe some 'dynamic' new product. But in some ways this is at least no more strange than the association of *revolution* with

VIOLENCE, since one of the crucial tendencies of the word was simply towards important or fundamental change. Once the factory system and the new technology of the late eighteenth century and early nineteenth century had been called, by analogy with the French Revolution, the INDUSTRIAL *Revolution*, one basis for description of new institutions and new technologies as *revolutionary* had been laid. (capitals and brackets in original)[1]

Revolution, in whatever sense it is used, implies movement, and in these developed usages, that means movement through time. The concept of the 'information revolution' is therefore in essence historical; and the critique of the concept offered in this book is also grounded in the past, a past limited to the particular circumstances surrounding the application, over the last two centuries, of science to the human communication process. We shall argue that there is nothing in this history to indicate that significant major changes have not been accommodated by pre-existing social formations, and that 'revolution' is therefore quite the wrong word to apply to the current situation. Indeed, it is possible to see this historical record as being regular enough, if the above premise of continuity is accepted, to serve as a model for all such communication technologies, certainly past and present and, probably, in the short term future too.

The pattern of change in telecommunications, although historical (which is to say, diachronic), can also be expressed as a field in which three elements – science, technology and society – intersect. The relationship between these three elements can be elucidated by reference to another conceptual model – one taken from Saussurian linguistics.

Utterance is, for Saussure, the surface expression of a deep-seated mental competence. In Chomskyan terms, each utterance is a performance dependent on this competence. By analogy, then, these communication technologies are also performances but of a sort of scientific competence. Technology can be seen as standing in a structural relationship to science – as it were – utterances of a scientific language, performances of a scientific competence. In the linguistic model the link between competence and performance is achieved by the operation of transformations which move the utterance from deep to surface level. These movements are rule-governed and it is the rules – grammar in language – which enable a speaker to generate comprehensible but unique utterances. In the model proposed for technological change these transformations are not claimed to be so regular as to be rule-governed. The notion of transformation in our model, on the contrary, allows the model to accommodate less predictable factors. Transformations address the operation of factors external to the actual performance of technology, factors which work to transform a scientifically grounded notion into a widely diffused device.

Phase one: scientific competence

The development of telecommunications devices can be seen as a series of performances ('utterances') by technologists in response to the first phase of the model – *the ground of scientific competence*. The centuries-old investigations of electromagnetic phenomena and photo-kinesics are the two fundamental lines of scientific inquiry which make up this first phase. The possibilities of using electricity for signalling march, from the mid-eighteenth century on, virtually hand in hand with the growth of the scientific understanding of electricity. Similarly, the discovery of photography involved knowledge of the different effects light has on various substances, a scientific agenda item from at least the Middle Ages on. The propensity of certain solids to conduct sounds seems to have been known in ancient times and was certainly a well-observed phenomenon by the late eighteenth century. The photoelectric responses of selenium were known more than a century ago.

Note
Transformations in general: the three transformations

In the model (see Fig. 1) phases, such as the phase of *scientific competence* just described, are acted upon and transformed.

(i) The first of these *transformations* moves the technology from the phase of *scientific competence* into the phase of *technological performance*. The first transformation (which will be designated *the ideation transformation*) thus moves from science to technology, its effect being to activate the technologist. The two subsequent *transformations* alter the work of the technologist, or, to use the terminology of the model, the on-going work of *technological performance*.

(ii) The second transformation (the transformation occasioned by *supervening social necessity*) pushes the work of the technologist from prototypes into what is popularly conceived of as 'invention'.

(iii) The third (a transformation which will be called *the 'law'* ' *of the suppression of radical potential*) moves from the *invention* of devices to their diffusion.

Each transformation takes the technology further from the realm of pure science and closer to the everyday world of actual generally-used devices.

We shall examine each of these three transformations in a little greater detail as they occur in the model, beginning with the *ideation transformation*.

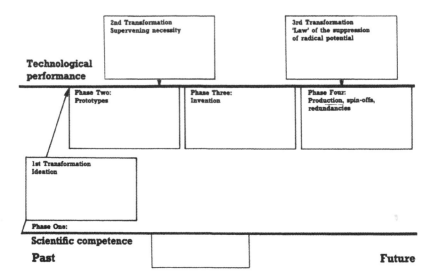

Figure 1 The model

The first transformation: ideation

The first of these transformations is the most local, which is to say it occurs within the laboratory. To continue with the linguistic metaphor *the ideation transformation* is akin to the processes whereby a transformation at the level of competence takes place, in the human brain, so that utterance, performance, can be generated. Ideation occurs when the technologist envisages the device – gets the idea, formulates the problems involved and hypothesises a solution. Those mysterious mental forces – creativity, intuition, imagination, what has been called 'the will to think' – are subsumed by *ideation*.

Although the technological idea will be grounded in scientific competence, it will not necessarily relate directly to science any more than a conscious understanding of linguistic competence is needed to generate utterance. Rather, just as in language a formal understanding of the deep structure of linguistic competence is not a prerequisite of utterance, so too a lack of formal scientific competence is no bar to technological performance. But the technologist will, at some level, have absorbed the science; just as a speaker, at some level, has absorbed grammar.

The *ideation transformation* interacts with the first phase and occurs concurrently with it; in telecommunications, transformations never precede science since the formulation of technological problems has always followed agenda set by scientific inquiry – and this, contrary to received opinion which sees the primacy of science in technological research as a recent development, has always been true of electrical telecommunications.

How the idea of television was first triggered by scientific advances has been outlined in the introduction. A Frenchman hypothesised the telephone in 1854, more than 20 years before Bell. A German thought of the telegraph in the last years of the eighteenth century, three decades before the first working device. Bell Laboratory workers began worrying about the transistor in the 1930s when solid state amplifiers had already been envisaged for a decade. Some of these thinkers went on to test their ideas technically; many did not. But more often than not their work was known to those who set about building devices.

> *Note*
> *Technological performances in general*
> *The three phases of technological performance*
>
> Before proceeding, we must now examine, in general, the last recurring element of the model – *technological performances*.
>
> *Ideation* transforms the processes of science into the testing of solutions – the building of devices which is the business of *technological performance*. This will go on until the device is widely diffused and even beyond, as spin-offs and refinements are developed. Performance is triggered by a transformation (*ideation*) and each successive transformation alters the nature of what the performance produces.
>
> (i) The second phase, after scientific competence, is designated the phase of *technological performance – prototypes*. During this phase, the technologists begin to build devices working towards fulfilling the plans which emerged from the *ideation transformation*.
>
> (ii) The third phase, *technological performance – invention*, is, from the perspective of the technologist, exactly similar but the operation of *supervening social necessity* (the second transformation) is catalytic. So within the laboratory the work continues as it did in the prototype phase, but the second transformation – *supervening necessity* – means the devices now produced are *inventions*. This third phase, then, shall be designated *technological performance – invention*.
>
> (iii) The operation of the next transformation, *the 'law' of the*

(Note. continued
suppression of radical potential, similarly affects the last phase of
technological performance. *Supervening social necessity*
guarantees that the *invention* will be produced. The above 'law'
operates as a constraint on that production. This final
transformation occasions a tripartite phase of *technological
performance – production, spin-offs and redundant devices or
redundancies*, which reflects the effects of the contradictions
which are at work.
These phases of *technological performance* are all discussed below
as they occur, beginning with the *prototype* phase.

Phase two: technological performance – prototypes

The four classes of prototypes

The solution to a problem raised by possibilities in the advance of
science has been proposed in the ideation transformation. Devices
must now be built.

Prototypes, such devices, can be of four distinct classes.

(i) The prototype can be *rejected* because a supervening necessity
has not yet operated and no possible use for the device is seen.
Rowland's demonstration of a working telegraph in 1816 would be
an example of this. The British naval authorities, understanding that
the semaphore was the only machine to use in long distance
signalling, simply refused to acknowledge the superiority of the
electromagnetic machine. Nearly every technology has its Rowland.

(ii) The prototype can be *accepted* because the early and
incomplete operation of a supervening necessity has created a partial
need which the prototype partially fills. The **daguerreotype**
photographic process which was widely used between 1850 and 1862
is among the clearest examples of this accepted group. It was
eventually superseded by processes which used negatives, the
essential mark of modern photography. Another example can be
found in the development by AT&T of non-geostationary
communications satellites. These were built and used a few years
before the introduction, by Hughes, of the current geostationary
device. The efficiency of Hollerith punch-card calculators,
introduced at the turn of the century but increasingly sophisticated
in the years after the First World War, can be said to have been so
well accepted that the development of the electronic computer was
delayed.

(iii) *Parallel* prototypes. These will occur when the device which
will become the parallel prototype is already in existence solving
another technological problem. Its potential use for a secondary

purpose is realised only after the operation of a supervening necessity. Various devices existed in the last two decades of the nineteenth century to demonstrate the validity of electromagnetic wave theory. Distinguished physicists such as Hertz and Lodge are associated with these demonstration machines. They were in fact a species of radio but were not seen as such. Their existence is, however, of importance in tracing the work of Marconi, Popov and others which led to radio. The cathode ray tube, before Rozing, would be another example.

(iv) Finally, in this second phase of *technological performance*, there can be *partial* prototypes which are machines designed to perform effectively in a given area but which do not. The telephonic apparatus developed by Reis in the 1860s and, arguably, Bell's earliest machines were of this type. Baird and Jenkins's mechanical televisions were also partial prototypes.

These then are the four prototypes of the second phase – rejected, accepted, parallel and partial. This classification is without prejudice to the efficacy of the devices. Except for partial prototypes which simply did not work very well, the other three classes of prototype in this phase all worked, more rather than less. The degree of their subsequent diffusion, though, depends more on the operation of the supervening necessity transformation than on their efficiency. An accepted prototype is a device which effectively fulfils the potential of the technology but, because the full power of the supervening necessity has not yet been called into play, there is still room for development. The rejected prototype might work just as well as the device eventually 'invented' but will achieve no measure of diffusion because there is no externally determined reason for its development. The parallel prototype is in similar case. The initial thrust of the technology is directed towards purposes other than those which eventually emerge. The effectiveness of this prototype in solving the problem for which it was originally designed has nothing to do with its effectiveness as a device in the second area. It is, in effect, a species of spin-off.

The second transformation: supervening necessity

The three broad types of supervening necessities

Just as ideation worked upon the ground of scientific competence to create these different classes of prototypes, so more general supervening social necessities can now work upon these prototypes to move them out of the laboratories. In the nature of the case this second transformation, being the social, is more amorphous than the first. There is no limitation on the forces that can act as supervening necessities; they can be the objective requirements of changed circumstances or the subjective whims of perceived needs.

(i) The least difficult group of supervening necessities to determine is occasioned by *the consequences of other technological innovation*. For instance, it was the railways that transformed telegraphic protoypes into a widely diffused device. Before railways there was no demonstrable need for such devices. Single track rail systems, however, required, as an urgent matter of safety, instantaneous signalling. Similarly the radio came into its own with the development of the Dreadnought battleship. Here, for the first time, naval battle plans called for ships to steam out of sight of each other, thus rendering the traditional signalling methods useless.

(ii) *Social forces working directly on the processes of innovation*, rather than being, as above, mediated through another technology, constitute a second, more vexed, group of supervening necessities. The rise of the modern business corporation created today's office, the architecture of the building that houses it and the key machines, telephone and typewriter, that make it function. Such offices were the site of the first telephones and typewriters. In the middle decades of the nineteenth century the possibility of the limited liability company was established for the first time in law. The legal development of the modern corporation thus mothers telephony. The growing urban mass and its entertainment needs are the supervening necessity for the cinema. The desire of all classes to emulate the ability of the rich to have images created of themselves and for their walls focussed the search for an effective photographic process.

(iii) *The commercial need for new products* and other commercial considerations would form a third group of necessities – less effective in guaranteeing diffusion and producing less significant innovation than the consequences of social change or other technological advance. Super 8mm film, Polaroid movies, 16rpm records and the compact audio disk (CD) can stand for the host of devices to which commerce makes us heir.

The action of a supervening social necessity does not account for the entire development and reception of a technology. Instead it transforms the circumstances in which the technologist labours, creating a fertile ground for effective innovation.

Phase three: technological performance – invention

The fifth class of prototype

It follows from the above that there must be the possibility of a fifth class of prototype, one which is either synchronous with or subsequent to the operation of a supervening necessity. The production of such machines is the business of further technological performance and leads to what is commonly called the invention. Since the difference between the devices now produced and the previous group of prototypes is the operation of a widespread transformation (social necessity), it is likely and, indeed history reveals, common that such creations will occur in a number of places synchronously. There is, therefore, no mystery in the synchronicity of invention. This third phase will be designated *technological performance – invention*.

Of all the phases of the model this is the best known. Herein are to be found all the heroes of communication technology's Hall of Fame – Bell and Shockley, Marconi and Land, Rosen and Hoff, Morse, Zworykin, Mauchly and Eckert.

The third transformation: the 'law' of the suppression of radical potential

The *invention* now moves into the market place. Yet acceptance is never straightforward, however 'needed' the technology. As a society we are schizophrenic about machines. On the one hand, although perhaps with an increasingly jaundiced eye, we still believe in the inevitability of progress. But on the other hand we control every advance by conforming its context to pre-existing social patterns. The same authorities and institutions, the same capital, the same research effort which created today's world is trying also to create tomorrow's. A technologically induced hara-kiri on the part of these institutions, whereby a business 'invents' a device which puts it out of business, is obviously impossible. But what is equally true, although less obvious, is the difficulty of inventing something to put other businesses out of business; and the bigger the threatened business the more difficult it is.

Progress is made while going down the up escalator (or, as optimists might argue, up the down). This jerky advance into the future can be seen constantly repeated in telecommunications history. Its daily cavortings can be read in the trade press. This is the measure of *the 'law' of the suppression of radical potential*. In the model this 'law' is proposed as a third transformation, wherein general social constraints operate to limit the potential of the device radically to disrupt pre-existing social formations.

Understanding the interaction of the positive effects of supervening necessity and the brake of the 'law' of the suppression of radical potential is crucial to a proper overview of how media develop. Constraints operate firstly to preserve essential formations such as business entities and other institutions and secondly to slow the rate of diffusion so that the social fabric can absorb the new machine. Such functions, far from atrophying this century, have increased. Whatever the general perception, there has been no speed up in the measurable rate of change. If anything, there has been a significant diminution in the cut-throat nature of the market place because the desire for stable trading circumstances, coupled with external restrictions and monopolistic tendencies, works to contain the crudest manifestations of the profit motive.

Two caveats must be entered as to the chosen designation of this third and crucial transformation. Beyond the proper and necessary caution required when postulating historical laws, *'law'* here is apostrophised to indicate that although the phenomenon under discussion can be found appearing in the histories of all telecommunications technologies it is not so regularly as to always manifest itself in the same form with equal force at the same point of development. It is recurrent enough to be a 'law' but not certain enough in its operation to be a law. Thus it is not a law, a universal hypothesis, in the Hempelian sense in that it does not assert

In every case where an event of a specified kind C occurs at a certain place and time, an event of a specified kind E will occur at a place and time which is related in a specified manner to the place and time of the occurrence of the first event.[2]

Secondly, *suppression* must be read in a particular way. As Lewis Carroll said, it is 'rather a hard word'. Here it is not meant to convey the idea of overt authoritarian prohibition or to indicate the presence of any form of conspiracy, conscious or unconscious; rather *suppression* is used in the more scientific senses given by the *OED*, viz:

to hinder from passage or discharge; to stop or arrest the flow of; (in Botany) absence or non-development of some part or organ normally or typically present.

It is possible that even with these caveats the word is still too 'hard' to cover the sense of a technology's potential simply being dissipated by the actions of individuals and institutions. However a 'law' of dissipation or retardation would be too soft to convey the strength of the forces at work in the process of technological innovation and diffusion.

The most obvious proof of the existence of a 'law' of *suppression*, then, is the continuation, despite the bombardments of technology, of all the institutions of our culture. To the conservative, such changes as have occurred loom very large; but any sort of informed historical vision creates a more balanced picture of their true size and scope.

The 'law' of the suppression of radical potential explains the delay of the introduction of television into the United States which lasted at least seven years, excluding the years of war. It explains the period, from around 1880 to the eve of the First War, during which the exercise and control of the telephone (in both the US and the UK) was worked out while its penetration was much reduced. It accounts for the delays holding up the long playing record for a generation and the videocassette recorder for more than a decade. It underlies the halt in the growth of all but through-air propagation of audiovisual signals for more than twenty years in most countries. It is far more powerful than the concept of 'development cycles' determining, through an examination of business alone, the factors and time involved in diffusing an innovation. This 'law' works in the broadest possible way to ensure the survival, however battered, of family, home and workplace, church, president and queen, and above all, it preserves the great corporation as the primary institution of our society. The 'law' of the suppression of radical potential is the 'law' by which Bernard Shaw's mighty company 'Breakages Limited' prospers.

> Every new invention is bought up and suppressed by Breakages
> Limited. Every breakdown, every accident, every smash and crash,
> is a job for them. But for them we should have unbreakable glass,
> unbreakable steel, imperishable materials of all sorts.[3]

'Breakages Limited' is but a poetic version of a phenomenon obvious to anyone studying technological change – that there are endless impediments to our forward progress and not all of these can be laid at the door of the innate conservatism of a culture which is, on the contrary, dedicated to change. More is at work than, as McLuhan put it, our driving into the future with our eyes looking in the rearview mirror.

Phase four: technological performance – production, spin-offs, redundancies

The 'law' of the suppression of radical potential is the final transformation and it occasions this, *Technological performance – production, spin-offs and redundancies*, the last phase of the model.

(i) Of the three distinct activities covered by this phase, the least problematic is that of *production*. The acceptance of the device is to a certain extent guaranteed by the operation of the supervening necessity. Much attention has been paid by economists to the symptomatic study of diffusion at both a macro and micro level with the result that the most scholarly literature available on innovation is skewed away from the processes previously described in our model in favour of a concentration on the production and marketing phases. The problems of moving prototypes into production and marketing will therefore be peripheral to this study.

But in the course of this movement the device can be modified, extended, refined; alternative solutions can appear as rival technologies. Such developments can themselves, as in the prototype phase, either be accepted or rejected.

(ii) If such a development is accepted, diffused, it is a *spin-off*. **Pacman**, for instance, is an accepted extension of microchip technology which was certainly not developed with that specific purpose in mind. Similarly the CD record, actually a video technology, appears to becoming publicly accepted. These are *spin-offs*, accepted products of technological performance synchronous or subsequent to the original device's diffusion.

(iii) If, on the other hand, the technological performances of this post-production stage are rejected, as, for instance, the Polaroid instant movie film was or as non-laser videodisk has been, they can be described as *redundancies* which then suffer the same fate as partial prototypes.

Necessities and constraints

'Revolutions' in telecommunications technologies cannot be readily accommodated in this model. But then neither can they be read easily into the historical record nor our present situation. Using the terms established above, the current 'revolution' can only be deemed to be occurring if the various transformations are ignored and the phases of technological performance are misread. Such ignorance and misreading leads to the systematic misunderstanding of media involved in the concept of the 'information revolution'.

Understanding the history and current position of telecommunications properly depends on an examination of the operation of social necessities and contraints, rather than on the performance of technology considered *in vacuo*. The above brief description of the proposed model references the history of telecommunications of the last two centuries. In the rest of this book the validity of the model will be established by treating this history in greater detail, firstly in the next chapter, by redescribing the development of television in the terms established here; and then in subsequent chapters the other technologies central to 'the information revolution' (computer, satellite and telephone) will be similarly dealt with.

This is not the place to inscribe an account of the debate as to the efficacy of what Popper has called 'historicism' or the propounding of 'historical prophecies'. Many, in seeking to understand the pattern of Clio's garments, have been tempted into predicting, on the basis of that understanding, what she will wear tomorrow. The present text moves in that company largely in response to a dominant tendency in the literature on electronic communication devices, both popular and scholarly, which uses an erroneous history, both implicitly and explicitly, as a predictive tool. Unfettered by much understanding of the past beyond the anecdotal, many currently propound insights into our future in the form of trajectories from our past. In this literature, the historical implications of the word 'revolution' are not denied; instead a supposed technologically determined transformational movement from 'then' to 'now' is celebrated. The purpose of this book is not only to explicate the 'then' by inscribing a more balanced account of what actually occurred in the telecommunications past but also to offer an interpretation, necessarily revisionist, of those occurrences. The model described above arises out of a consideration of the events of these histories.

This occasions a problem. What follows is an account of how a particular group of technologists, in history, interacted with their societies to produce a given set of devices. This enterprise is very close to Thomas Kuhn's project, which has received widespread attention, in the historical sociology of science. Kuhn's explanation of the nature of changes in scientific thought is schematic, just as our explanation of the developments of telecommunications is schematic. Further, Kuhn sees scientific changes as a species of revolution, just as many see technological changes. The problem is simply that Kuhn's schema has been extensively applied to fields well away from the history of science. It has become a species of 'supertheory'. The present work – although about science's sib, technology – is not Kuhnian except in that it offers a (different) schemata and is about (non-existent) revolutions; but nevertheless, before returning to television, it is necessary briefly to examine Kuhn's work and the points at which it touches our topic.

On Kuhn

Kuhn's *The Structure of Scientific Revolutions* – 'undoubtedly one of the more influential and controversial scholarly books to emerge in the last few decades' – proposes a pattern of cause and consequence relating to changes in the basic concepts governing scientific inquiry.[4] At the centre of Kuhn's schema is the 'paradigm'. A paradigm is generally accepted theory which rests on its ability to offer a better and more coherent account of the phenomena it addresses than do other competing theories.

> Paradigms gain their status because they are more successful than competitors in solving a few problems that a group of practitioners has come to recognise as acute.[5]

The particular consequence of the acceptance of such paradigms is that the paradigm must be:

> sufficiently unprecedented to attract an enduring group of adherents away from competing modes of scientific inquiry. Simultaneously [the paradigm must be] sufficiently open ended to leave all sorts of problems for the redefined group of practitioners to solve.[6]

This last is crucial –

> to be accepted as a paradigm, a theory must seem better than its competitors, but it need not, and in fact never does, explain all the facts with which it can be confronted.[7]

Explicating paradigms then becomes the business of what Kuhn calls 'normal science'. The practice of normal science is a sort of 'mopping up' operation to demonstrate the efficacy of the paradigm.

> No part of the aim of normal science is to call forth new sorts of phenomena; indeed those that will not fit into the box are often not seen at all. Nor do scientists normally aim to invent new theories, and they are often intolerant of those invented by others. Indeed, normal scientific research is directed to the articulation of those phenomena and theories that the paradigm already supplies.[8]

These two phases, the establishment of the paradigm and the work of normal science consequent upon it, account for the vast bulk of those activities we call science. But Kuhn goes further and offers an explanation as to the creation of paradigms. Paradigms emerge in two distinct situations. There can be, in any area, a pre-paradigmatic phase when random observations of phenomena are gathered and many varied theoretical explanations for them are offered. None of the latter is generally accepted until one of them, by virtue of the better account it offers of the phenomenon in question, attracts sufficient support to create 'a scientific community' which embodies a research tradition.

Thus the theory becomes a paradigm.

The second circumstance – and it would seem from Kuhn's examples the more characteristically modern – results from the practice of normal science. In that practice there is a slow accumulation of data not readily accounted for by the paradigm. These 'anomalies', as Kuhn calls them, are eventually of such weight that they occasion a period of 'crisis' among the scientific community. The grip of the paradigm is, as it were, loosened. It would seem that Kuhnian anomalies approximate in some measure to Popperian falsifications, except that for Popper such results – 'the refutation of our earlier theories' – constitute the heart and purpose of research whereas for Kuhnian 'normal scientists' they are the somewhat unwelcome (at best) result.[9]

The period of crisis is much like the pre-paradigm phase in that a universally accepted paradigm is required for the work to go forward; but it is different in that the agenda the paradigm must address is established by the anomalies rather than by a set of random observations. 'The resultant transition to a new paradigm,' Kuhn says of both these cases, 'is a revolution.' He goes on: 'Scientific revolutions are here taken to be those non-cumulative developmental episodes in which an older paradigm is replaced in whole or in part by an incomplete new one.'[10] Kuhn, then, offers a key to the history of science. Applying his schema to that branch of physics which most concerns communications technologies, electricity, he offers the following account.

Prior to the paradigm-establishing work of Franklin, a number of different theories were circulated accounting for the range of electrical phenomena then known. The most potent of these, articulated by the French physicist Dufay after 1733, held electricity to be of two distinct types, 'vitreous' as when glass is rubbed by silk and 'resinous' as when sealing wax is rubbed with fur. Franklin proposed the terms 'positive' and 'negative' for these phenomena and suggested that a single 'electric fluid' was at work but only flowed when positive. The fluid theory, which we still remember whenever we talk of electric 'current', was adequate to explain the basic performance of static electricity but, as experimental work in dynamic electricity progressed during the century from 1750 and especially after the introduction of the Voltaic battery in 1800, anomalies were increasingly exposed. The experimental work, in this instance as in others, nevertheless went forward. Faraday produced evidence that the electric fluid was made up of particles but those elements of the fluid paradigm most affected by anomalies were simply ignored.

For Kuhn the first half of the nineteenth century was a period of crisis in electromagnetic theory. The crisis concludes in the 1860s when the theoretical physicist Maxwell, working with the lines of force Faraday had envisioned surrounding his electrified or magnetised bodies, proposed a new paradigm which better explained these phenomena:

> The theory I propose may therefore be called a theory of the *electromagnetic field.* . . . The electromagnetic field is that part of space which contains and surrounds bodies in electric or magnetic conditions. It may be filled with any kind of matter, or we may endeavour to render it empty of all gross matter. There is always, however, enough of matter left to receive and transmit the undulations of light and heat, and it is because transmission of these radiations is not greatly altered when transparent bodies of measurable density are substituted for the so called vacuum, that we are obliged to admit the undulations are those of an ethereal substance, and not of gross matter. . . . We have therefore some reason to believe that there is an ethereal medium filling space and permeating bodies, capable of being set in motion and transmitting that motion from one part to another.[11]

Maxwell's concept of the undulating or wave nature of all electromagnetic phenomena is at the heart of telecommunications and related technologies. His mathematisation of these waves is, for our field, among the most important scientific advances of the nineteenth century. Maxwell's theory replaced Franklin's and thereby became the paradigm for the next half-century. But a central problem presented itself. If the waves or any other more substantial objects were passing through anything, even a medium as insubstantial as ether – the incorporeal stuff which Maxwell thought suffused the universe – they should be, however infinitesimally, slowed down; they should drag. On this basis the Michelson-Morley experiment of 1887 dramatically introduced an anomaly into the Maxwellian theory by failing to show any such drag on the motion of the earth. As the decades passed, the concept of ether became increasingly suspect. The anomalies accumulated as the business of normal science went forward. As Kuhn describes:

> Maxwell's discussion of the electromagnetic behaviour of bodies in motion made no reference to ether drag, and it proved very difficult to introduce such drag into his theory. . . . The years after 1890 therefore witnessed a long series of attempts, both experimental and theoretical, to detect motion with respect to the ether and to work ether drag in Maxwell's theory. . . . The latter [theoretical work] produced a number of promising starts . . . but they also disclosed still other puzzles and finally resulted in just that proliferation of competing theories that we have previously found to be a

concomitant of crisis. It is against this historical setting that Einstein's special theory of relativity emerged in 1905.[12]

And it is Einstein's paradigm that still holds sway, more or less, three-quarters of a century later. All of the technological advances in telecommunications this century have taken place during a period of what Kuhn would describe as 'normal science'.

The elegance of Kuhn's explanation of scientific progress has itself become something of a paradigm. The central notion of paradigm has been applied to various fields – political science, sociology, religion, economics and history in general. Such a wide degree of acceptance has been bought at the usual price. Objections, both major and minor, have been lodged and Kuhn has himself modified some of the elements he initially proposed.

Philosophers of science have had much difficulty with the concept of the paradigm. Kuhn has been accused of imprecision in his use of all his terms, especially the paradigm. One paper counts with a rigour that is positively scholiastic no less than 21 different senses of the word in his works.[13] Others have been uncertain as to the nature of the relationship between the wholly or partially rejected old paradigm and the new. Practitioners of 'normal science', it has been pointed out, are seldom, if ever, as *ad idem* as Kuhn originally suggested. There are further difficulties with the language of the schema. The term 'crisis' is, as Kuhn now allows, a little heady for many of the situations that normally precede a paradigmatic shift. In electricity, for example, the post-Franklin decades up to Faraday's time can scarcely be so described. Franklin's theory, throughout the eighteenth century, was less widely accepted than an alternative two-fluid explanation, even after Cavendish demonstrated in 1785 that positive and negative charges exerted the same force.

Similar objections can be raised to the term 'revolution'. There is an element of what has been called 'rhetorical exaggeration' in Kuhnian terminology which Kuhn has accepted. Rather than a complete break between one paradigm and the next a series of microrevolutions might be deemed, in Toulmin's view, to take place.[14] Most of all, objection has been raised to the idea, clearly inscribed in Kuhn, of a sort of anti-rationalism in science whereby any scientific community is doomed to be inevitably sawing the tree branch upon which it sits – privileging practice over logic in a way that reverses the model suggested by Merton for a sociology of scientific behaviour.[15] Thus Kuhn's account of scientific progress does require acceptance of a simple 'fallibilism' which argues for example that Einstein's paradigm will be overthrown as surely as was Newton's and therefore the truth about the physical world will remain for ever unknowable.

The widespread use and misuse of the paradigm as an intellectual conceit in disciplines far from the sociological history of science is, of course, not attributable to Kuhn. The attempted application of Kuhn to sociology is but the most obvious example of what has been called 'the misuse of the paradigm concept'.

> The result of this misuse is that the concept has come to be used in ways that Kuhn never intended. In some cases it has taken on attributes which he specifically disavows. Multiple interpretations of the term have had the effect of allowing sociologists to cite Kuhn as a source while, at the same time, they are not taking seriously the implications of his position.[16]

It is applications beyond the history of science that are making the Kuhnian schemata into its own paradigm. But as Kuhn himself has argued, history and other social sciences are 'protosciences' whose ability to generate exact hypotheses, a *sine qua non* of scientific paradigms, is limited. Such extensions of Kuhn, however, need not detain us. If his work is less a 'super theory' than might have been supposed in the second decade of its promulgation, its value is still considerable. It exerts an on-going influence on our perception of the nature of scientific inquiry.

However, two main objections to Kuhn can be sustained from his own field. There is a problem of circularity in the definition of the groups scientists form as communities to work out the implications of a paradigm. Kuhn in effect states that a paradigm is what is shared by a scientific community and a scientific community is defined by a shared paradigm. This masks a greater difficulty. Kuhn's project is in a sense to correct the somewhat hermetic tradition science enjoys. For Merton and others before Kuhn, the doing of science could be isolated in some way not only from the world, but also from the actual science in hand. Kuhn seeks to redress and indeed succeeds in redressing part of that bias. For him, what the science *is* crucially determines the professional, social relations of the scientists. But the relationships between these scientists and the world remain elusive. 'He treats the evolution of scientific thought and technique as if it were impervious to the influence of national cultural traditions'.[17]

In effect, Kuhn leaves the scientists inviolate in the laboratory and offers no account as to how social factors might determine the progress of paradigms. This, however, is easily remedied at one level at least, and more recent studies have highlighted connections between the scientist and the world. For instance the reception, in a Weimar Republic intellectually hostile to any determinism, of a theory of quantum mechanics which was acausally interpreted, has been pointed out. 'Goals and interests exist; they channel inference and judgement; they thus help to account for the emergence of a specific body of knowledge'.[18] Thus the failure to note this in the Kuhnian schema can be redressed; but only on a case-by-case basis. The schema itself cannot accommodate external factors in any systematic way. Kuhnian paradigms issue from the mind of science full-grown.

A second more serious objection to Kuhn can be found in the work of those attempting to apply the model to certain important paradigm shifts dealt with in some detail in *The Structure of Scientific Revolutions*. Of these, a most impressive critique of what Kuhn calls 'the scandal' of Ptolemaic astronomy before Copernicus has been mounted. Heidelberger claims, in contrast, that, 'There was no crisis in Ptolemaic astronomy'; and he suggests that in 1543 there was little to choose between Ptolemy and Copernicus, primarily because helio-centralism could not be proved on empiric observational grounds. As it was, Copernicus simply saved the phenomena of planetary regression rather more elegantly than did the developed Ptolemaic paradigm of his time; but he offered no more parsimonious model of the heavens because more epicycles were described in *De Revolutionibus* than were used in Ptolemy. Heidelberger suggests that all Copernicus did was harmonise a paradigm of observed heavenly motions (Ptolemaic astronomy) with a paradigm of theoretical causation of that motion traceable to Aristotle.[19]

Similar objections have been made to other major paradigm shifts – the Darwinian for instance – as being too simple to account for the historical facts.

It would seem, despite these critiques, that what is at fault in Kuhn is nothing more than too diachronous a schemata. Just as the language of 'crisis' and 'revolution' is not entirely appropriate, so too the sequentiality of the model is too limiting and strict. A greater degree of interaction between the various phases involved in paradigm shifts needs to be accommodated fully to explain the historical record. But there is a danger, too, that in denying, in detail, the exact ordering of a Kuhnian paradigm shift, the fact of the shift itself is denied. Let us recall that, 'Nevertheless, Copernicus's *De Revolutionibus* acted like a delayed time bomb.'[20] Heidelberger, like Kuhn, conducts the discussion of this paradigm shift as if no external forces were at work – as if the Turks were not at the gates of Vienna, as if Luther were not in full cry. Between the cowering and cowed Copernicus of Koestler's *The Sleepwalkers*, the serene Copernicus of Heidelberger and Kuhn's crisis-ridden Copernicus there still remain major gaps unaccounted for in theories of conceptual shifts in intellectual history.

This résumé of the debate on Kuhn is occasioned because our project is, self-evidently, no little informed by his model, at least as regards general purposes. It is something of a paradox that the history of technology – that history which is twin to the history of science – has not attracted Kuhnian interpreters. It is not the intention, however, that this book perform that function, for the imprecision of the community of technologies is not markedly less than the imprecision of the community of, say, sociologists. Technology does not work by the testing of paradigms and the application of a straightforward Kuhnian paradigmatic model would not be appropriate. Thus although the purpose of our proposed model is cousin to Kuhn's project, the terminology is different and specific to our purposes. It borrows nothing from Kuhn directly, although it is illuminated by Kuhn's account of the growth of knowledge about electricity. What is suggested in the first section of this chapter is a quasi-Kuhnian pattern which could account for the structure of technological change in one specific area – communication.

FUGITIVE PICTURES

Phase one: scientific competence

Monsieur Bequerel makes TELEVISION possible, 1839

Television depends in essence on the photovoltaic (or photoemissive) effect, that is the characteristic possessed by some substances of releasing electrons when struck by light. The observation of this phenomenon is credited to a nineteen-year-old, Edmond Bequerel, in 1839 but it seems that his father, the *savant* Antoine César, might have helped him to prepare his account for L'Académie des Sciences. The experiment demonstrated that a current passes between the two electrodes of a Voltaic cell when a beam of light falls on the apparatus. Bequerel, who made the first colour daguerreotype and whose son Henri was to discover radioactivity in 1896, stands in a line of electrical experimenters stretching back to Queen Elizabeth I's physician William Gilbert. He was the first Englishman to endorse heliocentralism and in *De Magnete*, the first book written in England on the basis of direct observation, he coined the phrase *vis electrica* to describe the property, noticed in antiquity, possessed by amber ('ηλεκτρον) and some other substances which, when rubbed, attracted light materials such as feathers.

What Franklin was to call positive and negative charges were demonstrated by the superintendent of the gardens of the King of France in 1733. In 1745 the first Leyden Jar was built and Cunaeus, a friend of the inventor Musschenbroek, got a serious electric shock from it. It prompted the beginnings of the theoretical discussion which led to the establishment of the Franklin 'one-fluid' paradigm. A parade of electricians, many of whose names are now immortalised in equipment or units of measure, elaborated, into the early nineteenth century, both the theory of and the laboratory apparatus for creating electrical phenomena. This is the tradition – that of Ohm, Volta and Galvani – into which Bequerel fits.

Parallel with this inquiry were investigations into the nature of light.

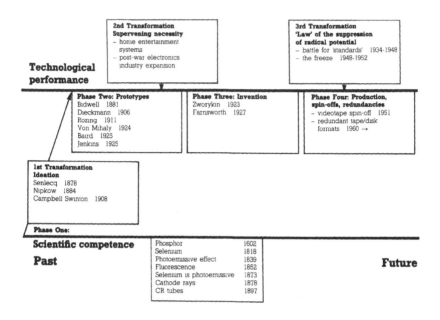

Figure 2 Television

For a century, a preparadigm phase, to use Kuhnian terminology, had existed. Newton's theory that light came in discrete bits, corpuscles, was pitted against the contradictory idea of Huygens that it was continuous. Dr Thomas Young was the father of physioptics. He was the first to explain how the eye focusses and to describe astigmatism and he also translated the demotic text of the Rosetta stone, did work on the heart and served on a committee to advise on the general dangers of introducing gas into London, a not untypical late-eighteenth-century range of interests. But his most important work was in proving Huygens right. The proof was to stand for the next 104 years.

In 1801, Young was studying the patterns thrown on a screen when light from a monochromatic source, sodium, passed through a narrow slit. Areas lit through one slit darkened when a second slit illuminated by the same sodium light source, opened. This phenomenon – interference – Young explained by assuming the light consists of waves and suggesting that interference was caused when the crests of the waves from one slit were cancelled by the troughs emanating from the second. He was able to measure the wavelengths of differently coloured lights, getting close to modern results.

In 1845, with the by-then dominant Franklin paradigm for electricity crumbling, Faraday, putting magnetic waves and Young's light waves together, observed that 'magnetic force and light were proved to have a relation to each other'. He added 'This fact will most likely prove exceedingly fertile.'[1] He had been working with electricity for more than a decade when he hypothesised that light waves might be transverse vibrations travelling along the lines of magnetic and electrical force he was studying. In this the great experimenter foreshadowed the unified electromagnetic wave theory of James Clerk Maxwell which was, a generation later, to bring all these strands together, replacing Franklin's paradigm and absorbing Young's observations. Maxwell's work and its experimental verification by Hertz in the late 1880s lie at the heart of all electronic communications systems. It was Hertz, following Bequerel, who observed a photo-emissive effect, that ultraviolet light striking a polished metal surface would cause the emission of what would come to be called electrons.

The photoemissive phenomenon is central to television. A Swedish chemist, Baron Berzelius, had isolated selenium in 1818, without noticing (in advance of Bequerel) that it had this property. Selenium is a non-metallic element of the sulphur group and was initially of interest because it had a very high resistance to electricity. Willoughby Smith, a supervisory telegraph engineer involved in the laying of the first successful transatlantic cable, was directed to selenium because of this property.

> While in charge of the electrical department of the laying of the cable from Valentia to Heart's Content in 1866 I introduced a new system by which ship and shore could communicate freely with each other without interfering with the necessary electrical tests. To work this system it was necessary that a resistance of about one hundred megohms should be attached to the shore end of the cable. . . .
> While searching for a suitable material the high resistance of selenium was brought to my notice but at the same time I was informed that it was doubtful whether it would answer my purpose as it was not constant in its resistance.[2]

May, an operator of Willoughby Smith's, noticed the correlation between the erratic behaviour of the selenium resistor and sunlight. Willoughby Smith investigated and made his results known in 1873.

> When the bars were fixed in a box with a sliding cover, so as to exclude all light, their resistance was at its highest, and remained

very constant . . . but immediately the cover of the box was removed, the conductivity increased from 15 to 100%, according to the intensity of the light falling on the bar. . . . To ensure that temperature was in no way affecting the experiments, one of the bars was placed in a trough of water so that there was about an inch of water for the light to pass through, but the results were the same; and when a narrow band of magnesium was held about nine inches above the water the resistance immediately fell more than two-thirds, returning to its normal condition immediately the light was extinguished.

I am sorry that I shall not be able to attend the meeting of the Society of Telegraph Engineers to-morrow evening. If, however, you think this communication of sufficient interest, perhaps you will bring it before the meeting.[3]

A decade and half later, in the year after Hertz had noted the photoemissive effects of ultraviolet light, Hallwachs produced a device specifically to create photoelectricity. Elster and Geitel, using first sodium and then potassium, perfected a practical and sensitive photocell to the point of commercial production by 1913.

Faraday's work in electro-optical effects was continued by Kerr in the 1870s and led, by 1905, to the direct modulation of an electric arc through the action of a light beam.

In addition, Guthrie, in 1873, observed that an electroscope would discharge in the presence of a heated metal ball, the underlying principle of what was to be known as 'The Edison Effect'. Edison, ten years later, noticed that within his electric lamp at certain voltages a reverse current between the two leads of the filament caused a bluish light to appear. Edison (and, independently, Elster and Geitel) discovered that these thermionic emissions, whereby a stream of electrons or ions are released because of high temperatures, occurred in vacuums.

A professor of electrical engineering at University College, Ambrose Fleming, in the years after the turn of the century, built a variation on the light bulb, a vacuum tube which, by specifically exploiting the Edison effect, turned an alternating current radio signal into a direct current capable of activating a meter or telephone. The patent for this **diode tube** rested with the British Marconi company to whom Fleming was consultant. In 1906 in Palo Alto, De Forest added a third electrode between the cathode and anode of the Fleming valve to increase its sensitivity. Edwin Armstrong, at Columbia, and von Lieben in Germany realised De Forest's **triode** would amplify weak radio signals and enable longer distances to be covered. In 1911 Armstrong patented the negative feedback circuit that was crucial to the amplification process and sold his rights to Westinghouse.

Eventually twenty years later the Supreme Court decided that, despite the essential contribution of Armstrong, De Forest had solo claim to the **triode tube**. The latter declared himself 'the father of radio'.

Thermionic valves were thus widely introduced by the development and expansion of radio and telephony. Photoelectric phenomena, however, although observed at the same time, did not lead to devices of such broad application. This was despite the fact that, for theoretical physics, photoelectricity was a central area of investigation.

Cathode rays had been described by Sir William Crookes in 1878. The first indication of the existence of sub-atomic particles came in J.J. Thomson's demonstration in 1897 that photoelectric phenomena depended upon particles and not on waves. He called them, using a term suggested by Johnston Stoney, 'electrons'. Thus was the belief in the indivisibility of atoms shattered and the dedicated, and fateful, investigation of the sub-atomic world begun.

Eight years later Einstein explained the mathematisation of photoelectric emission (photoemissive) phenomena, in effect using it as a proof and extension of Max Planck's quantum hypothesis of 1900 that 'radiant heat . . . must be defined as a discontinuous mass made up of units [quanta] all of which are similar to each other.' [our brackets][4] Maxwell's wave paradigm began to follow Franklin's one fluid theory into history since the experimental proof of Einstein's 'photons' demonstrated that the higher the frequency of the light the greater the speed of the emitted electrons – which would not be the case if light were continuous. Newton was right after all, but so was Huygens; our current picture of light is a paradoxical synthesis of both descriptions.

Crucial to Thompson's work which instigated the whole line of inquiry was the **cathode ray tube** (CRT), introduced the same year as his discovery by Braun. It made the streams of electrons visible to the eye, yet, unlike thermionic valves, it remained in the physics laboratory.

The ground of scientific competence for television also contains other elements. In 1602 a Bolognese cobbler and part-time alchemist, Cascariolo, found a mineral – a sulphide of barium known as 'Bologna Stone' – which would glow brightly after being exposed to light. This phosphor was of excellent value for the performance of tricks but it was so rare that a search was initiated for an alternative, that is for a stable phosphor which could be manufactured. In 1886 this was achieved.

Fluorescence was a similar phenomenon whose investigation is of equal duration. That certain substances will emit luminous coloured light was noticed in antiquity and much described in the eighteenth century. Sir George Stokes, however, was the first to offer, in 1852, a reasonable explanation of the effect and to name it (after the mineral fluorspar).

All these strands together constitute the ground of scientific competence in relation to which the technological development of television takes place. Photoelectric effects had been utilised in the production of light sensitive cells by 1913. Electro-optical effects had been refined to the point of practical modulation systems by 1905. Fluorescent paint was being manufactured. Phosphors were understood. The cathode ray tube existed. Knowledge of thermionic amplification had produced, by 1914, the practical basis for extensive radio and radio-telephony systems. By means of the facsimile telegraph, images had long been transmitted by electricity between distant points. And by the time of the First World War, men had been dreaming of television for forty years.

The first transformation: ideation

Monsieur Senlecq invents the TELECTROSCOPE, 1877

Following the announcement of the peculiar properties of selenium and the excitement generated by the introduction of the telephone, numerous notions for 'telescopy' were put forward. The television receiver was first imagined by a *Punch* cartoonist as a two-way interactive device whereby those at home could talk to those on the screen by telephone. The **telephonoscope**'s screen stretched the entire width of the mantelpiece. It is unlikely that the artist, in 1879, would have had much understanding of how such a device might work. Yet the basic principles which were to lead to television were already understood by the scientific community; although, even there, a number of factors were, typically, not accounted for in the earliest schemes.

Senlecq, a French lawyer, was the first to suggest how selenium might be used in a scanning system. He was, like most of these early thinkers, primarily concerned with a reprographic apparatus that would work telegraphically – telephotography or facsimile telegraphy in effect. Senlecq envisaged television, as did most of his peers, not as the instantaneous transmission of images onto a screen but rather as the transfer of a single image, perhaps a series of images, onto paper. He published a brief account of a **telectroscope** in an *English Mechanic* of 1878. The device would reproduce at a distance the images obtained in the camera obscura. The scanning notion was expressed as follows:

An ordinary camera obscura containing at the focus an unpolished glass (screen) and any system of automatic telegraphic transmission; the tracing point of the transmitter intended to traverse the surface of the unpolished glass will be formed of a small piece of selenium held by two springs acting as pincers, insulated and connected, one with a pile and the other with the line.[5]

Senlecq's receiver was simply a pencil operating on the same principle as the vibrating diaphragm in a telephone receiver, tracing its responses to the irregularity in the current (generated by the light hitting the selenium) on to paper. Senlecq refined the transmitter in a proposal two years later. Instead of a single moving block of selenium traversing the screen of the *camera obscura* he now suggested a mosaic of selenium cells, each to transmit by separate wires to a similar mosaic in the receiver.

His talk of 'fugitive pictures' stirred the imagination of others on both sides of the Atlantic. In a letter to *Nature* in April of 1878 Messrs Ayrton and Perry responded to the news that the great Alexander Graham Bell had apparently deposited a scheme for television at the Smithsonian. It will be learned below that Bell's reliability in such matters was less than it might have been, but in this instance there was simply a misunderstanding. Bell and a colleague, Sumner Tainter, were developing a device in which a voice-modulated beam of light fell upon a selenium cell which in turn modulated the carrier wave.[6] It was to be patented as the **photophone** and stands, in effect, on the line leading not to television but to modulated-light carriers of the sort just now being introduced into telephony. Ayrton and Perry, although confused as to Bell's line of inquiry, were quite clear that, public spiritedly, they wished to prevent him establishing patents in another major area.

The following plan has often been discussed by us with our friends, and, no doubt, has suggested itself to others acquainted with the physical discoveries of the last four years. It has not been carried out because of its elaborate nature, and on account of its expensive character, nor should we recommend it being carried out in this form. But if the new American invention, to which reference has been made, should turn out to be some plan of this kind, it may do good in preventing a monopoly in an invention which is really the joint property of Willoughby Smith, Sabine, and other scientific men, rather than of a particular man who has had sufficient money and leisure to carry out the idea.[7]

Bell obviously did not enjoy, in contemporary scientific circles in the UK at least, quite the heroic reputation popular opinion was to endow him with as the years progressed. Ayrton and Perry suggested again a mosaic of selenium cells as a transmitter. The image of the object to be reproduced was to be 'strongly illuminated' and focussed through a lens on to this screen. Individual wires were to run from each cell to the distant place. There a most ingenious apparatus was proposed. Electromagnets were to control the opening of a host of apertures. The strength of the magnetic pull would depend on the amount of electricity coming from the transmitter. This would in turn determine the amount of light falling on the selenium cell within the transmission screen and in this way the image of the object would be recreated.

To George Carey of Boston belongs the credit for the first completely described, that is fully drawn, system. He too opted for the selenium mosaic but at the receiving end suggested that the image might be impressed upon a piece of chemically treated paper. Again the facsimile telegraph lay behind the thinking. In a second scheme proposed at the same time, Carey suggested that a single selenium cell be arranged within the camera to traverse a spiral path, thus scanning the image. However, he does not seem to have realised the need to duplicate the rate and pattern of the scan synchronously in the receiver.

About the same time, Leblanc proposed a more efficient optical system for scanning which involved revolving mirrors. Another scientist attempted to overcome the slowness of selenium by mounting the cells on a steadily revolving ring fitted with mirrors. Again there is no evidence that the device, which was patented in 1877, was ever built. The difficulties of synchronisation meant that this idea was not taken up again until after Nipkow had patented the spinning disk. This was done in 1884. (Nipkow, who died in Berlin in 1940, worked all his life as an electrical engineer for a railway signal manufacturing company. In his seventies, in 1934, he assumed the presidency of the newly founded German Television Society. For the fifty years between the patent and the presidency he did nothing with television.[8]) In 1889 Weiller proposed a drum or disk mounted with tangentially set mirrors as the scanning device.

All these selenium-based devices were doomed to failure, at least for the transmission of movement, because of the slow responsivenesss (including lag) of the substance. As to their application in the field of telegraphy, other better facsimile systems existed and there was thus no supervening necessity for pursuing this line of inquiry. The basic principles of telephotography were outlined in a British patent of 1843 using chemically treated paper in the receiver and in 1847 a 'copying telegraph' relying on scanning was introduced by Bakewell. By 1862 a French divine, Caselli, sent a picture by wire from Amiens to Paris. Elisha Grey, the man who was beaten to the Washington Patent office by Bell, designed a device which could, with rheostats, electromagnets and styluses, copy a written document telegraphically.

A confusion seems to exist in the minds of many of these early television thinkers. They dreamed of the reproduction of movement, dreamed of it in advance of the cinema; but they addressed themselves to the transmission of stills, a species of almost redundant effort since other systems already existed for such purposes. The advent of moving pictures did nothing to increase the need they were addressing with these experiments.

But this general blindness was certainly not true of one senior British electrician, Campbell Swinton, who, in a letter to *Nature* on the 16th of June 1908, outlined the most significant of all the early schemes for television. The chain of events leading to this description of an all-electronic system appears to have been prompted by a public promise made by M. Armenguard, the president of the French Society of Aerial Navigation, that 'within a year, as a consequence of the advance already made by his apparatus, we shall be watching one another across distances hundreds of miles apart'. Shelford Bidwell, one of the few early pioneers actually to build an effective selenium device, knew of its limitations as well as anybody. Not conceiving of any alternative, he was prompted by the Armenguard announcement to put into print an account of his experiments which had taken place over a quarter of a century earlier. In part supporting him, Campbell Swinton was then moved to lay down the basic principles of modern television. He was dismissive of all mechanical scanning and multiple wire systems and suggested instead that the problem could:

probably be solved by the employment of two beams of kathode (sic) rays (one at the transmitting and one at the receiving station) synchronously deflected by the varying fields of two electromagnets . . . indeed so far as the receiving apparatus is concerned the moving kathode beam has only to be arranged to impinge on a sufficiently sensitive fluorescent screen, and given suitable variations in its intensity, to obtain the desired result.[9]

Three years later in his presidential address to the Röntgen Society he elaborated the receiver by suggesting a special cathode ray tube which would have, at its front, a mosaic screen of photoelectrical dots but he said 'it is an idea only and the apparatus has never been constructed'. In moving the vote of thanks Silvanus Thompson described the idea as 'a most interesting, beautiful and ingenious speculation'.[10]

In a way that can perhaps be categorised as rather typically British (though he was not to be as thoroughly inactive as Nipkow), Campbell Swinton carefully assessed the difficulties of an all-electric television system and determined it was not worth trying to build because of what he claimed would be the vast expense involved. Others were less hesitant. A full year before the 1908 letter to *Nature* a Russian, Rozing, had patented, in London as well as in Berlin and St Petersburg, an all-electric television receiver. In the year of Campbell Swinton's presidential address, this same Russian actually transmitted a signal to his receiver. But this is technological performance, the subject of the next section of this chapter. In Rozing's patent and in all these proposals the *ideation transformation* can be seen as it applies to the history of television.

Before it was called **television**, it was called **telephotography**, **telescopy** or **teleautography**. As late as 1911, a British Patent Official opened a new file on the matter as a branch of facsimile telegraphy, even though he called it **television**, a term first coined independently by Persky in 1900. By far the most interesting depositions in the new file were British Patents Nos 27570/07 and 5486/11 for a velocity modulation cathode ray receiver awarded to Boris Rozing.

Phase two: technological performance – prototypes

Mr Shelford Bidwell invents TELEVISION, 1881

The cold war casts a curious shadow across the history of television. The earliest published accounts of what is here being called *Phase two: technological performance – prototypes* were written from the mid-1930s to the mid-1950s by which decades the accident of Rozing's birthplace had assumed a significance it otherwise would not have had. British writers in the 1930s claimed that Campbell Swinton's analysis of the basic problem was, despite the fact he did not work on his proposed solution, superior to Rozing's more pragmatic approach.[11] And an American account dating from the 1950s described Campbell Swinton's 1911 proposal as 'a still more startling *invention*' than Rozing's work. (our emphasis)[12]

The Soviet treatment of technological history has become the butt of much Western humour, although Russian 'firstism' about aeronautics was substantially boosted by **Sputnik** (see chapter 5 below). Stalinist claims that Russians had invented both radio and television are no less – and no more – substantial and anyway compared well with British pretensions in these areas. Certainly Rozing's experiment is of a quite different magnitude from Campbell Swinton's musings, for all that the latter envisaged the more completely modern scheme eschewing mechanical elements and accurately forecasting the exact method of CRT utilisation that was to prevail three decades later. But Rozing's is an achievement that neither the chauvinism of others nor the rodomontade of official Soviet accounts should be allowed to taint – although, since his contributions all predate the revolution, the rhetoric of the latter remains faintly comic.

The Soviet account offers a whole parallel cast of characters to the one given above. Bakhmet'yev is credited with a single selenium cell scanning in a spiral to beat Senlecq's suggestion; mirrored drums and the like were perfected by Vol'fke in 1898 and Polumordvinov in 1899. More than that the Vol'fke patent proposed wireless transmission and a gas discharge tube at the heart of the receiver. And more than that, the Polumordvinov patent was for colour using a sequential system.

Boris L'vovich Rozing was born in St Petersburg in 1867, took his degrees at its university and taught, from 1893, at its Technical Institute. He was interested in the electric telescope (as he was still calling it in the 1920s), and noticing that the electron beam in the common cathode ray laboratory oscilloscope left complex luminescent patterns on the front of the tube, decided that this was his 'ideal mechanism'.[13] In 1907 he patented such a circuit using electron beam deflection to modulate the intensity of the beam, by altering its velocity, in the receiver.

Rozing's idea, innovative and important though it was, had its precedents. These do not detract from his achievement. They simply place it in context. His transmitter was built along the lines that were being commonly suggested before the turn of the century, that is, an electromechanical system using a mirrored drum of a known design. His more revolutionary receiver had been anticipated, too. A German patent of 1906 had already used the cathode tube specifically for 'the transmission of written material and line drawings'. This machine, designed and built by Dieckmann, stands with the other surviving apparatus of this early period – Bidwell's system of 1881. These two machines, although both are primarily facsimile devices, have claim to be the earliest *partial* prototypes of the first technological performance phase of television.

Shelford Bidwell demonstrated in 1881 apparatus much like that proposed by Carey but, for the first time, transmitter and receiver were synchronised. A single selenium cell was mounted in a box with a pin-hole aperture which was arranged within a frame so it could be cranked to rise and fall relative to a screen upon which was projected the image to be transmitted. At the receiving end a drum was rotated to match this motion and created a negative image on paper soaked in potassium iodide. The image was built up of a series of closely spaced brown lines. He demonstrated the device to a number of learned societies and donated the apparatus to the Science Museum.[14]

Dieckmann's device, which was also built and survives, used a standard Nipkow disk transmitter but here for the first time the image scanned was received on a cathode ray tube. Paradoxically, this breakthrough was counterproductive, for Dieckmann's intended purpose was facsimile transmission and the lack of a hard copy was a major disadvantage.[15]

These machines can be classified in two ways. As types of facsimile telegraphs, they are *redundant* devices. In television terms, though, they are the most *partial* of *prototypes* because, as transmitters of single images, only in the crudest sense can they be considered as television at all. They are, however, examples of technological performance since they were both built; and that is why, despite their primitive nature, they are included in this phase rather than in the previous ideation transformation.

It should not be thought that, since Rozing used techniques previously suggested or demonstrated by others, this flaws the claim made for his significance. Technology is not a field marked by 'eureka' discoveries. Building prototypes and 'inventing' are better described as 'systems engineering'. That is to say, it is less a question of solutions being born *ab initio*, fully formed in the mind of the technologist; rather it is a question of the technologist seeing in the available ground of scientific competence and among the ideas of his or her peers how to put old elements together in a new way. In this Rozing was doing nothing that other better-known 'inventors', before and since, had not done.

Given the general state of tube technology in this period, his proposed solution was at the very cutting edge of what was possible. What also distinguishes him is that he did not give up at the patent stage but actually built a series of partial prototypes including the one which, in 1911, transmitted a wireless image; but the rudimentary state of the cathode ray tube and electronic amplification meant that the line of development was not taken up by others. Even in the USSR it seemed as if the problem would be better solved by mechanical scanning using substances of greater sensitivity than selenium, so Campbell Swinton's suggestion and Rozing's experiments were not pursued during and immediately after the First World War.

The truth, which Rozing and Campbell Swinton saw but which most researchers ignored, was that while mechanical scanning and cathode ray devices were partial prototypes both working poorly at this point (and the tube was worse than the disk), the all-electrical system had the far greater potential. Despite this, the years from 1911 to the early 1930s were to be the golden age of the Nipkow disk and its variants.

Selenium lag was no longer the major problem with mechanical scanning systems – more responsive substances had been isolated and various sophisticated arrangements of spinning mirrors and the like had increased the sensitivity of the photoelectrical elements yet further. Although the number of scanned lines generated remained few, the essential problem was perceived to lie more in the difficulties of maintaining exact synchronicity between the disks at either end of the system rather than with definition.

John Logie Baird was born, a son of the manse, in Helensburgh, Scotland in 1888. He had a most varied career including the marketing of a medicated undersock, 'Osmo' bootpolish, Australian honey, and a soap bar, as well as a failed enterprise manufacturing jam in Trinidad. After these adventures, in 1923, almost penniless and recovering from ill health, he started to build a practical television system along Nipkowian lines. He sought help via the front page of the *Times*.

SEEING BY WIRELESS – Inventor of apparatus wishes to hear

from someone who will assist (not financially) in making working model.[16]

In 1925 he showed the image of a dummy in Selfridge's store in London. (Gordon Selfridge was an early backer, although the store's publicity stated: 'We should perhaps explain that we are in no way financially interested in this remarkable invention.'[17]) In January of 1926, he demonstrated to a scientific audience the transmission of the image of a living face. The earliest of these devices had relied on selenium but by this date a gas-filled potassium photocell was used which, with the complications of the scanning disk system and 'blinding incandescent light from batteries of 100 watt lamps at very close range', produced a 'faint and blurred' image made up of 30 lines.[18] Baird's associates applied to the Post Office for a broadcasting licence. By 1928 the Baird Television Development Company was building **televisors** (or receivers) for the home. BTDC was working on a 30 line picture scanning at 12½ frames a second.

Nobody in either BTDC or the Post Office was under any illusion as to the BBC's quite proper interest in these developments. Baird, who had been at technical college with BBC Director-General Reith, founded his Television Company the year before the Corporation received its charter, but the BBC treated Baird's enterprise as that of an unwarranted upstart. Captain Peter Eckersley, the man who built the BBC's centralised radio network against the technological odds and who can thus claim, with greater truth than Lord Reith (who fired him), to be the father of British broadcasting, was scathing about Baird:

> The advisers of the Baird Television Company believe that this apparatus is sufficiently well developed to have a public service value. They contend that the attitude of the BBC is obstructive and irrational. The advisers of the BBC believe on the other hand that the Baird apparatus not only does not deserve a public trial, but also has reached the limit of its development owing to the basic technical limitations of the method employed.[19]

The BBC has never found it difficult to adopt such a tone but in this instance it was fully justified. Baird was a very modern figure in the sense that his image was media sustained. His search for funding took him again and again to what were then called 'publicity stunts'. He himself, despite his background as an effective if minor businessman, was presented as the very model of the otherworldly mechanical genius, always good stereotypical copy. In short order, during these years, when the best he could produce was a low-definition barely synchronised head-and-shoulders image of pinkish hue on an envelope-sized screen, he also demonstrated, at the same level of efficiency, transatlantic transmission, videorecording, infra-red apparatus and a colour system.

The 'obstructive and irrational' play no little part in the processes of innovation, but Eckersley's response to Baird was entirely reasonable. For all that British print media were chauvinistically hailing every flickering advance as a triumph, Eckersley knew that for nearly half a century Baird-style television had failed to work for very good basic reasons grounded in the immutable laws of physics. Eckersley had the support of Campbell Swinton as the BBC maintained its stand in the face of a hostile pro-Baird press. The fact remained that Baird's was a *partial* prototype and, unlike Rozing's, it represented the end of the line, just as the BBC claimed.

Nevertheless, through a marriage arranged by the Post Office, the BBC did begin experimenting with television and evolved a working relationship with BTDC which was at times positively cordial. By 1930, 'televisors' were being sold at 25 guineas the set and in April sound joined pictures in the transmissions. The system still produced an oblong picture of only 30 lines definition, although it had by now improved sufficiently for actual programming to be undertaken. The BBC began serious exploration of the new medium, transmitting, in July of that year, the world's first 'upscale' television play – Pirandello's *The Man with the Flower in His Mouth* in cooperation with Baird.[20] General Electric had broadcast the somewhat less esoteric melodrama *The Queen's Messenger* from its Schenectady station two years before. The difference in style and ambition that today characterises British and American television culture can therefore be said to antedate the introduction of the all-electric system.

Whatever the excessive claims of the British press, Baird was far from being alone. Von Mihaly, an Hungarian pioneer, was reported in the British technical press as having transmitted shadows with a mirror drum system in March of 1924 and, as Baird was setting up shop in 1925, the American C.F. Jenkins, who had contributed to the development of the movie projector, demonstrated an elegant apparatus which used prismatic disks to scan the image. It could not work in reflected light and gave only a shadowgraph effect in the studio but it was capable of transmitting images from a film projector, exactly what Jenkins intended, since his project was the home delivery of movies. Jenkins, whose laboratories were in Washington, also conducted an experiment with the Navy, signalling weather maps to ships cruising down to the Caribbean.[21] In 1927 he applied for and obtained a licence to operate a television station from the Federal Radio Commission, the forerunner of the Federal Communications Commission, the body governing American broadcasting.

General Electric's E.F.W. Alexanderson had been experimenting with a similar system which in 1928 resulted in the start of experimental transmissions from the GE radio station in Schenectady. Not only did GE produce the first television drama but it also mounted a demonstration using a screen 7 feet square, erected in a local theatre.[22] Alexanderson, pursuing the line begun by Bell's **photophone**, went on to work with modulated light carriers for electrical signals.

Beginning in 1923 and finally scotching the rumour of 1880 that Bell had designed a television system, Herbert Ives and others at Bell Labs were instructed to commence television experiments. In 1927 they demonstrated, over 250 miles by wire and 22 miles by radio, a system identical in principle to these others. Two forms of apparatus were used, one giving a picture $2 \times 2\frac{1}{2}$ inches and the other $2 \times 2\frac{1}{2}$ feet, a multi-element water-cooled neon lamp as the screen.[23] They also used the apparatus to scan film taken either at the transmitter or receiver end, so that a better illuminated picture could be achieved. Over the next few years Ives and his fellow workers increased the lines to 72 and the frame speed to 18 per second. Since potassium cells were sensitive to blue light, Bell engineers imposed a blue filter into the scanning apparatus which again increased its effectiveness. In 1929 they introduced colour and by the early 1930s they had a two-way interactive system working in Manhattan. The **videophone** (that most beloved of all information revolution hardware) is thus over 50 years old.[24] Its 'inventor', this same Dr Ives, told a British visitor in 1931 that:

> frankly he has not the remotest idea whether the public want to see the fellow at the other end of the telephone line badly enough to pay a high price for the privilege. But when the AT&T started to develop the transatlantic telephone years ago, they did not know whether sufficient people would pay the necessarily high price to make a service profitable. But the transatlantic telephone does pay.[25]

The notion that great corporations behave more rationally in the matter of innovation than lonely private inventors is given the lie by AT&T's 50-year flirtation with this device. Its ardour for videophones remains uncooled. It remains as confused as ever by the distinctions between the unnecessary sight of 'the fellow at the other end of the telephone line' and crucial communication requirements like trans-oceanic links.

More seriously, the facsimile implications of these developments remained important, hence at least one strand of the telephone company's interest in the matter. AT&T could, though, at the time of these experiments, do far better with existing equipment than with television. It was using a standard well beyond the capacity of any mechanical scanning system – a 5 × 7 inch image divided into 350,000 elements taking seven minutes to transmit. Ives calculated that a television band to achieve similar detail would require the then rather unimaginable bandwidth of 3 million cycles per second, 7000 times the one being used.

By 1929 the FRC had licensed 22 radio stations to transmit pictures. The game with W1XAV, Boston, was for a group of MIT students to go to the station while another group, in the college, gathered round the scanning-disk receiver and tried to guess which of their friends in the studio they were looking at.[26] But in general, despite the proliferation of different firms and mechanical systems, the public was uninvolved in these experiments.

The Germans were responsible for the next stage of the Nipkow machine. Experimental transmissions had started in Berlin in 1935 using apparatus built by Fernseh A-G, a subsidiary of Zeiss Ikon and Bosch formed to exploit Baird's patents, which achieved, for film transmission, 180 lines and 25 frames per second. 'The success of this machine', a colleague of Baird's wrote in the 1950s, 'was to no little measure due to the micrometer precision engineering tools which the Germans had available for disk construction'.[27] The precision engineering of the system was taken a step further in that the disk was placed in a vacuum to reduce both interference and drive-power. Subsequently the Germans managed a mechanically created variation on interlaced scanning, thus achieving stability very close to the all-electrical systems.[28] This machine represents the mechanical scanning *partial prototype* in its final form.

Since this system, like Jenkins's, worked best when dealing with film rather than in the studio, Fernseh constructed a film camera with an attached developing tank, building on the proposal of Ives in 1927. It produced a photographic image in under a minute which was then mechanically scanned – the Intermediate Film (IF) system. The Germans also used an electrical system based on the Farnsworth patents, for which they had signed an agreement in 1935. As is explained below (page 63), Farnsworth had produced an alternative electric tube which also worked best when transmitting film; so the IF system was ideally suited for this configuration as well. As with many other *prototypes* the German Nipkow apparatus worked nearly as well as did the contemporary version of the *invention*. In 1938 the Germans built a videophone between Berlin and Nuremberg on a 180-line standard, but mechanical solutions were doomed, even here. Anyway, the Germans by this date were well into an extended public service using both Farnsworth/Fernseh equipment and a version of the RCA camera built by Telefunken. (See below, page 61.)

Baird's engineers, like their German opposite numbers, were also locked in battle with an all-electrical system (essentially the same as

Telefunken's and RCA's) developed by Marconi and EMI in concert. BTDC duplicated the earlier German mechanical feat and, in 1936, achieved their ultimate in mechanical scanning, 240 lines at 1/25th of a second. But it was a dead end. Despite Baird's dictum of 1931 that, 'There is no hope for television by means of cathode ray tubes', the receivers for these 'high definition' mechanical systems, both in Germany and the UK, were by now all-electric tubes with no trace of spinning disks.[29] It was only a question of time before the disks disappeared from the transmission end of the system, too.

Baird withdrew entirely from the activities of the company that bore his name and spent the next three years experimenting with a large screen, mechanically scanned (at 6000 rpm) colour television system with stereophonic sound! By then, Baird's collegues were using a mechanical scanner with IF for studio work and a purely BTDC telecine device, which observers felt was better than anything EMI had, for transmitting film. They also had Farnsworth's **image dissector** camera available, BTDC having signed an agreement with him, but they could not make it work. At the government's behest, the BBC was gearing up so that a final judgment between BTDC and EMI could be made – a sort of experimental run-off, as Asa Briggs points out, like nothing so much as a nineteenth-century competition between rival steam locomotives.[30] The game, though, had been fixed by physics as Peter Eckersley had somewhat brutally indicated nine years before. Each mechanical element had given way to an electronic equivalent. Electrical scanners inexorably drew away in terms of performance, ease of operation, reliability and general 'elegance'. Only when transmitting film was there any question of competition and even then the electronic potential was clear. In the words of one of the BBC pioneers required to produce programmes, turn-and-turn-about, using both systems: 'Working in the Baird Studio was a bit like using Morse Code when you knew that next door [in the EMI studio] you could telephone.' [our brackets][31] By 1936 the question was not, 'mechanical v. electronic?' but rather 'which electronic?'

The second transformation: supervening necessity

The US navy invents TELEVISION in Lancaster, PA, 1945

All the developments described above – up to and including the establishment of the German and British pioneering public television services – fall in the prototype phase. There was clearly enough interest in the new medium to sustain research and development, yet the final push out into the world of mass communications had not taken place by the eve of the Second World War. Despite the availability of the technology, television was still waiting to be diffused – waiting, in some sense, to be *invented*.

For technological performance to change from *prototype* to *invention* all that is required is the operation of an external *supervening necessity*. This formulation addresses directly the central question of television's technical history which is less why it happened in the way it did, but more why it took so long. Television, electronically scanned television, was a demonstrated and soundly grounded theoretical option by 1911. It was a mass medium by the mid-1950s. These forty-plus years yield one catalytic external *supervening necessity*, which is of greater moment than any single technological advance in the same period. It occurred after the conclusion of the Second World War.

The crucial factor which transformed television from toy to mass medium was the spare capacity of the electronics industry in 1945/6. In the two years from April of 1942 (according to the premier issue of *Television* magazine), defence spending had expanded the radio industry of America by between 1200 and 1500 per cent. More than 300,000 workers were involved. As the magazine put it:

> The question now arises what to do with these facilities after the War, for the demands of aural radio alone will not be sufficient to keep many of them going. *Only television offers the promise of sufficient business*. (Emphasis in the original)[32]

James Fly, then chair of the FCC, wrote that same year: 'I think it quite likely that during the postwar period television will be one of the first industries arising to serve as a cushion against unemployment and depression.'[33] There was precedent for this. After the First World War a similar situation had existed and radio had functioned similarly. Westinghouse had been making comparatively complex but compact SCR 69 and SCR 70 radio receivers for the Signal Corps. The servicemen who used these machines constituted the '100,000', the corps of the amateurs whose continued interest created the world of radio between the Armistice and 1920.

Frank Conrad had been Westinghouse's supervisor making the SCRs. He began broadcasting from his garage near Pittsburgh on a regular basis and a local store started selling readymade wireless receivers to members of the public who, not having had the benefits of a Signal Corps education, could not easily build their own. Conrad's boss became aware of this activity and removed him from the electrical switches production line on which he had been put. He was instructed both to increase the power and scope of his transmissions and to revive the assembly of SRC 70s.

Previously, in 1916, David Sarnoff had written to his general manager at Marconi's, anticipating sales of one million 'radio music boxes' at $75 the set within three years of the commencement of manufacture – $75 million. After the war, Sarnoff became an executive of the Radio Corporation of America. RCA had been the British Marconi Company in the United States. The new entity was created in 1919, at government behest, by expropriating the British owners. It was felt that, as a measure of national importance, radio ought to be locally controlled. In 1920, Sarnoff and a colleague visited an independent radio engineer who had perfected a uni-control radio that was simpler to operate than the SRC 70. The other executive was busy pointing out how useless the device was since the joy of radio was clearly to have a lot of knobs to tune and the enjoyment of the privacy of a headset. (Failure of vision, as much as unthinking enthusiasm for technology, can lead equally well to the misunderstanding of media.) Sarnoff, who understood that within the culture music was listened to collectively, contradicted his colleague, saying 'This is the radio music box of which I've dreamed'.[34]

Westinghouse, on the basis of Conrad's work, was invited to join in the ownership of RCA with General Electric, AT&T and United Fruit. RCA, by the third year of the radio boom, had sold $83 million of sets, $7 millions'-worth more than Sarnoff had predicted nine years before. Radio, by 1922, was already a $60 million a year business.[35]

Much the same thing was to happen with television. In 1945 television in America was still experimental, in that there was no widespread service. RCA had built a plant for the navy at Lancaster, Pennsylvania, which mass-produced cathode ray tubes. Immediately after the war RCA bought the plant back and within a year was manufacturing a 10 inch table-top television, the '630 TS'. It sold for $385, which compares well with Sarnoff's proposed 1916 price of $75 for a radio.

Even beyond the not inconsiderable effects of marketing the sets, RCA and others were strenuously arguing that the post-war economy needed enhanced levels of consumer demand in general which could be fuelled only by an effective new advertising medium. These considerations were finally to break the deadlock in the development of American television. The brake had been applied by various interests since at least the start of German and British public services (using, to all intents and purposes, RCA systems) in 1936. In the intervening decade the United States had been at war for only three years. The war cannot be claimed as the delaying factor in all the years of peace up to 1942. On the contrary, it was this war which provided, by creating a vast electronics industry, the final supervening necessity for television.

This necessity, though, would have been insufficient of itself had it not been for other long-standing drives at work in the society, drives which created and sustained the research interest in television. As in

Tolstoy's tale of the giant turnip, the efforts of the mouse are required at the end of the farmer's line of stronger helpers to get the monstrous vegetable out of the ground and into the house. The slowing cathode ray tube assembly lines are the mouse. The farmer, his wife and all the others are to be found, as it were, in those tendencies within society creating entertainment forms for the urban mass.

There had been important underlying supervening necessities discernible in various phenomena from the creation of the mass-circulation press through the refining of the organisation of the popular stage to the appearance of the cinema and radio. Across the one hundred and fifty years of the rise of mass culture two movements can be plotted.

First, there is the homogenisation of the consumers' experience. Mass circulation newspapers bring each and every reader the same thrill and the same *umwelt*; centralised booking and theatre ownership systems bring each member of the audience the same act, the same spectatorial experience. These nineteenth-century developments are massively aided by the twentieth-century technologies being discussed here. Not only that, but the rising dominance of the nuclear (as opposed to the extended) family and the provision of ever more comfortable accommodation for that family, reaching even into the working class, create a further movement to take these homogenised entertainments (except for people seeking mates) into the home.

Couple this with a second factor of yet longer standing – the addiction to realism which characterises all systems of representation within our culture. Twice in Western history, in Greece between the fifth century BC and the second AD, and again in Western Europe between the fifteenth century and the present, these systems, literary and pictorial, have moved from abstraction and idealisation to the particular, the concrete and the verisimilitudinous.[36] The rise of the realist tradition in the nineteenth century speaks eloquently to the current power of this addiction. Every mass medium has, in generating its codes of representation, eschewed all non-realistic possibilities. The camera's and the radio microphone's capacity for the fantastic remains largely unexplored. Realism rules.

These general cultural traits constitute powerful underlying super-vening necessities for television, but they are not of themselves strong enough in the period from 1936 (at the latest) to get a television system widely introduced to the public, either in the US, where such a possibility is denied by the government, or in the UK and Germany, where it is encouraged. They are in fact the sustaining elements behind the partial prototypes described in the previous section and the simultaneous research effort into all-electrical systems – the actual *invention* to be outlined below. Yet they cannot overcome the forces working to protect the status quo against the disruptive effects of the new technology. The result was a sort of stand-off. As with the telephone in the first years after Bell's establishment of AT&T, so with television. There was no public clamour for the new device, whatever form it took.

Paradoxically, this can be best demonstrated by the British experience in these years. The British, ignoring the supposed beneficence of the market place, inaugurated by government fiat a full-scale public experimental service in November of 1936 using two systems. In February of 1937 the London Television Standard of 405 lines at 1/25th of a second, interlaced, was adopted and the Marconi-EMI system was allowed to broadcast alone.

The public take-up of the service proves that it was premature. Barely 2,000 sets were sold in the initial year of operation, and this despite a reduction of about 30 per cent in their cost and the wonders of the first televised coronation. By the outbreak of war only about 20,000 sets had been sold. The problem was the programming; viewers' responses, audience research and the press all were largely favourable, although the level of repeats drew criticism.[37] More significant were the limited hours, worries about obsolescence because of the conflicting systems and an unawareness that the highly publicised low definition Baird picture was a thing of the past.[38] But these problems would have vanished if the demand had been there. Had the public pushed the market, as it was to do in the 1950s, the BBC would have responded as it did then, diverting resources to the new medium from the old, radio. But in the late 1930s the old medium had, like the BBC itself, barely come of age; and talking pictures were even younger than that. Both were prospering despite the depression. Public attention, and with it the public's purse, were elsewhere. And where the public might want television, the local entertainment industry stood ready to ignore the demand. For instance, in the US a 1939 poll estimated that 4 million Americans were eager to purchase television sets.[39] The FCC took no notice and these other media, because of the investments they entailed, constrained television development and thwarted such public appetite as existed.

The technological performance producing television prototypes had exploited the contradictions between the brake of established markets and the need to create, through innovation, new markets; but, as ever, those interests most threatened were exactly those most responsible for the prototypes. The corporate and government feuding occasioned by these somewhat schizophrenic circumstances inhibited the broader social necessities from working to diffuse the innovation. The final proof of this can be seen in the actual British experience, rather than in opinion polls. The BBC, having been forced into television, was, in the years before the Second World War, in effect operating a system without audience and without proper manufacturing base. In 1939 there were almost no sets.[40]

Television was *invented* in the straits between the Scylla of established industry and the Charybdis of innovation. The broad social supervening necessities, requiring highly iconic home-delivered entertainment systems, sustained technological performance so that research on the device continued through all this, but slowly. It would take the particular necessity of maintaining wartime levels of electronic manufacture to make the television receiver – in Britain and the US, as well as in the rest of the world – the shining centre of every home.

Phase three: technological performance – invention

Vladimir Zworykin invents TELEVISION, 1923

(i) A rose by any other name
Technological performances synchronous with or subsequent to the effects of supervening necessities constitute *inventions*. The technological performance that *invents* television is synchronous with the broad social necessities described above. A viable system of a totally modern design was available by 1934, more than a decade after it had first been patented. It was being used for both British and German services during 1936, and is not substantially refined before the action of that supervening necessity (spare post-war capacity in the electronics industry) which causes it, finally, to be diffused. From 1923 on the detailed principles of the device are known; and from 1934 the viability of their practical application is in no real doubt. The system remains essentially unchanged up to the present, which can be seen in the compatibility of the original device with all subsequent developments.[41]

Vladimir Zworykin emigrated to the United States after the Russian revolution, failed to find work and returned to Omsk in March of 1919, but came back six months later to a position with the Russian Embassy as a book-keeper. In 1920 he resumed his engineering career by securing a research post with Westinghouse.[42] He had studied with Rozing at the Institute in St Petersburg and in 1912 he had gone to the Collège de France to do graduate research in theoretical physics and X-rays. During the First World War in Russia he built tubes and aviation devices for the Signal Corps. In 1923 Zworykin patented a complete electrical television system including a pick-up tube, that is to say an electronic camera, along the lines suggested by Campbell Swinton. The camera was, as was Rozing's receiver, adapted from the standard cathode ray tube but of a very different design. The electrons were now directed at an internal screen which they were forced, by magnets mounted outside the tube, to scan in a zigzag pattern. Rozing had suggested that the speed of the stream of electrons could be varied in accordance with the intensity of the light, but Zworykin followed Campbell Swinton in varying the intensity of the beam itself to reconstitute the lights and darks of the scene before the lens. He also followed the Swedish researcher Ekstrom, who, in a patent of 1910, suggested that scanning could be achieved with a light spot.

Zworykin's camera tube, the **iconoscope**, used a sensitive plate of mica coated with caesium, a soft silvery element which has the distinction of being the first metal to have been identified with the aid of a spectroscope. (It had been, up to the nineteenth century, mistaken for potassium salts.) It proved to be more photoemissive than potassium and certainly much more than selenium. In the tube Zworykin built, the photoelectrons are emitted the entire time the screen is illuminated by the spot but the charge is stored until the spot returns to build up the next frame. Then the electrons are discharged. In all the competing systems, whether mechanical or electrical, the stream of electrons was created by a fragment of light striking the cell and being discharged immediately, The iconoscope's 'charge storage' system used the electrical consequences, as it were, of all the light in any one frame with an enormous increase in sensitivity.

Zworykin, writing in 1929 after his move to RCA, acknowledges other workers, French and Japanese who had been following Rozing's use of the cathode ray tube.[43] In this article, however, he describes a television/film projector interface (a telecine device) using an oscillating mirror – which is to say a mechanical (non-electrical) scanner. It seems as if the earliest results with his all-electric camera, in 1924, were of extremely poor quality and, in order to refine the receiver tube, Zworykin and his team at Westinghouse returned to signals created by Nipkowian mechanical scanning. When he moved to RCA he took up work on the camera tube again. In Germany of that year, 1929, von Ardenne (later Hitler's favourite scientist and the major figure in German radar research) had demonstrated an all-electrical system with 60-line definition.

By 1932 Zworykin had a camera that worked more effectively, at least using reflected light in a studio, than any other available. It produced 240 lines, matching the most advanced mechanical systems and it was demonstrated by a wireless transmission from New York to the RCA laboratory in Camden 80 miles away. This was still not quite an all-electric system since the synchronising pulses to stabilise the image were provided not by circuitry but by a spinning disk. Over the next two years electronically generated sync pulses were added. At the same time Zworykin designed the interlaced raster which scanned every other line in any one frame so that the frame time was halved to a 60th and picture stability further improved. Early in 1934 Zworykin wrote,

The present sensitivity of the iconoscope is approximately equal to that of a photographic film operating at the speed of a motion picture camera, with the same optical system. . . . Some of the actually constructed tubes are good up to 500 lines with a good margin for future improvement.[44]

Zworykin was fudging a little by not specifying a 16mm motion picture camera in the above quotation; for the ambition of all the television pioneers, mechanics and electricians, was to match the quality of the contemporary amateur, i.e. 16mm, picture.

The film industry defines film stocks in terms of lines per millimetre of film surface. The limiting resolutions of emulsions can be determined when the emulsion is used to photograph a chart upon which standardised blocks or lines of black have been printed. The film industry's norm was a 35mm film which could photograph between 30-40 of these lines per millimetre. To achieve the same density of visual information electronically something like 1200 lines on the cathode ray tube would be needed. This same normal stock formated for 16mm would, of course, also photograph the same number of lines per millimetre but to far cruder effect, since each line occupies more of the frame area. To match the resolving capacity of the 35mm norm, a 16mm stock would need to photograph around twice as many lines (i.e. 90 per millimetre). Conversely, the normal 16mm standard of 30-40 lines per millimetre can be matched on 35mm by 12 lines; and it was at that standard that the television researchers aimed since its electronic equivalent required only around 400-480 lines. In short, by the early 1930s, the whole thrust of the research was directed towards creating a 400-line picture, the equivalent of the contemporary 16mm film image.[45] This became 'high definition' television, the standard eventually adopted, in its 525/625 line guise, by most of the world. Today, 'high definition' is being redefined, as the world's television engineering community considers a 1200-line HDTV system. Such a standard, if adopted, simply matches the resolving characteristics of 35mm film.

The British contribution to the search for an electronic equivalent of 16mm film was much facilitated by corporate musical chairs in the late 1920s. RCA took over Victor Phonograph Inc. which in turn owned the British Gramophone Company. Thus RCA, born out of the ashes of British interests in the American radio market, came to own a slice of the British record industry. In 1931, the Gramophone Company was an element in the founding of EMI, and not only did RCA now have a share of an English laboratory but the director of research there, Isadore Schoenburg, was another of Rozing's ex-students.

Twenty years later an official EMI account suggested that the young 'believe everything of any importance in the [television] system came from the U.S.A.' and that this was a 'mistaken impression'. [our brackets][46] However, with Sarnoff on the original EMI board, it is easy to see the source of the 'impression'. All the crucial television decisions the company took between 1931 and 1934 can be considered rational only in the light of the American connection. This is not to say that further important advances enhancing the viability of the system were not made in the UK, but to all intents and purposes the **emitron**, EMI's earliest camera, was an iconoscope by another name. Even today British *amour propre*, to put it no more strongly, finds this difficult to accept.

Schoenburg, in 1934, decided to adopt a 405-line standard, which makes little sense if we are asked to believe he made this decision *in vacuo*. After all, the official requirement was at that time pegged to Baird's best – 240 lines.

> It was the most dramatic moment in the whole of television development. He [Schoenburg] said, 'What we are going to do, in this competition, we're going to offer 405 lines, twin interlace. And we're going for Emitron. We're going to give up mirror drum scanning, we're going along the lines of the electronic camera.'[47]

Sir Frank McGee, perhaps the most distinguished of Schoenburg's team, calls this 'the most courageous decision in the whole of his career', which it might well be; but it is difficult to believe that he took it without cognisance of what Zworykin had *publicly* announced, in January of 1934, to be possible, never mind what private communication might have passed between the two labs. Zworykin recalls that two of Schoenburg's engineers visited his lab late in the winter of 1933/4 for three months.[48] By 1934, 343 lines had been demonstrated on the RCA iconoscope, not quite sufficient to match the definition of 16mm amateur movies. Zworykin, as his January 1934 paper reveals, was confident that this could be achieved. Further, the majority opinion in the US was that 441 lines was the maximum that could be accommodated in the 6 megacycle channel the FCC had mandated for experiments. Schoenburg's decision to go for 405 lines and, therefore, an electronic system was without doubt farsighted; but it can scarcely be called the 'most dramatic moment in the whole of television development'. It is even more astonishing to find one of the engineers who visited RCA in 1933 claiming, in 1984, that 'McGee and EMI owe nothing to RCA and only in 1936 did the two companies sign an agreement for a complete exchange of patents and information.'[49] McGee protests in equally strong terms:

> It has been said that the Emitron (made at EMI) was a straight copy of the iconoscope (made by RCA) and that we at EMI were largely

dependent on know-how from the latter company. One reason advanced for this is that the tubes looked similar. Now I can state categorically that there was no exchange of know-how between the two companies in this field during the crucial period 1931 to 1936. And with regard to the similarity of the tubes it would be clear to anyone with a minimum of technical perception that, given the situation as it existed at that time when we did not know how to make an efficient transparent signal plate or photo-mosaic, there was no other way such a tube could be constructed.[50]

The case basically rests on the claim that 'know-how' was not exchanged between the two companies:

> Throughout the formative period of television R.C.A. and E.M.I. had certain commercial relations but these relations did not involve ownership of E.M.I. or control of E.M.I. policy by R.C.A., nor was there any exchange of research or manufacturing information on radio or television between the two companies. It is true that technical publications and copies of specifications of patent applications were exchanged, but 'know how' was not exchanged because R.C.A. was at that time only prepared to agree to such an exchange for a payment, which E.M.I. was not prepared to make. It is significant that in 1937, when the E.M.I. system was a proved success, a free exchange of 'know how' was agreed. Thus the position was that R.C.A. and E.M.I. were developing television on a basis of friendly but quite spirited competition, with neither party giving away any secrets to the other.[51]

What the significance of 'know how' is to persons with 'a minimum of technical perception' (which one must assume the British were) remains somewhat mysterious. History clearly suggests that the essence of both cameras' design was laid down in a patent of Zworykin's dated 1923. Zworykin worked for RCA. RCA owned a slice of EMI. EMI's research director shared a teacher in Russia with Zworykin. The teacher was a television pioneer. Contrary to some claims, EMI engineers did visit RCA. The Emitron looked, not like Zworykin's patent of 1923, but like his development of that patent, the iconoscope of the early 1930s. It was, in McGee's own words, 'fundamentally the same as the iconoscope.'[52] Why, half a century beyond the events in question, there should still be such a shrill insistence on primacy cannot be here addressed; suffice it to say, though, that in the long term it might drown out the real achievements of the British in this story.

The American iconoscope, in the early 1930s, was barely usable, being very noisy. Too high a ratio of interference to signal rendered results disappointing. Indeed, the tube produced more noise than picture and was a lot less impressive than the high definition mechanical scanning systems then coming on line. The reasons for this poor performance were properly analysed and ultimately corrected first at EMI. In 1933, a member of the EMI team perfected a technique to stabilise the DC component of the signal, thereby solving a clearly understood problem. More significantly, the then mysterious process whereby the electronic signal was derived from the tube was first elucidated by McGee. Secondary emission of electrons, within the tube, were found to be crucial.[53] Indeed secondary emission was a third way, in addition to photo and thermionic emission, to obtain, within a component, free electrons. As a consequence of this understanding, the team of Blumlein, Browne and White (the first two of whom were to die testing experimental radar equipment in flight in 1942) set about suppressing the unwanted signals.[54] The patented circuits which did this were passed back to Zworykin and incorporated into the RCA camera – the exchange of patents referred to above. RCA never signed patent agreements, always buying what it needed outright. This is one of only two occasions in the 1930s when this company policy was violated, obviously because of the closeness of the two organisations. (The other occasion is detailed below, page 65.) The importance of the British development should not be understated. McGee and Blumlein, in the Emitron, made the iconoscope work – but it was, all protests aside, nevertheless a development of the iconoscope.

The German experience is the proof of this. Baird Television Development Company, as we have seen, did not use its Farnsworth image dissector camera in the great TV systems run-off in the winter of 1936/7; but BTDC's opposite number in Germany, Fernseh AG, did. There a similar, if considerably more protracted, run-off had begun earlier in the year, during the Olympic Games. (Fernseh)/Farnsworth's rival was Telefunken, whose engineers had abandoned mechanical scanning at exactly the same time as Schoenburg took his 'dramatic' decision to do the same thing and for much the same reason. Like EMI, Telefunken was working, with RCA agreement, on a design derived from Zworykin. It yielded the **Ikonoscop-kamera**. During the Games a coaxial cable network linking five cities was established. Fernseh could use its Farnsworth camera directly for the indoor events only, such as swimming. Outdoors it used the intermediate film system it had perfected and sold to BTDC and even to Farnsworth's company in the US. Telefunken used the ikonoscop everywhere.

In the years that followed the systems were allowed to coexist. In 1938, Fernseh covered the *Parteitag* in Nuremberg. In 1939, Telefunken introduced its **Super-ikonoskop** which produced 441 lines and worked well in the most adverse lighting conditions. At the outbreak of the War Telefunken had Fernseh on the ropes. Like BDTC, the latter had only television with which to support itself. Telefunken, like EMI, was a major electronics firm, with or without television and Telefunken slowly demonstrated the superiority of its system. The Nazis decided to use the super-ikonoscop for coverage of the 1940 winter games. Fernseh had lost out. The German run-off had taken more than five years and was rather more to the point than the four-month British exercise. But then, had BTDC managed to get its Farnsworth image dissector to work, perhaps the British would have taken rather longer to decide in favour of EMI/(RCA).[55]

(ii)   First television camera

Philo T. Farnsworth's contribution to the development of television gives the lie to a number of assumptions about the necessary conditions for high-tech innovation in the twentieth century. The technical history of television has been used as a prime example of the industrialisation of innovation which is generally deemed to be such a prominent feature of modern life. Campbell Swinton encapsulated the implications for television of modern corporate research methods in a lecture before the Radio Society of Great Britain in 1924.

> If we could only get one of the big research laboratories, like that of the G.E.C. or of the Western Electric Company – one of those people who have large skilled staffs and any amount of money to engage on the business – I believe they would solve a thing like this in six months and make a reasonable job of it.[56]

In the event, with something less than the whole-hearted backing Campbell Swinton envisaged, it took the industrial researchers longer than six months to 'invent' television. The idea, though, that they generally worked in a way different from the old-style individual innovator nevertheless obtains an important boost in the received history. The position Baird occupies is crucial, for nothing is as powerful as the notion that he, an old-style private inventor, was limited to mechanical systems because he was unable to compete with the Marconi-EMI industrial lab in the more sophisticated area of the cathode ray tubes. In fact, Baird seems to have just had a thing about tubes and an obsession that his first thought was the best. Vic Mills, his earliest collaborator, alerted Baird to CRTs in 1924:

> I said you can't play about with those spinning discs and think you're going to get television. I told him to go ahead with cathode-ray tubes. I'd read about it in a book printed in 1919 and it made me want to take the long jump and avoid all this mechanical business. But if I knew only a little about the cathode-ray tubes, Baird,

apparently, knew nothing. He was simply not interested. He could comprehend the mechanical system but the idea of doing it all electronically appeared to be out of the question.[57]

These traits could just as well have manifested themselves under the aegis of a great corporation. Take, for instance and at random, the following: the almost universal failure to perceive a market for computers in the years before and after the Second World War, RCA's obsession with a non-laser videodisk technology, the decades of the imminent arrival of the videophone. And, as the role of Philo Farnsworth dramatically illustrates, the obverse is equally true: the lone innovator was not precluded from the cutting edge of experimentation. Farnsworth, the boy-wonder electrical genius who occupies Baird's place in the popular American understanding of television history, competed effectively with the radio industry, producing work every bit as sophisticated as that of the industrial teams.

He was a rural mid-Westerner who came from a Mormon home which only acquired electricity when he was 14. He learned his science from popular magazines and read up on mechanical television. At 15 he confounded his chemistry teacher at high school by describing, on the blackboard, an all-electric device which he thought might work better. He left school and had no further formal education.[58] In 1927, aged 19, he patented an electrical pick-up tube which operated on quite different principles from Zworykin's. Called an **image dissector**, it had the advantages of offering a more stable picture than the iconoscope, free of shading and saturation at the white levels. It used neither a scanning spot nor the storage principle. It worked by translating the image into a pattern of electrons which were then passed across an aperture. The disadvantage of the device was that, without charge storage, the image dissector was far less sensitive than the Zworykin design, but when Farnsworth patented the device this did not necessarily seem like too great a problem.

[A] picture produced by the dissector tube was of better definition and sharper contrast than that produced by the iconoscope. It became apparent that the Zworykin iconoscope was a superior instrument for use in studio and outdoor pickup where lighting was a problem. In motion picture work and bright sunlight, the dissector tube had an advantage.[59]

The image dissector could not work in a studio – but in the 1920s and early 1930s, the iconoscope did not work well enough in that situation to be real competition, and with the intense direct illumination in a telecine device, the comparative (but rather theoretical) disadvantage of the dissector was removed. Film transmission was seen as being central to television development in this period.

Farnsworth's achievement was that he built a more effective electronic camera earlier than RCA did. He was on to the secondary emission problem at the same time as EMI and indeed made, in 1937, what was to be his greatest contribution to electronics in general by designing an **electron photomultiplier** specifically to exploit it. He arranged a series of collector anodes with successively increasing voltages inside the dissector tube so that the weak electron output of the aperture would bounce between them and be exponentially increased. He was close enough, in fact, to worry RCA considerably. Both Sarnoff and Zworykin visited his laboratory and their professions of disinterest in his advances were not backed by their companies' moves against him. RCA mounted an interference action against him which he successfully defended (with the help of his chemistry teacher and at a cost of $30,000).

By then Farnsworth was not a man alone. Unlike many individual engineers, he conducted himself as well in the boardroom as he did in his laboratory. He had joined the Philadelphia Battery Company, Philco, a major rival of RCA's which had made its mark by specialising, from 1927 on, in car radios. Philco had begun by selling large domestic batteries to houses not on the electricity grid. As electrification killed this business, the company moved to radio where it was now outselling RCA nearly 3 to 1 and had over a third of the market. Rivalry between the two firms climaxed with Philco bringing an action against RCA for industrial spying, specifically charging that RCA operatives took some of Philco's female employees to Philadelphia where they

> did provide them with intoxicating liquors, did seek to involve them
> in compromising situations, and thereupon and thereby did
> endeavour to entice, to bribe and induce said employees to furnish
> them . . . confidential information and confidential designs.[60]

Protected by Philco's broad back, Farnsworth was able to resist RCA attempts to dislodge his patents of 1927. The dissector was sufficiently different from the iconoscope for Farnsworth to maintain his rights, but his receiver was close enough for him to pick up Zworykin's pictures on his apparatus.

He left Philco (amicably) and in the early 1930s forged his international links, assigning his patents to Baird in the UK and Fernseh in Germany. In 1937, upon the introduction of the photomultiplier, Farnsworth reached cross-patenting agreements with both AT&T and CBS, RCA's great rivals.[61] Farnsworth was too good a scientist not to realise that Zworykin's charge storage principle, despite the photomultiplier, was superior – was indeed television –

while his concept had produced, by however small a margin, a *partial prototype*; but he was also too shrewd a businessman not fully to exploit the very real contributions he had made. He and his partners poached RCA's head of licensing to be the president of the Farnsworth manufacturing company created in 1939. It was showdown time and, despite the fact that the iconoscope was by this time superior to the dissector even in film chains, he nevertheless still had an ace up his sleeve.

In 1935, after a three-year proceeding, Farnsworth had successfully won his own interference action against Zworykin and RCA. Farnsworth's device scanned an *electronic* image – in the word of his 1927 patent application, he had designed 'an apparatus for television which comprises means for forming an electrical image'.[62] Zworykin, in his 1923 patent, the patent official held, had suggested scanning an *optical* image – a significant difference.[63] With the start of RCA's service scheduled for the World's Fair opening in 1939, this became a critical issue. The improvements that had allowed the iconoscope to overtake the Farnsworth camera had been purchased by incorporating Farnsworth concepts. Indeed Zworykin now designated his machine the *image*-iconoscope because it scanned an *electronic* image, and that clearly infringed Farnsworth's patent. After five months of negotiations RCA was forced into a licensing agreement with him. This was an event with only one precedent in the history of the company – the EMI deal. In September the agreement was reached and Zworykin published the details of his latest camera.[64] Farnsworth's biographer claims that the RCA vice-president who signed the contract wept.[65] It is for this victory, one would like to think, as much as anything, that Farnsworth was commemorated by the United States Post Office in 1983 with a stamp bearing the somewhat suspect legend 'First Television Camera'.

Another stamp en bloc in this same issue remembered Edwin Armstrong whose career, although a contrast to Farnsworth's, is a further equally potent rebuttal of the received idea that the lone inventor no longer has a place in the twentieth century. After selling RCA some radio patents in 1922, Armstrong, as a result independently wealthy and married to Sarnoff's ex-secretary, set about obeying a wish of the great man that a radio system less subject to interference, noise, be created. Three years later he lost his triode tube battle with De Forest but continued to work at Columbia University until, in 1933, he produced a signal of preternatural clarity, at least to contemporary ears, by using a totally new system of modulation.

If an analogue of the original, be it sound or picture, is to be impressed on a carrier wave, there are in essence two ways only in which it can be done. Either the signal is modulated in strength or in speed. (Digitalisation raises different and more complex options.)

Both Zworykin and Farnsworth opted for strength – intensity – as the underlying principle of the television system, although they dealt with the electron stream very differently. It will be remembered that Rozing had proposed, albeit very vaguely, a frequency modulation system, whereby the greater the light the faster the stream of electrons. Exactly the same division was possible with radio.

The first radio system, like these first television systems, relied on amplitude modulation. Armstrong, in 1934, produced a signal created by frequency modulation but, at RCA, the great man's needs had changed. The noisy AM radios of the early 1920s now worked well enough for Sarnoff to be running a major enterprise. At the height of the depression, RCA shed less than 20 per cent of its employees and, in all the years of economic disaster, it made losses only in 1932 and 1933. Armstrong's device was no longer needed and, worse, it now posed a considerable threat. Sarnoff said FM was not just an invention but a revolution. In the event it was a revolution that RCA did not want. Armstrong was asked to remove his equipment from the Empire State Building.[66] The battle for FM radio became, because of frequency allocation issues, inexorably intertwined with the arguments about the development of television. One result of that battle is that FM is used as the standard for television sound.

Armstrong sued RCA for royalties and spent a full year in the witness box being grilled by corporation lawyers for his trouble. Unlike Farnsworth, he gave in and authorised a settlement. All his life, since 1911, he was embattled, first with De Forest and then with RCA. It proved too much for him. The money, $1 million, was paid to his estate, for he had stepped out of his apartment window before the deal was finalised. Armstrong's commemorative stamp bears the legend 'Frequency Modulation' and about that claim there can be no debate.[67]

Armstrong and Farnsworth made contributions equal in sophistication to those made by McGee and the other workers in the great laboratories. Baird lost out not because he was an individual but because he was wrong. The history of television is not as ready a prop for the industrialisation of innovation thesis as is commonly supposed.

Yet the greatest claim to the invention of television is undoubtedly Zworykin's, and it was in all essentials first built under the aegis of four large radio organisations, for every basic aspect of modern television systems conforms to Zworykin's original patent description of 1923 and the devices he built to refine and develop those ideas in the 1930s. But, in another crucial sense, the *invention* was still in doubt at the end of that decade because the device was not widely diffused.

This third phase of technological performance – *invention* – can take place either synchronously with or subsequent to the crucial supervening necessity which ensures diffusion. In the case of television, the *invention* phase is synchronous with some underlying supervening necessities, the general drives conditioning the development of popular entertainment, the addiction to realism in the culture, the supremacy of the nuclear family home. But it precedes the operation of the final supervening necessity – spare industrial capacity at the end of the Second World War – which ensures diffusion. Thus the third *invention* phase is not only synchronous with some underlying social necessities, it is also synchronous with the second phase of *prototype* performance – which is why the British could hold run-off between a *partial* mechanical prototype and the *invention* proper. Television, modern high definition all-electrical television, was 'invented' by 1934 at the latest. The man who thought of it in 1908 was convinced that it could have been produced nearly a decade before this. The pace of these developments, especially in America from 1934 on, is not determined by technological performances at all but rather by the operation of the next and final transformation, *the law of the suppression of radical potential*.

The third transformation: the 'law' of the suppression of radical potential

The FCC procrastinates about TELEVISION, 1934-1952

(i)   The common good of all, 1934-1948
The received explanation of what happened, in America, after television had been 'invented' is as follows:

The Radio Manufacturers Association, a trade group dominated by (but not wholly a creature of) RCA, had set up a television committee which, by 1935, was ready to set about establishing appropriate standards. RCA by this year had reached 343 lines in a interlaced raster scanning at 1/30th of a second, although Zworykin had made some tubes that could produce more than 400 lines.

In the following year, the FCC began the difficult business of making frequency allocations for television in what was then still called the 'ether spectrum'. The RMA prepared to give its evidence to the commission. On standards, a 441-line picture on a band 6 megacycles wide (the lines being thought the maximum accommodatable in that bandwidth) was put forward. Other matters, the polarity of the signal, the aspect ratio of the picture, the synchronisation standard, all these things were agreed within the RMA and sent to the commission. Although only three additional issues were to be decided subsequently (in the final agreement on standards reached six years later) the FCC, upon receipt of these recommendations, acted in nothing but the matter of bandwidth allocation, more or less following the manufacturers. As to everything else, the commission felt further experiment was in order. Within weeks of this determination, the Germans began broadcasting their version of the RCA system and within months the British version was established as the London Television Standard.

In America, two years of intense activity followed the FCC announcement, during which the RMA added refinements to the basic plan it had originally drawn up – such elements as horizontal direction of polarisation, a maximum on the amount of picture carrier that could be used for synchronisation. But there was one major difficulty. The possibility of a new transmission technique was opening up and this had implications for television bandwidth requirements.

Essentially, analogue electrical signals duplicate themselves on either side of a median line. If just one side of the signal is transmitted the receiver can create the mirror image and thus restore the original wave form. What is saved by this single side band (SSB) transmission is bandwidth, since the signal now occupies less space. The RMA took the imminent introduction of SSB on board when drawing up its revised proposed standards which it eventually got round to handing over to the commission in September of 1938. The FCC asked the RMA if it felt a formal hearing should be scheduled. The RMA said it did, but no action was taken.

Since the effective public demonstrations in Britain and Germany of what was its system (as amended by Blumlein and McGee) were about to celebrate their second anniversary, RCA announced that it would begin a public television service in April of 1939 to coincide with the opening of the New York World's Fair. Zenith and Philco felt that this was a premature move, but other manufacturers began to prepare for the making of sets (and the payment of royalties to RCA). The FFC were still contemplating hearings on the revised RMA standards. During that winter the implications of the SSB transmission were being worked through, so that NBC, the RCA subsidiary responsible for the announced service, had to change its equipment specifications as it was running down to the opening. The engineers met the deadline and the first American television programme specifically intended for the public, although broadcast on an experimental licence, was of President Roosevelt opening the Fair. The whole episode, for the FCC, was an example of the volatility that it wanted to protect against; but equally, the fact that NBC began broadcasting was proof that changing technical requirements could be accommodated.

In December of 1939, as NBC continued to broadcast a couple of hours a week, the Commission considered relaxing its ban on the commercial exploitation of the new medium and proposed that, if the monies accruing were used for further experiments, the stations could seek sponsors. A hearing was called on these new proposals for January of 1940, the matter of the standards still being left in abeyance. At this, the first public opportunity that had arisen to discuss national television policy (such as it was), there was considerable dissension within and without the ranks. The two agenda – commercial exploitation and technical standards – were conflated. Dumont Laboratories, an organisation outside the RMA, felt the proposed standards were too limited. Dumont was searching for a phosphor which would allow a very slow scanning speed without flicker, which in turn would permit more lines to be transmitted. The previous month Allen Dumont himself had shown the RMA engineers a tube built along such lines, but they were unimpressed. Philco, not the most satisfied RMA member, also complained at this hearing that the 441-line standard was too low and suggested that a slightly slower frame rate (24 frames) and an 800-line standard would be more appropriate. Since the firm had no devices to show, the proposal, and other subsequent and equally unfounded suggestions, can be seen as nothing but spoiler tactics designed to halt or at least slow down RCA. Thus the row confirmed the commissioners in their inaction over the RMA standards and they further decided to delay their own proposed introduction of commercialisation until September of 1940. They wrote:

> The commission therefore recommends that no attempt be made by the industry or its members to issue standards in this field for the time being. In view of the possibilities for research, the objectives to be obtained, and the dangers involved, it is the judgement of the commission that the effects of such an industry agreement should be scrupulously avoided for the time being. Agreement upon the standards is presently less important than the scientific development of the highest standards within reach of the industry's experts.[68]

Had the war in Europe not suspended transmissions, the British and German services would have been preparing to celebrate their fifth birthdays at the time of this decision. The London Television Standard, without SSB and with only 405 lines, worked well enough for there to be more complaints about repeats than about technical quality.

In America, RCA's response was to read the commission's inaction as a go-ahead. It announced, with considerable fanfare, increased production of sets and an extension of programming. The FCC, seeing these plans as a further attempt to freeze standards *de facto*, called another hearing for April. Dumont and Philco gave evidence to the commission that RCA's moves had caused them to abandon their work on alternative systems. Others, apart from RCA, claimed there was enough space within the RMA standards to permit refinements without rendering the public's receivers obsolete. RCA itself said it was only doing what it thought the majority of the industry wanted. The FCC abandoned its partial commercialisation plan and ruled that, 'As soon as engineering opinion of the industry is prepared to approve any one of the competing systems of broadcasting as the standard system, the commission will consider the authorisation of full commercialisation.'[69]

In this stand-off situation within the industry and between the industry and the commission, the FCC and the RMA agreed to the formation of a National Television System Committee, which would include all qualified engineering opinions whether within the RMA or not. In short order, at least by comparison with the previous pace of events, 168 people produced 600,000 words in minutes and reports, met for 4000 person-hours and spent a further 4000 hours gathering on-site evidence and watching 25 demonstrations. All this was done within six months. From the time of the initial agreement to proceed in this way to the acceptance, by the FCC, of the final report of the NTSC a mere 14 months elapsed. The monumental quality of this enterprise is somewhat reduced when it is remembered that virtually everything of substance had already been previously decided. The lines were increased to 525 because it was now known that this rather than 441 was the effective maximum for the 6 megacycle band. FM audio was chosen. The VHF spectrum was secured. But for the rest the NTSC endorsed the RMA proposed standards of 1939 which were a rerun of the 1936 suggestions.

On 27 January 1941, the chairperson of the FCC, James Fly, said,

This is another example of the best that is in our democratic system, with the best in the industry turning to a long and difficult job in an effort to help the government bodies in the discharge of their function so that a result may be achieved for the common good of all.[70]

How then can the commission's insistence on technical excellence be evaluated? It was by no means a given of regulatory policy, as are, for instance, the Independent Broadcasting Authority's stipulations regarding the financial stability of UK franchisees. The FCC was not, in other areas, so insistent on maximising technical quality. By ignoring FM it eschewed a commitment to the cutting edge of science when it came to radio. At this same time, in fact, the FCC finally granted the first FM radio licences, seven years after Armstrong's conclusive demonstration of the superiority of his system.

What had been gained technically during the period of the 1936-1941 delay was marginal. 525 lines were better than 441 lines but both were in the same range, the quality of the 16mm home movie image. At 525 lines that was still all that was achieved. As for the imposition of FM sound, it can be argued that the VHF band, being less subject to the sort of interference FM was designed to combat, did not require it.

Certainly, there was no comparison between the need for FM sound in television and the imperative created by the poor quality of the AM radio bands, yet such big business flourished in AM that the FCC did not even consider a mandated and disruptive radio move to FM.

More than all this, at the end of the day in 1941, the two major technical options which really did need serious consideration were ignored. The possibility of moving to the UHF band, which the FCC had begun to license for experiment in 1937, was left hanging over the future of television and FM radio (which would have occupied the vacated VHF channels). And, prior to the NTSC agreement, CBS had demonstrated a viable colour system which was also ignored. Indeed the uncertainties of the art in 1941, when the bullet was bitten, were if anything greater than in 1936.

If the official explanation of the agenda then makes no sense, what was going on? Simply, *the 'law' of the suppression of radical potential* was at work. The disruptions feared were many. Television had to be made to fit into a media system already accommodating live events of all kinds, print, films and radio. And the diversity of manufacturing and programming interests had to be continued so that the balances achieved across the entire mass communication industry would not be upset.

The contrast with radio is instructive. The conflicts between wireless and wired telegraphy had been resolved before the First World War when AT&T disgorged Western Union, which it had bought up, and promised the government to operate only in wired telephony (see below, chapter 6). Television threatened radio as radio had threatened telegraphy and telephony. Radio had though, by comparison with its potential victims, one major disadvantage – anybody could listen in. When, by 1920, this disadvantage had been converted into the raison d'être of the device, it threatened no other interest – not even the movies which then did not talk. But more than this, with radio the ground for exploitation was cleared in a far more effective way. RCA itself was owned by all the major players – Westinghouse, General Electric, AT&T. It was their creature, not their equal. And, parallel with this, there was an agreement among them and others to pool all patents.

AT&T, which had suffered concerted government attacks from the moment the telephone had established itself as an essential tool of our civilisation, was ever watchful about conflict of interest charges. Nevertheless, although it owned part of RCA, by 1923 it was eyeing radio hungrily:

We have been very careful [said the vice-president charged with radio responsibilities], up to the present time, not to state in public in any way, through the press or in any of our talks that the Bell

system desires to monopolize broadcasting, but the fact remains
that it is a telephone job, that we are telephone people, that we can
do it better than anybody else, and it seems to me that the clear,
logical, conclusion that must be reached sooner or later in one form
or another, we have got to do this job.[71]

But, apart from the ever-present threat of anti-trust action, radio was
proving a real bother – not a job for telephone people for all that the
station they owned in New York was the first to broadcast
commercials. (It did this because selling airtime seemed exactly like
selling telephone time.) It had to sack one of its own early radio stars
at the prompting of the Secretary of State because the broadcaster had
dared to criticise the politician. As one AT&T executive wrote at the
time, the phone company believed in a 'fundamental policy of constant
and complete cooperation with every government institution that was
concerned with communications',[72] and even, as the incident reveals, a
few that were not. AT&T's plan to treat radio channels like telephone
lines was seen by almost all other parties as an attempt to create
another telephone monopoly. 'Monopoly' and the threat thereof
became, in the debate about broadcasting that raged in the 1920s, a
code expression for the phone company and its schemes. It was only by
threatening to exploit the cross-licensing agreements and set Western
Union up as a rival radio network that the other members of the radio
manufacturing pool got AT&T to back off.

By 1926 AT&T was prepared to give up the imperial scheme for
radio. It had anyway determined that more money was to be made out
of leasing long lines to the radio industry than could be earned by
owning a few radio stations. The pursuit of both, having already
antagonised the other radio interests, would obviously create a conflict
of interest of exactly the sort the Justice Department would act against
without hesitancy. So it sold the New York station to RCA, at the
same time relinquishing its RCA stock. AT&T seemingly got out of
the radio business (except that networking depended on it completely).
Its continued interest in television was simply part of the on-going
work of Bell Labs.

AT&T had, more or less, decimated the telegraph companies
domestically by the First War. It was not going to let any other
organisation, brandishing a new technology, do the same to it. Bell
Labs, founded exactly to deal with radio, is in fact not an engine of
innovation but rather an expression of constraint, of the operation of
the suppression of radical potential. It is a real 'Breakages Limited'.
Bell Labs' primary interest in all new technologies has been to protect
the phone company. The best way of doing that is by controlling at
least some of the patents in any new area. By 1926 it was clear that the
phone business and the radio business were different. It was also clear

how AT&T, with its monopoly of long lines, could get a slice of the radio pie. The protection afforded by the ownership of radio patents and RCA stock was no longer needed. On the contrary, the limitations on AT&T's area of manoeuvre by still being that involved were considerable.

In 1930 the more direct pressure of consent decrees pushed General Electric and Westinghouse out of RCA. The Justice Department had found it increasingly uncomfortable to have them, at the manufacturing level, in supposed competition with themselves. RCA was no longer a creature of the industry but was now a corporation like any other. It did not dominate radio manufacture nor did it dominate programming, but none of its rivals in either of these areas could match its total range of operation. Its corporate style was a consequence of this. RCA was a strapping and o'erweening bullyboy when it came to television. It commandeered all the primary patents without there being any general understanding that this power would be used with restraint. Sarnoff was even prepared to take on the phone company: 'The ideal way of sending messages is to hold up a printed sheet that will be immediately reproduced at the other end; facsimile transmission and television are about ready [for that].'[73]

Some viewed this belligerency with approval. RCA owned a telegraph company and in 1937 was seen to be trying to extend its business in this area by buying up Western Union. Film industry elements thought that the Justice Department might be deliberately encouraging the creation of a giant capable of tackling AT&T which Hollywood would cheer because AT&T had found its slice of the movie pie, too. 'We are the second largest financial interest in the motion picture industry,' wrote J.E. Otterson of AT&T in 1932. Otterson was executive vice-president of Electrical Research Products (ERPI), the AT&T subsidiary set up in 1927 to exploit the Westrex sound system which AT&T owned. (ERPI also held the Ives television and film/television interface patents registered that same year.) Within five years Otterson could report:

The motion picture industry in the United States owes us about sixteen million dollars and our expected revenues from the industry for the next ten years is about sixty-five million dollars. This is a large stake and establishes our interest in the welfare of the motion picture industry.[74]

A year later, Otterson evaluated the position in these terms:

It is true today, as it has been for three or four years, that the Telephone Company can control the motion picture industry through ERPI without investing any more money than it has now invested. . . . The industry is in crying need of the kind of strength and character that could be obtained through the influence of the Telephone Company.[75]

But the 'Telephone Company' had to maintain as low a profile as the ultimate Hollywood Tsar as it had tried to do previously as radio overlord. It signed up every major studio, bar RKO which was owned by RCA, and insisted from the outset and well before any all-electrical television system was in the offing that all sound films shot using its equipment (the majority) were not licensed '[F]or any uses in or in connection with a telephone, telegraph or radio system or in connection with any apparatus operating by radio frequency or carrier currents.'[76] The circumlocutions are a prophylactic against the Justice Department. AT&T meant, in 1927, that no films, recorded on its sound system, were to be sold to television. The 'strength and character of the Telephone Company' (which, as we have seen, scarcely encompassed the idea of free speech) did not, of course, express itself in loyalty to the motion picture industry.

AT&T was hedging its bets. In the next decade, it maintained its interest in the videophone. It did a deal with Farnsworth over patents which in turn gave it a lever over RCA, but it also responded to a legal threat from RCA by dropping all restrictions and limitations in the contracts and royalty agreements for sound film. Out of these manoeuvres grew the mutually beneficial, from the legal stand-point, general cross-patenting agreement between the two giants.[77] Most importantly AT&T determined that the one essential which television would need, when it finally got started, was long lines with which to build networks. By 1936 Bell Labs had developed an effective wire capable of carrying the enormous bandwidths that video signals demanded. In 1937 AT&T began to build the national network of these coaxial cables with a link, costing $580,000, between New York and Philadelphia.[78] From the phone company's point of view radio and television were essentially the same pie and AT&T, having made its peace with RCA, was ready for television even if the rest of the players were not.

Thus the ostensible agenda of the debate between 1936 and 1941 was comparatively meaningless. The decision made in 1941 introduced a system only marginally more refined than that proposed in 1936. All subsequent advances in that period could have been made compatible – just as colour was to be. The hidden agenda was to hold down RCA and that worked.

The British, grappling with the same problems, were quite specific as to what had to be guarded against – 'any monopolistic control of the manufacturing of receiving sets'. The Selsdon report goes on:

The ideal solution, if it were feasible, would be that as a preliminary to the establishment of a public service, a Patent Pool should be formed. . . . We have seriously considered whether we should advise you [The Postmaster General] to refuse to authorise the establishment of a public service of high definition Television until a comprehensive Patent Pool . . . has been formed.[79]

Lord Selsdon and his committee decided this was impracticable and gave the nod to the service anyway. In America though, this is what, in effect, the FCC by its prevarications sought to do, masking a process of mandated diversification, not just of manufacturers but of programming institutions as well, beneath a rhetoric of technical concerns. The commission was well aware of the underlying patent problem. In May of 1939, just after negotiations between RCA and Farnsworth had begun in earnest, the commission's patent department noted that, should RCA buy out these patents, it would acquire complete control over the technology. The following year, perhaps as a result of having failed to do this – only a licence was secured – General Sarnoff told the Senate that, beyond RCA, only the Farnsworth patents mattered. The NTSC standards meant in effect Farnsworth and RCA licensing for all equipment. The patent staff, at the end of the war, pointed this out to the commissioners, yet – like the Selsdon Committee – they went ahead anyway, not allowing the question to be discussed. In 1947, RCA finally managed to achieve company policy with regard to the Farnsworth patents – it bought them outright. Its monopoly position was confirmed with the supremacy of the **image iconoscope**, a camera combining Zworykin and Farnsworth elements.

Despite this, RCA was contained. When television finally happened, it was not alone. There were rival networks, many manufacturers, diverse programme suppliers. Its containment does not imply that it did not profit mightily from its television investment. It is simply to say that it did not profit as mightily as AT&T had done from the telephone – that is, with a virtual monopoly. The market place itself was not enough to ensure that result. On the contrary, since no 'inventor' came up with a system as viable as that of RCA's 'inventors', the unfettered company would have cleaned up; or, as with early American telephony, chaos would have ensued with many minor competing systems needing to be absorbed. Hence the interventionist role of the FCC, unconcerned about the public but working most effectively to keep the industry stable through a period of threatened upheavals. The 'law' of the suppression of radical potential ensured that all the major radio industry players and AT&T should remain in the new game – and they did.

In this the commission was further aided by the war. The FCC authorised the commercial operation of television stations in accordance with the 22 NTSC standards on 1 July 1941. On December 7th the United States entered the war. The war prevented the creation of the mass medium of television but it also allowed the manufacturers to regroup for the new product with a minimum of disturbance. They ceased making domestic radios, more or less, and worked on material for the armed forces. When that stopped, they were ready to take up television. Indeed, their need to make television receivers was the supervening necessity that finally halted debate.

However, if the end of the war found AT&T and the radio industry agreed about televison, AT&T's clients on the West Coast were not yet accommodated into the emerging order.

Received opinion has been that the film industry was caught napping by television and the Hollywood studio system was destroyed, but it is now generally agreed that the consent decree of 1948 forcing the studios to divest themselves of their cinema chains is of far greater moment than the arrival of television's first big year. At best television was a third blow to Hollywood, which was not only suffering under this enforced reorganisation but also enduring the beginnings of McCarthyism.

The movie moguls knew all about television. In the 1920s, not only was AT&T telling them they could not sell the new medium their product but the very first television demonstration in America, by Jenkins, was specifically designed for the home delivery of movies. The Academy of Motion Picture Arts and Sciences in its report on television, in 1936, stated 'there appears to be no danger that television will burst upon an unprepared motion picture industry'.[80] By then radio shows, either with stars or edited soundtracks or both, had been produced in Hollywood for years and the movie community's understanding of that world influenced its basic attitude to television. It was neither ignorant nor disdainful but rather imperialist. The moguls tried to usurp electronic distribution of video signals. They failed but, for all that their forces were distracted during a crucial period of television's development, the 'law' of the suppression of radical potential saved them anyway. Television became the dominant medium and it was owned by the radio interests yet, despite that, Hollywood (albeit changed and regrouped) nevertheless became its major production centre. Exactly how that was achieved constitutes one element in the operation of the 'law' of the suppression of radical potential in the next era of television history – from the end of the war to the mid-1950s. Maintaining stability among the radio production interests during this same period is another element.

(ii) Freeze, freeze thou bitter sky, 1948-1952
By 1948 television was finally poised to cover the nation. It had not been making much money but there were 4 networks, 52 stations and

nearly a million sets in 29 cities. In those communities, at least according to popular press reports, all other entertainments were suffering. Then, in September, the FCC ceased to process licences. The official account justifies this action because stations, at 150 miles minimum distance, were found to be interfering with each other. Originally, in 1945, the commission had determined that stations using the same channel should be 200 miles apart. This meant that whereas Chicago was assigned five channels, New York got only four and Washington and Philadelphia three each. The industry protested and the FCC dropped the distance to 150 miles against the advice of the engineers. The engineers were right. The confusion that followed caused a four-year freeze on new stations. A further 400 applications were simply held up.

Since it is clear that a competent radio engineer with a good map could have solved this problem in something less than the 43 months it took the FCC, interference is scarcely a convincing explanation for the length of the delay. Adding the Korean War (which halted nothing else in America but which is cited as contributing) hardly helps explain the length of 'the freeze'.

The period between 1948 and 1952 saw the refinement of the *de facto* deal made in 1941 within the radio industry. Some major players were found backing technical options which had closed off. CBS, for instance, pushed for a move to UHF and its own colour system. It was losing out on both and it used the freeze period effectively to regroup for a battle with NBC on the VHF waveband with RCA colour.

The main threat to the stabilised diffusion period, which the 1941 agreement on standards had made possible, was caused by colour. By 1949, RCA engineers had produced what Sarnoff said they would, a colour system compatible with the NTSC 1941 standards, which is to say that the colour signal would appear in black-and-white on a pre-colour monochrome receiver. Abandoning a semi-electronic system, the RCA Laboratory concentrated on a method which used green, blue and red filters to sensitise three separate pick-up tubes within the camera. When the three resultant signals were superimposed on each other, via a system of mirrors, a full colour signal was created.

To receive it, RCA followed up on a concept of a German engineer, W. Flechsig, who had thought of a colour cathode ray tube in 1938, in which triads of colour phosphor dots, red, green and blue, were to be activated by a mesh of fine wires. (Discrete trichromatically sensitive specs of material had been pioneered as the basis of a colour photography system by the Lumière brothers in 1907. These **auto-chrome** plates were the first colour stocks to be successfully marketed.) Keeping the triads of phosphor dots, RCA engineers had modified Flechsig's otherwise difficult, if not impossible, proposal by suggesting that, instead of wires, electron beams could be used passing through a mask drilled with holes – hence 'shadow mask tube'. When Sarnoff demanded a compatible colour system in 1946, this was the prototype to which H.B. Law of the lab turned.

The rival CBS system, although clearly a less sophisticated device, looked attractive to the commission exactly because it would break the patent monopoly RCA had achieved with its outright acquisition of the Farnsworth patents and which it was hoping to extend with its dot sequential colour television. CBS had been experimentally broadcasting in colour, under the direction of Peter Goldmark, since 1940 transmitting 343 lines. This system has the honour of being the very last in the line that starts with Nipkow in 1884, for Goldmark used a spinning disk, both in transmission and at the home end, albeit not for scanning but to create colour with the three primary filters built into the disks.[81] Goldmark was in a considerable tradition, even leaving the pre-twentieth century proposals by the Russian Polumordvinov aside. Baird, using trichromatic filters in a disk, had transmitted a colour picture in July of 1928.[82] After his retirement from the race of 1936 he went back to this work. In 1929, Dr Ives, of the Bell Labs' videophone, although presumably no more certain as to the utility of the device than he had been two years earlier, nevertheless revealed a colour version.[83] It used sodium photo-emissive cells that were sensitive to the full range of visible colours.

In 1941, the NTSC ignored both this tradition and the CBS experiments when making its recommendations; but it did not believe colour developments would necessarily result in compatibility with the monochrome standards it was establishing. On the contrary, in a questionnaire issued at that time, the NTSC advisory panels concerned with this matter voted 28 to 7 against a compatibility requirement.[84] Anyway, contrary to the received history, by the late 1940s the CBS system was compatible with RCA monochrome receivers, if they had tubes smaller than 12½ inches and once a simple a tuning bracket was added to the set. In the light of this the FCC adopted the CBS system in the middle of the freeze, in September of 1950, but the system was never introduced to the public. The manufacturers, tied to RCA for their black-and-white business, refused to accommodate the CBS bracket or make spinning disk receivers. A lawyer, involved in a congressional investigation of the FCC's apparent failure to implement its stated anti-monopolistic policies, wrote: 'We do not know whether any pressure was brought on them [the manufacturers] by their licensor [RCA]; but we do know that their refusals effectively "killed" the CBS color system which the FCC had adopted.' [our brackets][85] A Senate report of 1950 evaluated the CBS/RCA systems (and a third system in a yet more experimental state) finding, across some 18 measures of utility, efficiency and effectiveness, the RCA system had eight better performance characteristics than CBS, four as good and six worse. In all CBS's colour fidelity was deemed to be better, but its sets could not, because of the spinning disk, produce pictures bigger than about one foot across.[86]

It might, perhaps, be thought that this hidden battle about colour, rather than the public débâcle over station distances, was the reason for the freeze. In one sense this is true, since the colour issue was part of the whole question of the continuing stability of the radio industry during the period of television's diffusion. But colour in a more direct sense cannot be the reason. It came to the fore after the freeze had begun and was resolved, in favour of RCA, after it was over – at the end of 1953. Anyway, rather like the British introduction of monochrome service in 1937, colour was premature. Americans no more rushed to buy colour sets than had the British to buy black-and-white televisions in the late 1930s. Sarnoff predicted in the mid-1950s that: 'by 1963, all of America would be blanketed by color, and each and every home will be receiving its entertainment in full color.'[87] In the event, although NBC – alone – ran a full colour schedule in 1964, it took longer than that. That year 1.4 million colour sets were sold, enough to begin to affect the ratings. In 1965, when 20 per cent of television stations were still monochrome, CBS followed NBC; in 1966 ABC became a full colour network. (The Europeans, who had seemingly delayed decisions on colour for a decade, turned out to be colourising their networks at almost the same time as the Americans. The BBC's second service, inaugurated in the spring of 1964, was a full colour channel by December of 1967.)

American network television went fully to colour exactly fourteen years after FCC authorisation, when sales of colour receivers in the US were level pegging with sales of black-and-white sets for the first time. However sales did not really take off until the early 1970s. By 1973, twenty years after colour had been authorised, 43½ million US homes, around 2/3rds, had colour. The result of American pioneering in the early 1950s was the adoption of an RCA system markedly inferior in quality to the systems developed for European broadcasters.

The commission, in opting for RCA in 1953, abandoned the rhetoric it had used to constrain television in the late 1930s and early 1940s, the argument that delay was necessary because technical improvements could be expected. Further the RCA decision was made despite the policy of resisting monopoly in all aspects and at all levels of American broadcasting. FCC lawyers warned of the serious consequences of proceeding with the RCA system without considering the patent situation. However, the commission had seen that the dangers of RCA's monopoly were less disruptive to the industry than an insistence on the CBS system for the ideological sake of diversity.[88]

Beyond the fracas about colour and signal interference, the major truth about the freeze is that the industry was not 'frozen' by it. The number of sets increased to 15 million and more than a third of the population had them. Television was bringing in 70 per cent of broadcasting advertising revenues by 1952. The 'freeze' worked to suppress television as an area of exploitation for new interests. The owners of the first 108 stations, in effect the radio industry, were able to bed down and sew up the new structure. The shape of American television behind this protective wall was established with the minimum of disturbance, despite the internecine disputes. Advertising revenues, programmes and personnel were transferred from radio to television with ease. NBC had been encouraging its affiliates to obtain television licences for years. The two dominant radio networks, NBC and CBS, transferred their hegemony to television. In 1948, Mutual, Philo and Paramount (the film production company) abandoned plans for networks. The third network, ABC, barely survived and the fourth, DuMont (part owned by Paramount Theaters, the cinema chain) was dead by 1955.[89] Some went so far as to suggest that the colour inquiry was reopened, after the wavelength assignments had been cleared up, specifically to consolidate the protection the major networks had received.[90]

The freeze concluded with the issuance of the FCC's *Sixth Report and Order* which institutionalised these results. With AT&T's help (via its coaxial rates structure, again agreed by the commission), the minor networks were struggling. *The Sixth Report*, which opened up a huge range of UHF stations, most of which were never assigned and all of which (until the coming of cable rebroadcasting) were economically marginal, mopped up the possibility of an extensive fourth network. By 1952, in utter obedience to the 'law' of the suppression of radical potential, the broadcasting industry had metamorphosed from radio to television and nearly every audio-caterpillar had successfully become a video-butterfly.

And something else, less remarked upon, also happened during the 'freeze'. In 1948 the top programme in the television schedule was Milton Berle's *The Texaco Star Theater* on NBC, a variety show. It was soon joined by Ed Sullivan's *Toast of the Town* on CBS. Both of these were live productions from New York. In 1952 the top show was filmed in Hollywood, *I Love Lucy*. This is not to say that in these four years production moved from the East to the West Coast. It is simply to point out that the television industry's structure, which looked at the outset of this period to be essentially live and New York, looked at the end of it also to have a place for film and Hollywood. The implications of this move would take most of the 1950s to work out, but by the end of the decade, the era of the live New York productions in the prime-time schedule was largely past and people were already referring to it as 'the Golden Age of American television'.

Hollywood's first idea about television was to incorporate it. All the early pioneers had shown large screen as well as small formats and the possibilities of theatre television looked as real as domestic options. At the Swiss Federal Institute of Technology Professor Fischer shared Hollywood's general view of television's potential as a theatrical entertainment form and, in 1939, he developed the **Eidophor** which was to become the standard device for large screen television projection.[91] In 1941 RCA demonstrated a 15 × 20 foot screen which was installed in a few theatres.[92] By 1948, Paramount's own system was installed in their Times Square showcase and the heavyweight boxing bout was established as a staple of the distribution system. The 'freeze' allowed enthusiasm for theatrical television to blossom, and over 100 cinemas were equipped for it, but within a year of the freeze being lifted, expansion stopped. The FCC, in 1953, denied the theatre owners' request to be allocated their own UHF bands and, without the networking possibility that would have made the system economically viable, the theatres could not match the burgeoning domestic television industry. The large-screen equipment, especially the Eidophors, only survived, beyond being used to relay the occasional prize fight, as back projection devices inside television studios. Another potentially disruptive element had been contained.

A further option was explored during the 'freeze' and took a longer time to die, if indeed it is dead yet. Subscription television, for which (by the late 1940s) there was an ample range of hardware, began in Chicago on a trial basis in 1949.[93] In 1953, Paramount's test in Palm Springs was closed in response to a threatened legal action. The charge was that as the producer of the films it was showing on its system, it was once more, and illegally, engaged in exhibition. Paramount's theatre wing, now United Paramount Theaters, had been ordered by the consent decree not only to split from the production company but also divest itself of half of its cinemas. In 1953 it bought itself into ABC and became a major part of the raucous campaign against pay television which, for a time at least, united the cinema owners and the broadcasters. The FCC regulated these tests, but made it clear that it thought its duties lay in protecting the existing system from unexpected competition. Pay TV languished.

Hollywood's way forward into the television age did not lie with alternative television distribution systems but with the radio broadcasters. The freeze had shown two things: firstly, that short-form Hollywood series could be as popular as anything produced elsewhere, and, secondly, that old movies had appeal. The studios were quicker to respond to the former possibility. They had difficulty in establishing telefilm operations because to do so would have upset their theatrical clients – just as, today, the American networks cannot offer popular programme services to the cable industry without upsetting their broadcasting affiliates. But despite this, RKO did set up a subsidiary for telefilms in 1944. Slowly the deals were done; ABC with Hal Roach Jr.; the first Disney special; *Warner Brothers Presents*. After the freeze the trickle of Hollywood prime-time product became a flood. By 1955, telefilm raw stock consumption was ten times greater than that of the feature side of the industry. Of course, there were many new production entities involved in this but, despite the rise of some smaller entrepreneurs, it was essentially old Hollywood. A few of the players had regrouped under new banners but most were still manning the same stores.

And after the freeze, with this beachhead in the schedules firmly held, beginning with RKO in 1956, the majors sold their libraries, often for shares in television stations. By that time, New York had come to call, with both CBS and NBC building major production facilities in Hollywood, consolidating the tradition begun with radio. New York was left with news, sport, documentaries, variety and the day-time soaps, and Hollywood got the dramatic staple of the schedule. In 1949 none of this was clear. By 1952, the mould of American broadcast television was setting fast.

The 'law' of the suppression of radical potential worked against the supervening necessities to hold television in a limbo just outside widespread diffusion throughout the 1930s and the 1940s. After the supervening necessity of spare electronic industry capacity made its introduction inevitable, the 'law' worked to contain its disruptive forces. From 1936 to 1946 it kept television in the lab and from 1946 to 1952 it kept it in an, albeit extensive, ghetto. For the technological determinist these delays can make little sense. The device was, without any doubt, viable in 1934 and working by 1936. It would probably, had the social forces been arrayed differently, have been viable the better part of a decade earlier. The slow introduction of television is the perfect refutation of such determinism as well as being a model for telecommunications developments in general.

This is true not just of the United States where the regulatory process existed in such an uneasy harness with the potential excesses of the free market – in the sense that, without the FCC, America could well have had more than one system on more than one standard, as had been the case with telephony in the years between the lapse of the Bell master patent and the imposition of order by the US government. In the UK, where government fiat had encouraged the premature television service, post-war progress matched that of the United States, although conducted far more rationally. Here, and throughout most of the rest of the world, the 'law' of the suppression of radical potential worked straightforwardly through austerity. By the end of 1952 there were still less than 3 million receivers in the UK. Thus when, by 1952 (the year when America finally got the go-ahead for national television), the BBC completed its network of major transmitters to reach 78 per cent of the population, it had less to do with the shakedown of warring elements within the industry and more with basic economics. Yet there were such elements, and, as had been the case with radio, the BBC was allowed to develop a new technology without competition from commercial broadcasters. Only when the system was established and the BBC had successfully transformed itself from a radio into a television entity did the advertising lobby in Britain finally get into the act, winning, in 1954, the right to broadcast commercials and, necessarily, also the programmes to put round them.

Elsewhere, with improving economic conditions, 1952/3 marks the true start of television diffusion. Italy began a five-year plan to cover the nation. North Germany's network was completed. The French added Lille to Paris and started work on three other transmitters. In Canada, CBC began programming. By the early 1950s, and only by then, could it be said that television had finally arrived. And, in every nation, the arrival did not displace whatever interested parties pre-existed the *invention*. Everywhere, radio manufacturers and producing entities switched to television. There were no casualties. There were few new faces.

Phase four: technological performance – production, spin-offs, redundancies

RCA produces TELEVISION SETS, Lansing, PA, 1947

The analysis offered above accounts for the success of the *production* phase of television. It will not be further discussed; but the concepts of *spin-offs* and *redundancies* in the model need to be illustrated.

(i) A spin-off – Bing Crosby invents videotape, 11 November 1951
At the outset of this book, it was suggested that of all the new devices now available the **videocassette recorder** might prove to be the most significant. Overshadowed by the enormous public relations exercise mounted by the world's cable interests, VCR growth has been, by comparison, almost unnoticed, especially in the US; yet by 1984 upwards of 40 per cent of British homes and 17 per cent of American (15,000,000) had the device. 80 million blank videocassettes were being sold, in the US, each year. The prerecorded cassette business collected $475 million in revenues in 1983 from this market alone and is projected to be worth $5.7 billion in 1987.[94]

Of course, such projections are a commonplace of the 'information revolution' but there is reason to believe the VCR has better potential to fulfil at least some of this promise of profit than do many other telecommunications developments. Firstly, it fits – that is to say, it meshes with abstractions such as the ongoing drive to put entertainment in the home as well as practicalities such as the fact that it displays its signal on the cathode ray tube of the domestic television receiver. It is even packaged like the hi-fidelity audio systems that preceded it as part of the growing 'home entertainment centre'. Secondly, it offers a real add-on advantage to television since it breaks the tyranny of the programme scheduler, allowing owners to time shift and save programming of their choice. Thirdly, after initial purchase or including rental fees, it is inexpensive – it gives families the opportunity to enjoy theatrical films at a fraction of the present cost for such a family to visit a cinema. Whereas 10 per cent of American VCR owners hired prerecorded material in 1979, 70 per cent did so in 1983. It is competitive with the cable home movie channels which have, like broadcasting, fixed schedules and, unlike broadcasting, a fixed monthly charge. (Cassetted movies are also more pornographic than those shown on cable!) None of these factors might of themselves seem enough to distinguish the VCR's potential from, say, cable's, the perceived front runner in 'information revolution' distribution systems. But there are significant structural differences in the attractiveness of the two technologies and their patterns of growth..

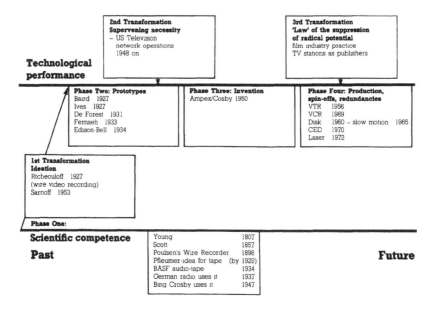

Figure 3 Videotape recordings

Although cable expansion has been fast (see chapter 6 below) it has been hedged with ambiguities. In America, its development was occasioned by the poor quality of the broadcast signal and that has continued to be a more important factor in its growth than is generally acknowledged. Whether cable is bought primarily for reception rather than for the add-on services it provides remains an open sort of question. Elsewhere, cable's diffusion has been helped by a population's desire for transnational signals, as in Canada and Belgium. Such effect as cable has had is of extreme significance for the firms that have been successful in its diffusion but, overall, the impact on the underlying economics of the broadcasting industry has been contained. ABC and CBS now both have interests in cable programming services while CBS operated, under a special dispensation from the cross-ownership rules of the FCC, cable systems. NBC's major interest, via its parent RCA, is in the operation of satellites to distribute signals to cable systems. The 'law' of the suppression of radical potential has worked. Contrast the nascent US television industry with around one-third of homes in 1952 already commanding 70 per cent of broadcast advertising dollars with cable's perforamnce in 1984. With more than 40 per cent of homes on the wire, cable secured $400 million of ad revenues – about a quarter as much as American radio billed ($1.5 billion), never mind network television's $8.4 billion.

VCR's growth has been less ambiguous – the thing is bought for itself – and steadier, following such established patterns as the demand for colour TV receivers in the 1960s. Indeed, the colour receiver market is closely bound up with the VCR's diffusion, not just as a model for the pattern of growth. In part, colour TV sales worked first to constrain the diffusion of VCRs in the early 1970s. There was then a considerable pause for, although VCRs were freely available from 1974, the market did not begin to register, in the US, until 1979 when, as in 1962 for the colour TV sets, nearly half a million units were sold. By 1981, the 1.3 million VCR sales matched the 1.3 million TV set sales of 1964. And again, in 1983, more than 4 million VCRs marched with the nearly 5 million colour sets sold in 1966. In 1984 nearly 8 million units went into American homes – far more than the number of television sets sold in 1967, colour TV having peaked as a consumer item that year since more than three-quarters of all homes owned one. Projections suggest that there will be, in the US, 51 million VCRs by 1990 – more than there will be cable boxes.[95]

The diffusion of the VCR has already taken some decades and the full implications of its arrival are still being worked out. On copyright for instance, in October 1981, a US court held no infringement was involved in home taping of broadcast signals. Sony had successfully resisted an action brought by United Artists and Disney. Hollywood in general, though, was not overly concerned since sales and rentals from prerecorded cassettes are a source of revenue. As for American cinema owners, they perceive the cassette user as being quite distinct from their audience and otherwise lost to the movies. In other countries there have been more overt attempts at suppression. In France, for instance, cinema owners managed legally to restrict the speed with which distributors could release cassettes of films still popular in the theatres.[96] As regards broadcast television, in Britain and in Germany cassettes were blamed for a diminution in audience size in the early 1980s. But these arguments, which are far from over, do nothing more than illustrate, against an already slow pattern of diffusion, the further workings of the 'law' of the suppression of radical potential. The impact on all these areas, film exhibition, copyright, broadcast scheduling, is quite gradual. As did television itself, the spin-off techniques of recording an electronic video signal have developed at a dilatory pace.

A television camera is a transducer converting light into an electrical signal. That signal can be stored, just as audio signals were, on disks. The problem is that the video signal is vastly more complex and is carried at far greater frequencies, millions as opposed to thousands of cycles per second, occupying enormously enlarged bandwidths.

Magnetic recording, using a wire although other media were suggested, had been introduced by a Dane, Poulsen, in 1898. The antecedents of this *invention* can be traced through Berliner's **gramaphone** (i.e. disk) and Edison's **phonograph** (i.e. cylinder) to Scott's **phonautograph** of 1857 and description by Thomas Young (of the light interference experiment), in 1807, of a method for recording the vibrations of a tuning fork on a drum.[97] The whole history of sound recording and the development of television itself constitutes the ground of scientific competence for videorecording.

The ideation transformation occurs in a British patent, filed by B. Rtcheouloff in January of 1927, suggesting the Poulsen **telegraphone** could be adapted for television. The device was almost certainly not built.[98]

Technological performance of the prototype phase begins with Baird who, because he was working with 30 lines at 12½ frames per second, was not overwhelmed by bandwidth and created, in September of 1927, the **phonovisor** using ordinary gramophone industry audio equipment to impress the signal on a wax disk. He tinkered with it for three months, recorded some images, publicised it, and moved on.[99]

Other systems had been tried, yielding more *partial* prototypes. Lee De Forest, of the triode valve, demonstrated a device in 1931 which allowed for the waveform of a television image to be impressed on 35mm film coated with pure silver, using a series of needles mounted on a revolving disk. It did not work. (Ten years later he was still arguing for the advantages of telerecording over coaxial cable distribution):

A coaxial line from New York to Los Angeles represents about $15,000,000 . . . but they can have all this in defiance of all time zone difficulties. . . . How? By the very simple and very sensible method of making television films and sending them from city to city in tin can containers.[100]

In 1934 a device which, like De Forest's, took the waveform of the videosignal – this time to modulate a light – was demonstrated by Edison Bell (UK) Ltd. Since films were made with optical sound-tracks, the system proposed, as it were, an optical optical track. Ordinary 35mm film was exposed to a fluctuating light which was responding to the videosignal, just as in the 'sound camera' a light responded to the audio waveform of the microphone's output. The film therefore recorded the videosignal as if it were a film soundtrack. The film could then be fed back into a television receiver system to recreate the original image. Although theoretically very elegant, the device also did not work well enough for further development to be then warranted.

High-definition television – in the 1930s sense – produced signals beyond the ability of the recording industry of the day and, from the late 1920s, the most widely used retrieval systems, in so far as they were developed, all depended on putting the television image on to film, essentially by pointing a film camera at a television screen. There were a number of technical problems, such as frame roll, whereby the camera's shutter action allows the film to record gaps in the electronic picture unnoticed by the human eye; but, basically, these were simpler to deal with than the bandwidth problem.

In 1927 Ives made the conceptual jump from television devices which played back existing films to an apparatus which exposed film specifically for television scanning (see above, page 49). Bell Labs wanted to improve its poorly defined television image with the direct illumination that a film/television chain could provide, but there was a side benefit to this film/television marriage. If the television picture was *filmed* at the receiving end, the film image could then be conventionally projected. By 1933 Fernseh A.G. demonstrated a variant on its intermediate film process which allowed for the projection, via a Nipkow disk, of a 180-line picture 10 × 14 feet. The technique survived to be seriously considered by Paramount for its theatre TV system in the late 1940s. In the Paramount version the camera exposed the electronic picture, presumably to be received over the UHF theatre network had that been allowed by the FCC, and 66 seconds later the developed film dropped by chute straight into the gate of the cinema projector. (The Eidophor was also in use in these experiments, as was mentioned above, but, although it produced a large screen image, it did so directly from the received electronic signal and was not a retrieval device.) As we have seen, theatrical TV was not a meaningful supervening necessity and the experiments were terminated. However, by the early 1950s, more straightforward telerecording or kinescoping techniques using a variety of shutter systems were in use in the broadcasting industry. At that point, with television about to become a mass medium, these *accepted* prototype film systems were proving inadequate, especially in America.

Without a national system of coaxial cables, programmes had to be, as De Forest had suggested and to use today's terminology, 'bicycled' to stations outside the AT&T network. As the net was built, another problem arose – time differences between the two coasts. This led to the era of 'hot' or 'quick' *kines*, requiring that the telerecording be produced in under three hours to enable the other coast to see it, via coaxial cable, at the same point in the evening schedule.[101] The growing industry insisted on a national audience; it promised the delivery of no less to its sponsors and advertisers. This need – which Lee De Forest can be forgiven for not anticipating – created a supervening necessity for a quicker system.

Beyond the particular difficulties caused by the size of the American republic, television producers everywhere faced the alternatives of either live transmission which, in the evenings, incurred high labour costs, or film programming, again involving high cost. So there was a further supervening necessity, within the industry internationally, which also required a more appropriate retrieval system than film.

On 11 November 1951, videotape recording (VTR) was *invented*. On that date, the electronics division of Bing Crosby Enterprises demonstrated a monochrome recorder using specially configured tape. Crosby himself had bought into the audiotape process, a German technique which had been on the market since the early 1930s, when a few fledgling firms were trying to exploit it in the US immediately after the war.

Magnetic tape was first produced by BASF (Badische Anilin-und-Soda-Fabrik) in 1934. Fritz Pfleumer, an independent industrial consultant of considerable creativity, had developed the idea in the late 1920s. He had been asked to help a cigarette manufacturer prevent the gold in the then fashionable gold-tipped cigarettes from coming off on smokers' lips. Pfleumer embedded the metal in plastic rather than in paper and the problem was solved. He also developed a device which could sense, electrically, whether or not all the metallised tips were at the top of the packet. He was a music lover dissatisfied with the fidelity of disks available, and it occurred to him that this technique of embedding metal into a plastic might be used, if the metal were magnetised, to record sound. Physical systems of filing the metal produced particles too uneven to give a good recording result so BASF, industrial chemists, became involved. Chemical precipitation of ferric oxide proved to be viable and, in 1934, magnetic recording tape was ready.

However, there was no supervening necessity for it and since it recorded music, just as the widely diffused record did, it was *redundant*. In 1937 AEG began to use the tape, now made by IG Farben which Pfleumer had joined, in its wire Poulsen recorders, **magnetophons**. During the war these machines were much improved by researchers in the German State Broadcasting Service. But the 'law' of the suppression of radical potential operated to protect the record industry throughout these decades. No wire recorder – not even for office dictation use – was manufactured in America between the collapse of Pulsen's original firm before the First World War and 1937. In that year Bell, ever watchful across the whole field of these technologies, introduced a high quality steeltape recorder for its recorded telephone weather service but, essentially, all these techniques remained, outside of specialised uses, a solution looking for a problem.[102] (The Nazi leaders, at Goering's insistence, had their wartime speeches recorded on magnetophon tape.)

The American broadcasting industry in general, although heavily reliant on disks, refused to use tape, steel or plastic. Network radio, in particular, was live radio, even when whole shows had to be repeated for the other coast three hours after the original transmission. Hence Bing Crosby. When he wished to record his *Kraft Musical Theatre* show but NBC Radio refused to allow it, Crosby was popular enough to defy Sarnoff. He took his show to ABC, then only in its third year of operations, having been, until the government ordered otherwise, NBC's second network. However, the disk recordings Crosby made were indeed less than perfect and the show's ratings began to suffer. The Crosby organisation laid down the specification it required for a recording medium. Only two firms were working on the German magnetophon, having acquired patent rights from the US Alien Property Custodian and in September 1947, one of them, Ampex, demonstrated for Crosby's technicians a prototype improved American version of the German machine which met their needs. The reel-to-reel audiotape recorder was *invented* and the radio industry fell in behind Crosby. His firm became sole West Coast distributors of the machines.[103]

For the public at large, the problem of creating a portable system to play prerecorded material in cars eventually broke through the record industry's objections to tape – which were well founded, being based on fears that it would allow home duplication of its product – and a supervening necessity for the innovation had at last been found. The young augmented the need for mobile music by demanding it wherever they went and so the audiocassette player made Pfleumer's bright idea of four decades earlier a useful technology for everyone (except the record industry), after all.

In video, the need both to create national networks and to avoid labour expenses incurred by live prime-time transmissions during unsocial hours had the same effect. Videotape, too, became a necessity.

To solve the bandwidth problem, the tape could be moved at high speed past the recording/playback heads, which were stationary; or various arrangements could be made whereby the heads spun as the tape passed them, thus reducing its speed. Just over two years after the Bing Crosby demonstration, RCA showed a recorder that used ¼ inch tape for monochrome and ½ inch for colour. It ran at 30 inches per second, recording 4 minutes of programming at 3-megacycles (about half the allocated broadcasting bandwidth).

The Ampex Corporation, which began work on the problem at this time, was to establish, by 1956, the broadcasting industry standard for the next three decades. Its machine used a magnetic tape, two inches wide, driven at 30 inches per second past a drum containing 4 magnetic heads which rotated at 14,400 revolutions per minute. On 30 November 1956, CBS transmitted the first videotaped programme (backed up by Kinescopes). RCA, having exchanged patents with Ampex, produced a colour recorder along broadly similar lines the following year. Ampex in 1960 created a device which allowed the VTR to be locked into external sync signals produced in the studio or elsewhere. Its output could therefore be cut into a programme or dissolved to without any disruptions. In 1964 Ampex, for the BBC's second channel which was eventually to be in colour, produced a high band recorder which could successfully deal with the new 625 line signal.[104]

Spinning the heads created problems and in the early years it seemed as if a direct tape path might be a better solution. RCA, in 1956, produced another ½ inch colour machine using this longitudinal scan configuration which travelled at a finger-munching 240 inches per second. In 1958, the BBC demonstrated ½ inch **Vera** (*Vision Electronic Recording Apparatus*) which ran at 200 inches per second and produced a 15-minute tape on a 20-inch reel. These machines, although not as practical for everyday operations as traverse scan devices of the Ampex type, were nonetheless worth exploring because they accommodated broader bandwidths on narrower tapes. In experimental devices demonstrated in the late 1960s, tape speeds reached 4000 inches per second and the bandwidth capacity of 50MHz, enough for many TV signals, even of what is now being called 'high-definition' (1000 lines +).

The third possibility, helical scanning, allowed the tape to move more slowly. Here the head spins in a drum, recording each field in one long path stretching diagonally across the tape. For professional use, before the introduction of electronic videotape editing (in the late

1960s), this tape path created a considerable disadvantage. Traverse scanned tapes could be, with some difficulty, physically cut; helically scanned tape could not. But there were compensations. The picture could be constantly displayed in fast motion, forwards and backwards or as a still frame. The tape speed and size could also be reduced. In September 1959 Toshiba demonstrated a machine using the helical technique and 2 inch tape on which the tape speed was halved to 15 inches per second. The first 1 inch machine appeared in 1962.

From the very beginning of this line of development, some had seen its potential as a home entertainment system. In 1953 General Sarnoff had pronounced:

> Magnetic tape recording of video signals should make possible simple means by which a TV set owner can make recordings of television pictures in the home. And they can be 'performed' over and over through the television receiver just as a phonograph record is played at will.[105]

By the late 1960s the expectation was growing that, with these less expensive techniques, videotape would be able to provide 'home libraries of films and television programs'.[106] Sony put 1 inch tape into a cassette in April 1969. The tape width was reduced to ¾ inch seven months later and the speed was now 3.15 inches per second. That same month Panasonic introduced a machine using ½ inch tape cassette which Philips had developed which ran at 7½ inches per second. Philips revised this design the following summer, keeping the ½ inch tape but reducing the speed to 5.6 inches per second.[107] 1969 saw one other addition to this proliferation of mutually incompatible boxes – JVC introduced a square cassette using ½ inch tape. It is no wonder that 'videocassette' was noted as a neologism at this time.

Eventually this ½ inch tape size in a number of formats was to prevail. The world market is dominated by the Beta and VHS configurations of Sony and JVC, with the former having the edge as the emerging professional standard, while the latter dominates the home market.

By the early 1970s videocassette recorders were a readily available consumer durable, nearly two decades after their industrial cousins were introduced. It was to take about five years for them to create space for themselves on retailers' shelves and the better part of another decade for them to achieve the status of major consumer durable. During this entire period, the only significant technological advance was the transistor, but this was not the enabling technology. Valves would have functioned adequately and did. The first transistor-ised broadcast standard videotape recorder was demonstrated by RCA in 1961.

The VCR, however it fares against established media, cable and other electronic distribution systems, is clearly a major *spin-off* from television. It has also attracted a considerable number of *redundant* technologies. *Redundancy* is, in part, a matter of timing, although it in part relates to function. It occurs, like *spin-offs*, when technological performance continues beyond the point of *invention*. With redundant performance, all that is produced are devices that duplicate the essential functions of the *invention* or its *spin-offs*. Their efficiency (as with prototypes in the second phase) is not the crucial factor. Redundant devices can work as well as those they seek to replace or sometimes they can work less well. Such levels of performance will not disturb the diffusion of the previous technology. Even superior performance will not be sufficient, necessarily, to sweep away established *inventions* and *spin-offs*. In fact, superiority can be nothing but a snare and a delusion for a redundant device is, in essence, a retarded spin-off. Timing is everything.

(ii) Redundancies – the non-laser videodisk, 1960; 8mm videotape, 1980

Many manufacturers have been seduced into *redundant* videotape formats, looking to overturn the hegemony established by Sony and JVC by finding the electronic substitute for the amateur 8mm movie camera. In the 1970s, Akai marketed a ¼ inch monochrome system to this end which did not survive. Sony introduced, in 1980, a tape format that corresponds to the size of 8mm film.

This prototype, with all its comforting backward-looking connotations, emerged from an international industry-wide agreement – itself a suspicious event – and it is difficult not to believe that Sony, consciously or unconsciously, has engineered something of a Machiavellian plot. The 8mm alternative has been seized on by Kodak and a variety of manufacturers excluded from the Sony/JVC VCR hegemony. But they should take warning from the fact that, despite the consensus, Sony and JVC have taken steps to nullify 8mm video's one great advantage, its smaller size. JVC first introduced a small recorder which takes a cutdown compact version of its VHS cassette and Sony has gone even further by producing a colour camera of amateur film dimensions, the **Betamovie**, which records, within the body of the device, on to a full-size Beta cassette. Its only current disadvantage is that it cannot playback through the **Camcorder** (as they are being called) but needs a full-size Beta machine for this purpose. Now JVC has a camcorder using an ordinary VHS cassette, and unlike the 8mm machines, it is compatible with the millions of VCRs already sold.

Kodak announced 8mm video cassettes would cost $12.99 for 30 minutes and $15.99 for 90 minutes. VHS and Beta alternatives were available in the US at this point (1985) at around $6 for up to 9 hours playing time. Perhaps Kodak marketing experts are expecting the familiar yellow box to be a big draw.[108]

This path of redundant formats has been well trodden by the firms who made the welter of cassettes in the late 1960s, and even, in the case of Philips, the 1970s. Since many of them are the same people, 'retreading the path' would be a better way of putting it. Philips has been especially persistent in seeking to avoid the client status of the licensee. Having lost round one with its square video box, it not only introduced another format, V2000, which also has made little impact but even, like JVC and Sony, developed a compact version. Faith is a wonderful thing. The 8mm video lobby was talking, in 1984, of lopping-off 20 per cent of the American VHS/Beta market.[109]

Eastman has reason to be alarmed by video's growth; and the simplest solution, to join the bandwagon and become a licensee of the Japanese, was not company style – until the winter of 1984. Then the company marketed VHS videotape promising the Kodak colours the world was used to for its new product – which claim must border on the very edge of truth-in-advertising since all its film stocks are subtractive colour processes while video colour systems are additive. But, for the better part of a century, Kodak has been used to owning the entire wagon – more or less – and it faces the video challenge with both the cushion of the lion's share in the gargantuan film stock trade, and a tradition of innovation.

For instance, in 1955, Eastman, clearly seeing in videotape a threat to its considerable telerecording business, revived a 16mm film with verticular lenses embossed into it which acted as prisms. This was first suggested in 1908 by Gabriel Lippmann, professor of physics at the Sorbonne. Kodak and Paramount had worked on it in the 1930s, failing however to produce an effective 35mm version. NBC, who could not have been displeased at having an alternative to the Ampex videotape its rival CBS had been supporting, used the film from September 1956. Two months later CBS transmitted tape for the first time, but NBC persisted with Kodak until February of 1958, when it gave up and went to videotape.

Although all telerecording developments have been rendered redundant for the broadcasting industry's primary needs, other kinescope systems persisted into the 1960s and are finding renewed possibilities in the fluid tape/film environment of both the theatrical film and television production processes in the 1980s. Production is moving ineluctably towards completely electronic processes except at the point of distribution and exhibition where film's quality and adaptability are not matched in video. Even if high definition 1250-line TV (HDTV), which would match the resolution of 35mm film, is widely diffused (as is likely) celluloid would still retain its advantage as a universal medium not subject to the vagaries of the local electrical supply and television standard. (HDTV was being pressed by the Japanese broadcasting authority, NHK, in the early 1980s. In 1985, non-Japanese technicians, in true obedience to the 'law' of the suppression of radical potential, were intent on determining the proposed system's compatibility with the 525/625 standard rather than its merits so that its disruptive potential could be contained. They were also talking about 'enhanced' (i.e. better) 525/625 TV to 'reduce', as two senior BBC engineers put it in a technical article, 'unfavourable comparisons with HDTV'. 'In 20 years or more,' stated another senior British technician at an international symposium on the matter, 'the NHK system may be obsolete. It would be much better engineering practice instead to match the three links of production, transmission and reception more closely by using enhanced television practices.' In effect, and in total obedience to the 'law' of suppression, the new possibility is being denied until it can be absorbed.[110]

As early as November 1969, Technicolor announced that it was developing techniques to use a television signal of 2000 lines, well beyond the million pixels (picture elements) required to duplicate the quality of a 35mm frame, and a laser-based conversion method to produce a top quality film negative, 35mm or 70mm.[111] Such announcements were premature because only during production was the film process matched by tape, the one substituting, as the recording medium, for the other. The post-production phases of editing and sound mixing were unavailable for video and had yet to be developed. Electronic editing was just coming on stream and computerised random access editing systems (where tape displays a real advantage by allowing instant access to any frame as opposed to film which has to be spooled) were at least five years into the future. And even during taping, although studio cameras were used outside the studio to obtain an image for the videotape to record, they were bigger and less mobile than comparable 35mm film cameras.

In fact, a decade and a half later the industry is still foundering in a mess of different systems, none working to as sophisticated a standard as that proposed by Technicolor in 1969. Tape's radical potential to disrupt the film industry – which could have been realised by the rapid development of field production and post-production facilities to match its capabilities – is being contained. Technicolor, not untypically, indulged in a species of technological synecdoche reading videotape as a whole system instead of seeing it as just one element. Because of the slow emergence of electronic substitutes for each and every one of the devices used in filmmaking a typical stabilised change-over period is occurring within the industry. The slowness, it must be added, has not been due to any technological constraint since every electronic device now being introduced, from smaller self-contained camera/recorder packs through computers to sound mixing systems was more or less waiting to be pressed into service.

Polaroid's response to instant video was to produce an instant film for the 8mm market. This failed to provide video's off-air recording capability, produced no library of material such as theatrical films and did not match, in price or quality, 8mm film. It was thus redundant not only in terms of tape and but also of amateur film and was rapidly withdrawn.

There have also been repeated attempts to revive the line begun by Edison Bell's *partial prototype*, described in the previous section, and, in various ways, to impress the waveform of the videosignal on to various media. General Electric in 1959, Eastman in 1961, RCA in 1964 and 3M in 1965 all produced systems which recorded the television picture after its conversion into an electron beam on to film (Eastman/RCA/3M) or melted it on plastic (GE). The most strenuous attempt to get this technology into the market was mounted by CBS which developed EVR in the late 1960s.

The acronym stood for *E*lectronic *V*ideo *R*ecording, a considerable misnomer since it was actually an electron beam process using 8mm film in cassettes,. The other systems all produced variants on the recording capabilities of VCRs of interest to professionals and, therefore, these research efforts were sustained. The 3M machine, for instance, was marketed in 1967 and produced a picture resolution of better than 600 lines with far less noise than the contemporary videotape image. The General Electric process looked as if it might have potential in the tape duplication field. But EVR was designed to challenge the emerging domestic videocassette head-on. It was smaller, which did not matter, and it lacked the recording capacity of tape, which did. The attempt to defeat the VCR was futile. In the UK the professional film technicians union (ACTT) organised the workforce at the EVR cassette production plant just in time to negotiate redundancy payments for them.

The capacity to record, as distinct from playback, was to prove an important element in the VCR's attractiveness to consumers. The audiodisk analogy does not seem to work for video. Unlike radio, where the first retrieval technology, disks, did not allow for home recording, with video the first retrievable home system, VCR, did. Time-shifting TV and personal off-air archives have turned out to be major uses which render the sustained attempt to market videodisks, which were all prerecorded, as futile as the efforts to diffuse EVR. It must be said, though, that CBS's commitment to EVR was terminated with a great deal more dispatch and foresight than RCA exhibited pushing the disk ten years later.

Of itself, disk offers two advantages over tape. It can be randomly accessed to display instantly any frame: and, if utilising laser-based technology, it is the most durable means available for recording audiovisual signals. It was the first of these capabilities that led to the earliest developments. Philips, by 1960, and various other players, major (like Sony) and minor (like MVR Corporation of Palo Alto), produced disk devices which stored instantly accessible single picture frames. Typically the disk held only a few hundred pictures, a few seconds. The supervening necessity that brought these hand-built machines into the studios was the sports programmes' need for careful, considered, not to say academic, analysis of plays. (Supervening necessities must not be subjected to value judgments along utilitarian lines of the greatest good for the greatest number. They can be found in the small as well as the big. For producers and watchers of sports shows, the need for 'instant replays' was obviously real and pressing enough for these devices to be built and sold.) In August 1965, an MVR unit was used in a CBS football game transmission. The world witnessed the wonders of 'action replay' for the first time. Two years later MVR produced a slow motion machine.

RCA introduced a domestic variant of this device which used a physical system to read the signal on the disk. This had been first demonstrated by Telefunken/Decca in 1970 where an 8 inch disk was shown holding only 15 minutes of material. RCA extended the capacity but with so fragile a surface that the player had to be designed to extract the disk from its cover which it otherwise never left. There was a problem when children or fans jumped up and down in excitement near the machine but eventually the thing was able to work in an everyday setting and was, with considerable hullabaloo, marketed. It gave a picture of sharper resolution than VCR, but since this was still bound by the line standard of the receivers, such improvement was not enough to make the device a necessity. The same was true of its ability to deliver stereophonic sound.

Other advantages proved just as illusory. Disks were cheaper than prerecorded cassettes, but the majority of VCR users did not buy prerecorded cassettes – they rented them or used the VCR's recording ability to copy off-air, either ploy making cassettes cheaper than disks. That the disk could randomly access frames, and alter running speeds, etc. was of course of no interest since, within the culture, that is not, by and large, how audiovisual messages are viewed. (The ability of VCRs to do many of these same tricks is an excuse for manufacturers to increase cost, such facilities being equally pointless toys. The only real requirement is for an effective multi-channel, multi-day timing device and this genuinely useful addition to the VCR, actually nothing but a simple time-switch, commands premium prices.) RCA persisted with the marketing of the videodisk, despite its long history of involvement with videotape, a domestic version of which they had introduced as a consumer standard in 1969. Finally in the spring of 1984, having endangered the entire corporation (network operations at NBS also at this time performing poorly in ratings and profits), and with at most half a million players sold, RCA gave up.

The RCA/CED disk must not be confused with laser disks which are virtually indestructible – a life of six centuries without any special storage is claimed – and are clearly needed if the world's audiovisual archive is to be preserved. They also have uses as data-storage devices for the same reason. They are being marketed, however, like the CED disks, as redundant alternatives to records and tapes, both audio and visual.

Philips were the first to bring a practical laser application to the audiovisual image storage problem. In 1972, they produced the VLP disk. On this the information was stored in a series of micrometer pits arranged in a spiral pattern which were read by a laser-generated spot. The disk held up to 45 minutes' worth of material.[112]

The ground of scientific competence that leads to the **laser** goes back once more to Dr Young. The experiment establishing the wave hypothesis of light also demonstrated that certain light sources were coherent – with waves in phase both temporally and spatially – and others, natural light say, were not. In 1913 Niels Bohr made an assumption about the nature of atoms which was to impinge most fruitfully on Young's observation of coherent light. Atoms exist in a ground – low-energy – state or can be excited into higher-energy states. They do this by absorbing one of Einstein's light particles, photons, of exactly the frequency required to reach the excited state. Conversely the atom can decay from the excited state into the ground state by emitting a photon of this frequency. Atoms do this spontaneously. In 1917 Einstein demonstrated that atoms could be induced to release photons by the external presence of a photon of the right frequency. The external photon and the one released by its stimulation would be exactly in phase, coherent in terms of Young's experiment. Once begun, the stimulated emission of photons would be like an avalanche. But a problem arose. As the emitting atoms decayed they reabsorbed some of the emitted photons and the avalanche petered out. To prevent this, a *population inversion* is required whereby a greater number of atoms must be in the excited state than there are atoms in the ground state.

The search for a substance and technique in which this would happen constitutes the ground of scientific competence for the LASER, *Light Amplification by Simulated Emission of Radiation*. The ideation transformation occurs in a seminal paper, written by Charles Townes and A.L. Schawlow (of Bell Labs) in 1958, which outlined the general principles whereby population inversion could be avoided and suggested potassium vapour as the substance.[113]

Townes already knew how to achieve avalanches of emission but

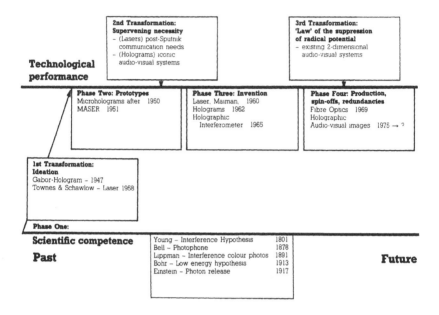

Figure 4 Lasers

with microwaves, comparatively long invisible wavelengths. In 1951 he and his colleagues at Columbia built a MASER – *M*icrowave *A*mplification by *S*imulated *E*mission of *R*adiation – using ammonia. The same year a Russian, Fabrikan, had the same idea. Proposals along these lines were also being made, in 1953, by Weber and, in 1954, by the Russians, Basov and Prokhorov.

All this activity clearly indicates a strong supervening necessity – masers were of interest to the telecommunications industry as a means of amplifying very weak signals, such as those coming from space or passing through a transoceanic cable. Lasers would be of similar value and the idea of using light as a carrier goes back to Bell's photophone and the experiments of Alexanderson at GE. Sputnik, by inaugurating the space race, reinforced the general supervening necessity for all aspects of high technology created by the Cold War and specifically helped lasers which, like masers, were potentially useful in spaceside communications.

Lasers are another device whose development is said to offer a contrast between the supposedly serendipitous nature of pre-twentieth century innovation and what today's technologists do. Yet here again the historical record does not quite sustain this picture. The pace of development up to the point of ideation was leisurely: science fiction dreams of deathrays did not speed them even during the Second World War and, before the Cold War took on its extraterrestrial aspects, the concept languished. Now, in the *invention* phase, looking for a substance that would allow for laser transition led to a search, albeit limited by a sophisticated understanding of the atomic structures of the substances examined, but not too unlike the procedures of not only the nineteenth century 'amateur inventors' but also the mediaeval alchemists.

The gold at the end of this rainbow turned out to be a ruby and it was identified in 1960 by Theodore Maiman of the Hughes Aircraft Company. Bell Labs understood the potential of rubies for lasers but had abandoned them as being too poor to give any hope of success and had concentrated instead on gases. Maiman had persevered with the stone, or rather a synthetic version of it, and in 1960 had achieved laser action. Once this was done a helium/neon laser was made and in the years that followed hundreds of different laser transitions were demonstrated. (It was, though, the Russians, Basov and Prokhorov, who got the Nobel Prize.) Four decades had passed since the theoretical possibility of the device had been determined and the ground of scientific competence laid. Rubies had been around a lot longer.

Apart from having a myriad of major industrial and military uses, lasers are attractive to communications engineers exactly because they are an ideal carrier wave for any electronically encoded data. We are still on the threshold of their diffusion. The 'law' of the suppression of radical potential is working to contain, not to say divert, the laser from its potential place at the centre of all our electronic communications systems. A quarter of a century after its introduction, the laser stands waiting to take over telephony. Laser-based moving holographic image-making systems, as is detailed below, have yet to be *invented*. In the meantime, communication laser applications are being relegated, outside of the military, to redundancies like the videodisk.

Marketing the laserdisk as a consumer durable in some sense endangers its survival as an archival tool. Archive requirements are not, of themselves, a supervening necessity, although there is some growing awareness, at least in industrial and academic circles, that the stock of audiovisual images held around the world is in danger of deteriorating beyond the point where information can be retrieved. Colour films, especially, are subject to degradation and it is possible that much early imagery will be lost. Nowhere, though, is a systematic attempt being made to transfer this material to laserdisk and thereby gain the promised 600 years of durability. Instead, in 1978, Philips marketed a variant for computer data storage; the disks' interactive capability was being exploited by the US military for education and training purposes (just as 16mm film had been used forty years before); and in 1984 in America, Pioneer, the domestic manufacturers, ran an advertising campaign to sell the players which featured, uniquely in a commercial for a visual device, the blind musicians Ray Charles and George Shearing, stressing the wonderful sound the laserdisk produced.

Perhaps this was in response to Philips' attempt, begun the previous year, in agreement with Sony (another suspicious event), to market a 5 inch laser videodisk upon which nothing was recorded except sound. The plan is to have the population which is currently addicted, in various degrees, to the present system of musical reproduction, called stereo, jettison this in favour of the new system, compact disks (CD), within ten to fifteen years. The disks are compact because Sony and Philips have immediately eschewed one advantage that would accrue from using the developed VLP for this purpose, to wit the complete recordings of long musical works on a single platter. Apparently the record industry believes that it and its need for profit, and not, say, Wagner, should dictate the 'natural' package of a piece. The supervening necessity behind this development is crystal clear. Leaving aside the raging and arcane debate as to the merits and distortions of the CD system, one thing is obvious – CD disks cannot be copied in the home. If they are rerecorded on ordinary stereo they lose their acoustic advantage. *Cui bono* from this inability? The record industry.

It has seen its worst fears of tape realised. Audiocassette recorders are now being manufactured which not only record records but also have a second head for recording other audio cassettes. In Britain the audiodisk business peaked in 1978 – by 1983 sales of singles had fallen 13 per cent, and albums 27 per cent. Significantly, prerecorded cassette sales were up by 75 per cent but this increase of 15 million units was not enough to balance the loss of nearly 32 million albums and 15 million singles. In this half-decade, the British record industry, manufacture, distribution and retail – some 12,000 workers in all – lost 1000 jobs a year.[114] CD thus represents the long-term salvation of the industry and, therefore, it will be a technology harder fought for than some of the other redundant devices here discussed. Sales projections were not being met initially. In the UK less than 30,000 CD players were sold in the first year and by 1984 there were fewer than a million worldwide. Sony introduced a shoulder-strapped portable which was of necessity at 5 inches a good deal bigger than its own audiocassette Walkman, needed an external battery pack, cost about 6 times as much and could not be jogged up and down.[115] But cheaper players began to appear in 1985 and sales started to pick up. The disks' indestructibility made the technology attractive to boat and car owners and the adventurous affluent began to bite the bait.

(iii)   A.N. Other invents the holographic moving image, ?1995?
To conclude this re-evaluation of the technological foundations of our televisual culture we will consider a device yet to be invented – holographic television.

In 1947, Dennis Gabor, an English industrial researcher, was looking to improve the performance of electron microcopes to the point where atomic structures would be revealed. He felt that, with a coherent light source, he could correct the electron picture and achieve this result. He used a high pressure mercury lamp and eventually produced miniature holograms of microphotographs.

A hologram is a photographic record of the difference between a reference beam of light and the pattern of interference created in that beam by an object. It allows for a true stereoscopic representation of the object to be recreated – true, that is, in the sense that the hologram allows the eye to look round the sides of the object just as it could in reality. This is unlike a 3D system using, say, coloured filters and double images to create an illusion of depth where every person sees the same illusion and where head movement will not reveal new facets of the scene.

It was the development of the laser in the early 1960s that gave holographers their basic tool.[116] The laser delivers exactly the strong coherent light source needed. By splitting a laser beam, an arrangement could be made whereby half passed across the object to be recorded while the other half, via a mirror, bounced round it. The photographic plate registers the beam as it has been disturbed by the object and also the undisturbed part as a reference. When a laser is once more pointed at the plate, the object reappears in front. (This system is analogous to the colour film process invented by Gabriel Lippmann, of the lenticular stock, in 1891 where a special fine grain film was exposed as mercury was poured against it. The interference caused between the incoming lightwaves and those being reflected back from the mercury enabled colour information to be registered as standing waves in the emulsion. Lippmann **Photochromes** remain among the few indexically accurate (i.e. independent of dyes and filters) colour photographs ever made, but can only be played back by exactly positioning a strong light behind the plate and the viewer before it.)

Dr Gabor was to receive his Nobel Prize in 1971, twenty years after he had pioneered the holograph. With the coming of the laser Gabor's technique prospered. In 1963, Emmett Leith and Juris Upatnieks showed the first laser-based holograms, marking the start of the prototype phase of this *'invention'*. In the last two decades, holograms have recorded live subjects (with pulsed lasers, since continuous ones cannot cope with any vibrations); in the USSR, since 1962, Denisyuk has produced holograms which would 'playback' in ordinary incoherent light; in 1965 a holographic interferometer was introduced to industry as a way of studying stress by comparing 'before' and 'after' holograms of the object; in 1968, Stephen Benton of Polaroid demonstrated a white light transmission (i.e. lit from behind) holo-gram which removed the need for a playback laser. Holograms achieved full colour in the 1970s.

What use is holography? Supervening necessities are currently limited to certain industrial applications and fine art. Curious uses are not, however, hard to envisage. The military value of having a true three-dimensional image of terrain to be taken – a Belfast terrace, for instance – is self-evident. It would be the ultimate map, a Borgesian conceit corresponding at all points with reality.

But the real question is whether or not holography will acquire movement. It has been slower coming. A conventional film of a person sitting on a rotating plinth – to take an early commercial example of 1975 – was hologrammed to produce a few seconds of stereoscopic movement. It is, like television in the 1920s, just around the corner. But, as with television, as with videotape, as with the devices discussed in the following chapters, the question with holographic moving images is less how than when. Obviously the immense investment in audiovisual reproductive processes is a main factor in suppressing the radical potential of holography. The full effects of videotape are still being absorbed. Beyond what it would have to amortise, the industry of the moving image also has a number of ideas as to what should be consumed next – first component television delivering stereo sound, then home video movie systems and flat display screens breaking the hegemony of the cathode ray tube in receivers with solid-state devices and replacing the pick-up tube in cameras; and, when these have saturated the market, 1200 or more lines high-definition television will follow. After all that, the question will still remain – does our culturally determined addiction to realism constitute a sufficient supervening necessity to ensure moving holo-graphy gets to be *invented*?

'INVENTIONS FOR CASTING UP SUMS, VERY PRETTY'*

Phase one: scientific competence

Monsieur Descartes renders the COMPUTER unthinkable, 1644

In the last chapter, the unrevolutionary development and absorption of television is accounted for in terms of a suggested model for all such histories. It is obviously crucial for the success of the schema herein proposed to demonstrate that it applies to the other central technologies of the information revolution. We will continue with the computer.

At first sight, the model seems singularly inappropriate. How can the pattern of delay and constraint, established above in connection with television, be meshed with the sudden ubiquitousness of keyboards, monitors, diskettes, drives and printers in these last few years? The answer is that the misperception which saw television as an explosive arrival in the 1950s is at work again in the 1980s with the computer. The received history of the computer selectively downplays the lateness of its development and the comparative slowness of its diffusion. For, even by the most conservative of measures, we are now entering the fifth decade of the so-called computer age.

> As is now realized, we had the technical capability to build relay, electromechanical, and even electronic calculating devices long before they came into being. I think one can conjecture when looking through Babbage's papers [nineteenth century], or even at the Jacquard loom, that we had the technical ability to do calculations with some motive power like steam. The realization of this capability was not dependent on technology as much as it was on the existing pressures (or lack of them), and an environment in which these needs could be sympathetically brought to some level of realization. [our parentheses][1]

* 'but not very useful': *Samuel Pepys Diary*, 14 March 1668.

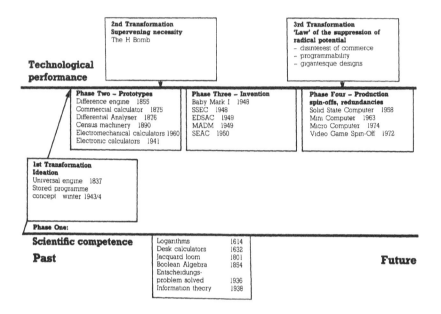

Figure 5 Computing

Thus Henry Tropp, the Smithsonian's historian of the computer pioneers, seeks what in this essay is termed *a supervening necessity*. Nicholas Metropolis, one of the earliest users of **ENIAC** (the machine popularly thought to be the first electronic computer) and the builder of **MANIAC** at Los Alamos, is more specific:

> It is interesting to contemplate how different even global events might have been if some commercial computer company had undertaken the design and construction of an electronic computer in, say, the ENIAC class, that would have been available before World War II. The electronic components used in ENIAC were certainly available.[2]

There is with the computer, then, exactly the same mystery as with the *invention* of television; the puzzle has less to do with why the device appeared when it did but rather why it did not appear very much earlier.

And if the computer's beginnings conform to our schema, so too does the pattern of its diffusion. The 'law' of the suppression of radical potential, as shall be shown below, operated to delay various stages of development for significant periods, exactly as the schema would suggest and, above all, diverted research effort away from the small desktop device that represents the full radical potential of the computer.

The computer – *machina ratiocinatrix* – is, in terms of the philosophical underpinnings of our culture for these last three centuries, an unthinkable instrument. It offends, fundamentally, the Cartesian duopoly of mind and matter. Boyle speaks for all the West when he claims 'engines endowed with wills' are men and cannot be anything else – not even, in those years, other animals.[3] Although from the first objections were raised to this dichotomy and although much of Descartes' view (that the pineal gland was the seat of the mind, for instance) has long since collapsed, nevertheless it is still to a large extent true that '*Nulla nunc celebrior, clamorosiorque secta quam Cartesinorum*'.[4] Certainly the arguments around artificial intelligence are cast in terms which Descartes and his 'school' would have no trouble in comprehending.

These same seventeenth-century *savants* were also intent on enshrining mathematics as the 'queen' of all sciences and the clearest evidence available of the glory of God's creation. The dominance of the empiric scientific method which began in this period stressed observation and measure and thereby, to a degree, encouraged the production of calculating devices both mental and physical. This work was enormously aided by the appearance of printed tables, freed from the inevitability of scribal error and capable, through repeated editions, of incorporating corrections.[5] All this led to a tradition of contradictory attitudes to machines that 'think': Western scientism it can be claimed requires, on the one hand calculators while, on the other, philosophically denying the possibility of what are today described as computers.

Crucial, then in developing the ground of scientific competence for the computer is the removal of the mental roadblock against the *machina ratiocinatrix* erected during the first phase of the scientific revolution. This, despite some earlier musing and theorising, finally occurred in pure mathematics in 1936.

Alan Turing was a scion of empire, the son of an Imperial civil servant and, as such, doomed to a Kiplingesque round of English foster homes, somewhat reduced in term by his father's early retirement. (He numbered Stoney of the 'electron' among his forebears.) After Sherborne, he went up to King's College, Cambridge where he studied under Max Newman. Following a brilliant under-graduate career, Turing was elected fellow, at the age of 22, in 1935. In 1936 he published a paper – 'On Computable Numbers, with an application to the *Entscheidungsproblem*' which dealt, elegantly, with the Cartesian obstruction.[6]

The agenda Turing addressed was at the heart of advanced pure mathematics. By the late nineteenth century, in the wake of the creation of non-Euclidean geometrics (among other developments), mathematicians were becoming, for the first time since the Greeks, increasingly concerned about the consistency of the axiomatic systems they used.

> [T]he creation of non-Euclidean geometry had forced the realization that mathematics is man-made and describes only approximately what happens in the world. The description is remarkably successful, but it is not the truth in the sense of representing the inherent structure of the universe and therefore not necessarily consistent. . . . Every axiom system contains undefined terms whose properties are specified only by the axioms. The meaning of these terms is not fixed, even though intuitively we have numbers or points or lines in mind.[7]

It was against this background that Bertrand Russell coined his famous epigram:

> Pure mathematics is the subject in which we do not know what we are talking about, or whether what we are saying is true.[8]

Despite this there were those, led by David Hilbert, the greatest mathematician of his generation, who in the early decades of this century insisted on the primacy of the axiomatic method. Against these assertions stood, ever more starkly in relief, a set of highly technical problems which can perhaps be most simply instanced by drawing 'the cork out of an old conundrum' viz the ancient 'liar paradox'. This paradox can be classically expressed in the sentence, 'This sentence is false.' For twentieth-century mathematics, dealing with mathematical equivalents of the 'liar paradox' meant confronting the problem of consistency across an increasing range of topics. At first, although at the cost of developing a variety of methods and schools, many of the paradoxes or antinomies (strictly – a contradiction in law) were resolved; as, for example, in the system proposed in Whitehead and Russell's *Principia Mathematica*. Because of this work, well into the 1920s, Hilbert continued to assert: 'that every mathematical problem can be solved. We are all convinced of that.'[9] In 1931, this time-honoured approach sustained the most telling attack yet. It was contained in a paper by the Czech Kurt Godel – 'On Formally Undecidable Propositions of *Principia Mathematica* and other systems'. In this Godel demonstrated that it was impossible to give a proof of the consistency of a mathematical system 'comprehensive enough to contain the whole of arithemetic'.[10]

> Godel's second main conclusion is even more surprising and revolutionary, because it demonstrates a fundamental limitation in the power of the axiomatic method. Godel showed that *Principia*, or any other system within which arithmetic can be developed, is *essentially incomplete*. In other words, given *any* consistent set of arithmetical axioms, there are true arithmetical statements that cannot be derived from the set. (italics in original)[11]

Godel's incompleteness theorem highlighted a number of subsidiary problems, chief among them, from our point of view, the question of decidability. If there were now mathematical assertions that could be neither proved nor disproved, how can one determine effective procedures in such cases? This was the decidability or decision problem – *das Entscheidungsproblem*. Just as Hilbert had declared that mathematical systems had to be consistent and complete, so too had he insisted upon the discovery of effective procedure as a necessary part of mathematics. Godel's attack on consistency and completeness rendered the decidability problem moot.

It was Turing, five years later, who dealt with *das Entscheidungsproblem*. Turing had been struck by a phrase in a lecture of Newman's where Hilbert's suggestion that any mathematical problem must be solvable by a fixed and definitive process was glossed by Newman into 'a purely mechanical process'. Turing, in his paper, found a problem that could not be so decided i.e. solved – in Turing's language 'computed'. It involved an artificial construct known as the Cantor diagonal argument whereby 'irrational numbers' could be created. (Cantor was one of those nineteenth-century mathematicians whose work set the stage for the crisis in axiomatic methods.) To dispose of the decidability problem, Turing constructed a mental picture of a machine, a conceit, and demonstrated that it could not compute certain numbers. Therefore there were mathematical problems which were not decidable; but, Turing wrote, 'It is possible to invent a machine which can be used to compute any *computable* sequence.' (our italics)[12]

Because of this conceit of a machine, 'On Computable Numbers' had, beyond its immediate significance in pure maths, broader and perhaps more far-reaching implications. The proof involved a machine which read, wrote, scanned and 'remembered' binary numbers inscribed on a unidimensional tape. It might not be able to compute the irrational numbers of Cantor's trick but it could, in theory, deal with a vast, enormous range of other computations. The very disposal of Hilbert's problem required no less. Turing had conceived of a tremendously powerful tool. He was not overstating the case when he christened it (in homage to Babbage) a 'universal machine'. Of course, he had no intention of building such a machine. Later it would be said of him – perhaps unfairly in the light of his wartime experience as a practical electronics engineer – that although 'He was intrigued by devices of every kind, whether abstract or concrete – his friends thought it would be better if he kept to the abstract devices.'[13]

In Cambridge in 1936 he was still concerned only with the abstract. When he wrote 'computer', he meant, as did all his contemporaries, a person who performs computations:

> The behaviour of the computer at any moment is determined by the symbols which he is observing, and his 'state of mind' at that moment. We may suppose that there is a bound B to the number of symbols or squares which the computer can observe at any one moment. If he wishes to observe more, he must use successive observations. . . . Let us imagine operations performed by the computer to be split up into 'simple operations' which are so elementary that it is not easy to imagine them further divided. Every such operation consists of some change in the physical system consisting of the computer and his tape.[14]

The human computer and his tape were to become the machine computer and its programme. 'To each state of mind of the computer corresponds a[n] . . . configuration of the machine.'[15] It was to be a *machina ratiocinatrix*.

> His argument held out the possibility of building an actual machine to do all the work that can be done by any human computer; and he extended the same model to provide a new analytic account of states of mind and mental operations. At their intellectual or rational core (which was a central preoccupation of René Descartes' arguments) all such states and operations rested on the kinds of procedures that his universal machine could in principle perform; and, from this insight, . . very general conclusions followed. First, the idea of a 'thinking machine,' which the seventeenth-century philosophers had regarded as a contradiction in terms, was now after all an admissible idea.[16]

Of course, with mathematicians all over the world attempting the *Entscheidungsproblem*, it was almost inevitable that Turing would have competitors. In mid-April 1936 he presented his paper to Newman in Cambridge. On April 15th, Alonzo Church, of Princeton, sent his demonstration of a different unsolvable proposition away for publication. Church's solution was close enough to require Newman's intervention on Turing's behalf to ensure that the younger man's work could be published. Turing equated Church's idea of 'effective calculability' with his own notion of 'computability' and offered, in an appendix to 'On Computable Numbers', a proof of their equivalency, admitting Church had reached 'similar conclusions' about the *Entscheidungsproblem*.[17] In October of 1936, Emil Post, a mathematician at the City University of New York, submitted a paper to Church suggesting a mechanical device – a 'worker' – for demonstrating Church's proposition along lines close to, but less ambitious than, a 'universal machine'.[18] Post nowhere mentioned 'states of mind'. Church, in a footnote to Post's paper, writes: 'The reader should compare an article by A.M. Turing, *On computable numbers*, shortly forthcoming in the *Proceedings of the London Mathematical Society*.'[19] In a subsequent comment he went further, acknowledging the power of Turing's approach by coining the phrase 'Turing machine' (or now, 'turing machine') as a synonym for 'universal machine'.

These men stand in a line of mathematical logicians traceable back to the self-taught nineteenth-century English *savant*, George Boole. In 1854, Boole faulted the Cartesian foundation by demonstrating that logics can be expressed algebraically. Boole's investigation was designed to explicate:

> the fundamental laws of those operations of the mind by which reason is performed; to give expression to them in the symbolic language of a Calculus, and upon this foundation to establish the science of Logic. . . . There is not only a close analogy between the operations of the mind in general reasoning and its operations in the particular science of Algebra, but there is to a considerable extent an exact agreement in the laws by which the two classes of operations are conducted.[20]

Turing, as a child at prep school, complained that his first arithemetic master gave 'quite a false impression of what is meant by x', presumably because he did not sufficiently indicate Boolean possibilities.[21] For most of us the xs and ys of elementary algebra 'stand for' unknown numbers, which was after all what François Vieta, the seventeenth-century mathematician who introduced them, intended; for the pure mathematician, though, these letters can be manipulated to symbolise the entire world of mind and matter. 'Pure mathematics were discovered', states Russell, 'by George Boole in his work published in 1854.'

The most powerful of Boole's concepts, so far as the scientific competency needed for the computer is concerned, is a 'special law to which the symbols of quantity are not subject':

this law in effect is that $x^2=x$ for every x in his [Boole's] system. Now in numerical terms this equation or law has as its only solution 0 and 1. This is why the binary system plays so vital a part in modern computers: their logical parts are in effect carrying out binary operations.[22]

The law turns, then, on the use of a binary system of notation which, although arguably the most primitive conceivable such system, dates in its modern mathematical form to Bacon and to Leibniz, the latter using it for – among other things – the creation of a mathematical proof of the existence of God. Boolean algebra, by reducing certain types of thought to a series of on/off states, is the means by which a turing machine can be said to think, make judgments and learn. That Boole and all the other pure mathematicians, including Turing before the end of the Second World War, built no such machines does not detract from their centrality in preparing the ground of scientific competence which could be transformed by technology into the computer.

In the third decade of this century there was much going on in mathematics which would help to translate activities popularly considered as uniquely human into forms that would be 'machine readable', if any such machines had existed. In 1938 Claude Shannon, whom Turing was to meet during a wartime visit to the United States, published his MIT master's thesis, 'A Symbolic Analysis of Relay and Switching Circuits' in which the insights of Boolean algebra were applied to telephone exchange circuit analysis.[23] This produced a mathematisation of information which not only had immediate practical applications for his future employers, Bell Labs, but also plumbed another part of modern computer science's foundations. Information Theory, as Shannon's work is called, defines information as the informational content of signals abstracted from all specific human information. It concerns not the question, 'what sort of information?' but rather, 'how much information?'[24] 'The word *information*, in this theory, is used in a special sense that must not be confused with its ordinary usage. In particular, *information* must not be confused with meaning.'[25]

In a telephone exchange, design requirements dictate that there be less concern about the content of messages than about the accuracy with which the system will relay them. Information becomes reified, quantifiable so that it can be treated as a measure of 'the probability or uncertainty of a symbol or set of symbols'.[26] By how much does the transmitter's message reduce uncertainty in the receiver? – by that much can the informational content of the message be measured and capacity of the channel of communication be determined.

Say we were awaiting the result of a race between four horses which would be signalled to us by flag – the code, as is necessary in any information system, having been previously agreed between transmitter and receiver. The waving of two differently coloured flags would suffice to convey the information. (Horse 1, red/red; horse 2, red/blue etc.) In an eight horse race the same two flags could be used but each horse would require three waves. (Horse 1, red/red/red; horse 2, red/red/blue, etc.) Each wave of the flag is a binary digit, a bit – red being, let us say, the equivalent of '1' or 'on' and blue being '0' or 'off'. The four-horse race requires a channel with a two-bit capacity, the eight-horse race requires three bits. Sixteen horses would require four waves or bits and so on. Each wave, each transmission of a bit, reduces our uncertainty as to the outcome of the race. The 'bound B to the number of symbols or squares which the computer can observe at any moment' (of which Turing writes) can be expressed as the capacity of the computer, human or mechanical, to address a discrete number of bits.

The quantification of information in Information Theory parallels and perhaps determines the reification of information which is so crucial a part of the 'information revolution'; that is to say, the rhetorical thrust that has the production of information in so-called 'post-industrial societies' substituting for the production of physical goods depends upon such reification.

The implication of all this work was not immediately apparent even to those most able to understand it, the mathematical community. Pure mathematical logic was so pure that few human activities could be considered further from putting the bread on anybody's table. But many were to get rich at Turing's feast, once they understood him.

> To illustrate the difficulty of deciding when something is really understood, consider that many mathematicians have the experience of knowing a theorem and a proof for several years and then suddenly 'understanding' the theorem in a new way. Occasionally there may be several of these episodes of suddenly understanding what one had apparently known all along.
> The example I shall take to illustrate the difficulty of measuring when an idea was first understood is the idea of the computer as a *symbol manipulator* rather than as a number cruncher. This was in a sense one of the decisive steps in the history of computing. . . . One could claim that Turing, when he proved that the universal computing machine could do anything any computer could do, must have understood the idea that computers are symbol-manipulating machines.[27]

Here then is the importance of 'On Computable Numbers'. By moving from number-cruncher to symbol-manipulator Turing threw the first plank across the Cartesian chasm between human being and machine. The thinking of those who were to design the first computers enlarged this bridge. John von Neumann, a student of Hilbert's and a mentor of Turing's, was one of the fathers of the American computer and a mathematician of enormous range – from game theory to nuclear devices – who dominated the field in the first decade after the war. He wrote, in 'First Draft of A Report on the EDVAC', the document that contains the original master-plan for the modern computer:

> Every digital computing device contains certain relay-like *elements*, with discrete equilibria. . . . It is worth mentioning, that the neurons of the higher animals are definitely elements in the above sense. They have all-or-none character, that is two states: Quiescent and excited.[28]

The seductiveness of the analogy between human neural activity and digital symbol manipulators has proved irresistible. In his last work, von Neumann wrote:

> In terms of the number of actions that can be performed by active organs of the same total size (defined by volume) in the same interval, the natural componentry is a factor 10^4 ahead of the artificial one . . . the natural componentry favors automata with more, but slower, organs while the artificial one favors the reverse arrangement of fewer, but faster, organs.[29]

'The natural componentry' is the human brain, the 'artificial' a bulky late-1950s computer.

Von Neumann does not appear to be mounting a case for artificial intelligence *per se* but simply drawing some parallels. Such efforts have been a characteristic of Western thought throughout the modern period, beginning with Lamettrie's *L'Homme Machine* in 1750. Seeing humanity in the image of whichever machine most dominates contemporary life is what might be called *mechanemorphism*. It first privileged the clock, and then the combustion engine as suitable models. Freud thought electromagnets a metaphor for the brain, and today this tendency is finding its most extreme expression with the computer. Mechanemorphism has conditioned not only our overall attitude to computers but also the very terminology that has arisen around them. To take one example: what crucially distinguishes the computer from the calculators that precede it is its capacity to store data and instructions. Calling this capacity 'memory' and the codes that control its overall operation 'language' facilitates the mechanemorphism of certain computer enthusiasts.

The potential confusion of human and machine has never been

greater; and, now, understanding human beings in terms of the computer implies, for some, more than just the acknowledgement of the binary nature of the synapses in the nervous system. It calls into question the fundamental privilege of biological beings, so that it is a further small step from von Neumann's parallels to cybernetics, 'which treats of the principles underlying the common elements in the functioning of automatic machines and of the human central nervous system.' For Norbert Wiener the most important commonality is that both automata and nerves are subject to the Second Law of Thermodynamics:

As entropy increases, the universe, and all closed systems in the universe, tend naturally to deteriorate and lose their distinctiveness, to move from the least to the most probable state, from a state of organisation and differentiation in which distinctions and forms exist, to a state of chaos and sameness. . . . But while the universe as a whole, if indeed there is a whole universe, tends to run down, there are local enclaves whose direction seems to be opposed to that of the universe at large and in which there is a limited and temporary tendency for organisation to increase. Life finds its homes in some of these enclaves.[30]

Since the definition of life is, for Wiener, simply a measure of entropy, there is no duopoly left and machines can just as well inhabit the enclaves of life as any biological being.

Life, (he writes,) purpose and soul are grossly inadequate to precise scientific thinking . . . if we wish to use the term life to cover all phenomena which locally swim upstream against the current of increasing entropy we are at liberty to do so.

There has followed from this nearly forty years of predictions that any day now there will be an 'electronic brain'. The claims were originally quite restrained. Lord Mountbatten described such a brain in 1946 as one which would perform functions 'analogous to those at present undertaken by the semi-autonomous portions of the human brain'. Some of these would 'exercise a degree of memory, while some were being designed to employ those hitherto human prerogatives of choice and judgment'. In stating this view, Mountbatten was following what has developed as a goal for one section of the computing community. The 'electronic brain' has been a most eloquent theme lodging itself firmly in popular consciousness and as such it is another of the crucial underpinnings of the 'information revolution'. At its highest level this theme is now encompassed in cognition studies. These have as their topic artificial intelligence and have replaced cybernetics, at least in its most catholic Wienerian manifestation, as the main arena of academic mechanemorphism.

Turing, given that he had discussed 'states of mind' in his original paper, was prepared, by 1950, to be bold enough to predict a machine, with a 10^9 storage capacity, which could fool an 'average interrogator' into thinking it was human '70 per cent' of the time. It could do this only in the circumstances of what Turing called 'the imitation game', which limited the interplay of human and machine to question and answer and allowed no more than five minutes of interrogation. Further, he still had in mind as a model for his 'thinking' machine a human computer who 'is supposed to follow fixed rules; he has no authority to deviate from them in any detail. We may suppose that these rules are supplied in a book, which is altered whenever he is put onto a new job.'[31] Despite these extensive caveats, six pages later Turing claimed:

> In considering the functions of the mind or the brain we find certain operations which we can explain in purely mechanical terms. This we say does not correspond to the real mind: it is a sort of skin which we must strip off if we are to find the real mind. But then in what remains we find a further skin to be stripped off, and so on. Proceeding in this way do we ever come to the 'real' mind, or do we eventually come to a skin which has nothing in it? . . The original question, 'Can machines think?' I believe to be too meaningless to deserve discussion. Nevertheless I believe that at the end of the century the use of words and general educated opinion will have altered so much that one will be able to speak of machines thinking without expecting to be contradicted.

But from the beginning there were those who resisted the attractions of the mechanemorphic view. Both Douglas Hartree and Sir Charles G. Darwin, those in charge of computer development (and Turing) at the British National Physical Lab (NPL), immediately saw fit to decry Lord Mountbatten's prediction in the pages of *The Times*. 'The term "electronic brain" is misleading. I hope this term will be avoided in the future,' wrote Hartree.[32]

The hope has not been fulfilled. The end of the century fast approaches and the original computers and three further generations of machines have been developed displaying the prerogatives of choice and judgment but still appearing to be manifestly inhuman to the average interrogator. The predisposition to contradict Turing's view continues unabated; the philosopher J.R. Searle, for example.

> No one supposes that computer simulations of a five-alarm fire will burn the neighbourhood down or that a computer simulation of a

rainstorm will leave us all drenched. Why on earth would anyone suppose that a computer simulation of understanding actually understood anything? It is sometimes said that it would be frightfully hard to get computers to feel pain or fall in love, but love and pain are neither harder nor easier than cognition or anything else. For simulation, all you need is the right input and output and a program in the middle that transforms the former into the latter. That is all the computer has for anything it does. To confuse simulation with duplication is the same mistake, whether it is pain, love, cognition, fires or rainstorms.[33]

Searle's position is currently bolstered by the fact that no machines come close to the capacities supposedly needed for displays of artificial intelligence. Apparently the fifth and next generation of computers will be true *machinae ratiocinatrices*. If this proves to be the case, Searle will still be able to deny the machines a 'mind'; and conversely, if the prophecy proves false the faithful, in the manner of Jehovah's Witnesses, will wait, hope undiminished, for the sixth generation. For the moment one can only agree with M.V. Wilkes, the man, contrary to received history, who built the first fullsize working computer and who wrote three decades ago: 'I see a bright future for machines in executive capacities, but I do not see them behaving like human beings.'[34]

All this has little or nothing to do with Turing's grasp of the difference between a number-cruncher and a symbol-manipulator. Symbol-manipulation was to be of ever-increasing significance over the twenty years from 1936 and, although not clearly perceived at first, it was inexorably to transform schemes for building number-crunchers into the first computer designs.

The first transformation: ideation

Richard Clippinger invents the COMPUTER, January 1944

(i) Exactly the same sort of memory device
A symbol-manipulator, as opposed to a number-cruncher, must have an in-built set of instructions which can be varied by the operation of the machine itself as a result of its actual processes of manipulation; and it must therefore have an extensive data-store. In practice, meeting these requirements most likely means that the symbol-manipulator must be entirely electronic. This is a computer; everything else, even if entirely electronic and massively fast, is a calculator.

The confusion in the popular account between the earliest computers and the powerful electronic calculators that immediately preceded them is because the line between number-crunchers and symbol-manipulators was, in the 1940s, a thin one. The very same people were involved with both sorts of machine. These teams, created by many wartime combatants to assemble the ultimate calculators, used that experience to launch themselves into the designs of full-scale computers; as they built the one they began to draw the other. The two classes of machines looked very similar, performed similar functions and were, largely, assembled from the same parts. Programmability and 'memory' factors can be surprisingly difficult to determine, for is a machine unprogrammable if it can be rewired, however laboriously, to perform different tasks? And since the biggest electronic calculators did store and retrieve numbers in the course of their operations, how much 'memory' constitutes an effective data-store? Finally, all these developments were shrouded in the secrecy of the two wars, the hot one with Germany and the cold one that followed it almost immediately.

The American team building ENIAC, which, in this book and contrary to received history, will be designated an all-electronic calculator, was housed in the Moore School of Engineering in the University of Pennsylvania. The history of this machine, a prototype computer, is detailed below (page 134) but it is here necessary to consider the moment in that history when the idea of the computer proper first emerged.

The ENIAC team had been at work since the spring of 1943. The design had originated with John Mauchly, a professor at the school who had a varied background as engineer and physicist. The money for its development, provided by the Ballistics Research Laboratory, had been found by Lieutenant Herman Goldstine, who was in the Moore School as the naval officer commanding the Laboratory's Philadelphia substation. He was using the advanced mechanical calculator the school owned. These two and a young engineer, Prosper Eckert, headed the team designing ENIAC.

ENIAC's progress, though, was very slow. Eckert and Mauchly began to generate ideas for improvements, but these new ideas could not be incorporated into the ongoing machine, for that would have slowed the pace even more. By August of 1944, Goldstine began lobbying for a new machine to utilise the advances being suggested. In September he put on paper the superior lines of enquiry he thought ought to be followed:

there are two further directions in which we should pursue our researches. The switches and controls of ENIAC [which are] now arranged to be operated manually, can be positioned by mechanical relays and electromagnetic telephone switches which are instructed by a Teletype tape. . . . [With this arrangement] tapes could be cut for many problems and reused when needed. . . . The second direction to be pursued is providing a more economical electronic device for storing data than the accumulator. Eckert has some excellent ideas on a cheap device for this purpose.[35]

This note contains suggestions for both programmability (the Teletype tape) and 'memory' (a device for storing data). There is a strong possibility that the team's thinking was in advance of what Goldstine thought proper to put down on paper. The 'memory' suggestion represents the point of ideation reached that autumn but the programme suggestion is still crude, giving no indication that the machine might alter its own functions in the light of its actual operations – that is, have the read-write capacity which is the final distinguishing mark of the symbol-manipulator. However, the solutions to the 'memory' problem being proposed by Eckert contained within them also a solution to the programmability problem and one that went well beyond tickertape to read-write.

It was all very well building devices to take advantage of electronic speeds, but 'memory' required that the electron stream be stopped and held in some way. William Shockley, a Bell Labs scientist who will figure again in this story, had developed an elegant device which could achieve this. It allowed electronic pulses to be converted into a physical form and then reconverted into electrons. Using this development, Eckert designed a 'memory' for the calculator in which mercury-filled tubes, 4½ feet long, would be arranged so that the pulses (representing the digits in the calculator) would activate a quartz crystal at one end of the tube. This would create an ultrasonic wave in the mercury. The wave would travel infinitely more slowly than the pulse. The slow-down would act like a memory, storing the pulse. At the tube's far end, another crystal would convert the ultrasonic wave back into a pulse. Place the mercury tube in a loop and a stream of pulses, digits, could be made to circulate continuously until needed by the computer. The machine would 'remember' the digits. Mauchly outlined the implications in these terms:

[L]et us consider some of the fundamental characteristics of this type of machine, in particular those points which differ significantly from present machine design [i.e. ENIAC-type calculators]. Of these, three have a definite bearing on the handling of problems: (1) an extensive internal memory; (2) elementary instructions, few in number, to which the machine will respond; and (3) ability to store instructions as well as numerical quantities in the internal memory, and *modify instructions so stored in accordance with other instructions*. [our brackets and italics]

By extensive internal memory is meant a memory accessible at electronic speeds, of the order of a thousand numbers, each of perhaps ten digits. [our brackets and italics][36]

The evidence is that, long before August of 1944, the implications for programming in the mercury-tube 'memory' solution had already been realised. Richard Clippinger, of the Ballistics Research Laboratory, had suggested that programming instructions might be encoded into the mercury as well as the data that needed to be stored. Prosper Eckert had written a paper in January of 1944, 'Disclosure of a Magnetic Calculating Machine', in which various storage systems were explored, including magnetic disks or drums. The paper, which was not actually typed up until February of 1945, contained subsidiary suggestions such as neon gas display panels and envisaged a binary machine,[37] but it also suggested: 'An important feature of this device was that the operating instructions and function tables would be stored in exactly the same sort of memory device as that used for numbers.'[38]

For whatever reason the idea of the stored programme was not fully articulated in Goldstine's suggestion that tickertape be substituted for ENIAC's manually set switches and plug-board. Clippinger has yet to receive full credit for his insight and Eckert's claim to the idea has been compromised by delay in publicising the 'Disclosure of a Magnetic Calculating Machine'. The paper was untyped from January of 1944 until February of 1945 and then it still remained secret. By the time Eckert and Mauchly published their account of the progress of the stored programme machine idea, in September of 1945, it was too late. By that date the nascent world of computing knew of it from another source.

In August 1944, John von Neumann, to whom Goldstine had introduced himself on a railway station, first visited the ENIAC team. A series of meetings ensued wherein von Neumann, with Eckert, Mauchly, Goldstine and the others, explored the problems of logical control that the storage lines would create. Von Neumann organised the memoranda of these meetings into a report, issued on 30 June 1945, 'First Draft of A Report on The EDVAC'. **EDVAC** stands for *E*lectronic *D*iscrete *V*ariable *A*utomatic *C*omputer, where the crucial term is *Variable*. The 'First Report' was issued by the university, it has been suggested by his partisans, without von Neumann's approval, since it fails to credit the others. Its publication marks the earliest comprehensive written outline for the modern computer, the point of ideation, the transformation from the ground of scientific competence. It also marks the beginning of what was to be an increasingly acrimonious relationship between von Neumann and Goldstine on the one hand and Eckert and Mauchly on the other.[39]

That Goldstine, over a quarter of a century later, called this paper 'the most important document ever written on computing and computers' and in his book on computing history systematically underplays every contribution but von Neumann's, is nothing but an indicator that the new machine did not bring forth new men.[40]

Goldstine claims that, 'von Neumann was the first person . . . who understood explicitly that a computer essentially performed logical functions, and that the electrical aspects were secondary.'[41]

It is obvious (writes Goldstine) that von Neumann, by writing his report, crystallized thinking in the field of computers as no other person ever did. He was, among all the members of the group at the Moore School, *the* indispensable one. Everyone there was indispensable as regards some part of the project – Eckert, for example, was unique in his invention of the delay line as a memory device – but only von Neumann was essential to the entire task. (italics in original)

Goldstine grounds this somewhat hyperbolic claim on von Neumann's 'writing his report' in which he finds five major contributions. The third is the crucial one laying out von Neumann's stake in the stored programme concept.

Third, in the famous report he proposed a repertoire of instructions for the EDVAC, and *in a subsequent letter* he worked out a detailed programming for a sort and merge routine. This represents a milestone, since it is the first elucidation of the now famous stored program concept together with a completely worked-out illustration. (our italics)[42]

Goldstine introduces a letter rather than the 'famous report' because the latter, a general and quite accessible document, is actually somewhat vague about the importance of the stored programme. Section 2.4, for instance outlines why EDVAC needs a considerable memory and the various tasks it would have to undertake, including the storage of instructions, but it does not privilege this function.[43] To maintain his claim for his hero, Goldstine cannot in fact rely on the 'First Report'. He needs 'a subsequent letter' to be 'the first elucidation of the stored program concept'; but his partisanship must not be allowed to obscure the crucial contribution that von Neumann made to EDVAC and the modern computer. To von Neumann's credit, after all, must go the supply of the logic of such storage mechanisms so that he 'was the first to . . . *exploit* the fact that when orders or instructions are stored in a high speed READ-WRITE electronic "memory", they can be manipulated arithmetically and modified by the machine itself'. (our italics)[44] Eckert and Mauchly, as stated above, issued their own progress report on EDVAC exactly twelve weeks after the 'First Draft'. Their report, and this is the only part quoted by Goldstine, thanks Dr von Neumann for his availability as a consultant.

Leaving the claims and counterclaims aside, Eckert had a way to make a 'memory' and Clippinger had seen its potential as a stored programme device. Mauchly points out that the ENIAC design already incorporated branching and looping requiring 'electronic stepping switches and counters that contained program information which was altered by the program itself'. Mauchly then claims, 'Clearly, we were already providing "stored program" in the most important phases of control' – which is again a hyperbolic statement since the 'control', if it could be so called, was of a very limited and physical kind.[45] ENIAC was not a stored programme computer in any modern sense. These (and other) Americans were eventually to take such authorial differences to the patent court.

Moreover, there are other claims – an 'indistinct conception' of the nineteenth-century pioneer Babbage that he might arrange a set of punched cards to cause other cards to be punched or a 1936 patent application of the German Konrad Zuse which envisaged 'a computational plan' mixing data and instructions.[46]

Nevertheless, and despite these, it was the 'First Draft', whatever its degree of originality, that received wide circulation. Turing used it to be read 'in conjunction' with his own proposal made at the National Physical Laboratory to which he had gone in October of 1945 specifically to build the **ACE** computer. The Manchester University team that Turing joined on leaving the NPL which had made the first electronic computer, the **BABY MARK I**, to run a stored programme (on 21 June 1948); the team led by M.V. Wilkes, that made **EDSAC**, at Cambridge University, the first fullsize stored programme machine actually to work on a real mathematical problem; the team that made EDVAC and its descendants (the Edvacides as they might be termed); von Neumann's own machine, the IAS computer, at the Institute for Advanced Studies at Princeton (and the Iasides, its descendants); all these, although differing considerably in their capacities and architecture, were in line with the 'First Draft'.

It was as if, in the early 1890s, somebody had specified, in some detail, the fundamental design elements of an automobile – and that this document acted prescriptively to limit the range of options available to engineers in the decades following. From this point of view, von Neumann's influence cannot be overstated. Positioned at the heart of the military-industrial complex, he planned a generation of enormously complex machines primarily useful to government-funded applications. All other potential exploitations of the stored programme idea, including most significantly the implications of the small size of the very first working computer, remained unexplored.

But the computer's birth was a late one. Such limited privilege as can be accorded von Neumann, in this ideation transformation, must reflect this. It is less that he was the obstetrician facilitating a difficult birth; rather he was the adjacent doctor, say in a traffic jam on the way to the hospital, who finds the baby delivered into his lap. The actual parentage is still much in dispute, although it is clear that the full credit must go to the ENIAC team in one way or another. But the thinking of that team did not issue from the mind of Zeus full grown. The tradition in which the idea of the read-write store occurred was at least 110 years old.

(ii) Executed by steam

By 1883, the English mathematician Charles Babbage, whom we have already had occasion to mention, was abandoning work on a complex calculator which had occupied him for the previous decade. Instead he now envisaged a device which could tackle, mechanically, any mathematical problem. This **universal analytic engine** (memorialised in Turing's **universal engine**) was never built, but the design embodied a series of concepts which were to be realised a century or more later.

The machine was to be programmed by the use of two sets of punched cards, one set containing instructions and the other the data to be processed. The input data and any intermediate results were to be held in a store and the actual computation was to be achieved in a part of the machine Babbage designated the mill. It would have been able to perform conditional branching operations, basic logical steps, by hopping backwards and forwards between the mill and the store. It was to print out its results automatically.

All of this is so close to the modern computer – the mill as the CPU, the operational cards as the ROM and the store as the RAM, etc. – that Babbage has been hailed as its father. Wilkes wrote: 'Babbage was moving in a world of logical design and system architecture, and was familiar with and had solutions for problems that were not to be discussed in the literature for another 100 years.'[47] Nevertheless, the possibility that the analytic engine could alter its programme in the light of its computations remained an 'indistinct conception' in a late notebook and therefore, for all his prescience, his thought remained one crucial step away from the computer proper. Like ENIAC, and like all the machines from ENIAC back to the analytic engine, the device was a calculator of a most advanced type. (It should equally be noted, though, that attacks claiming Babbage was forgotten and therefore unable to influence twentieth-century computer pioneers are unfounded. References to Babbage's work occur regularly from 1889 when his son, Major General H.P. Babbage, published *Babbage's Calculating Engines*.[48])

Babbage's interest in automatic calculation sprang from the common root – boredom. Like Mauchly, Zuse and Aitken, three major computer pioneers a century later, and like Leibniz, Pascal and Napier before him, Babbage disliked the computational aspect of mathematical work. In one story he reportedly muttered at a colleague, the astronomer Herschel, 'I wish to God these calculations had been executed by steam.' Herschel is said to have replied: 'It is quite possible.' In another account, Babbage remembers himself, 'in a dreamy mood', dozing over a table of logarithms in the rooms of the Analytical Society at Cambridge (of which he was a founder member) and when questioned by another member as to what he was dreaming about, he replied, 'I am thinking that all these Tables (pointing to the logarithms) might be calculated by machinery.'[49]

By 1823 Babbage had convinced the British government of the viability of his plan to do calculations, if not by steam, then certainly mechanically. The navy's need for nautical tables – a need dramatised by shipwrecks supposedly caused because of errors in the *Nautical Almanacs* – was the supervening necessity that set Babbage to work.[50] The creation of error-free mathematical tables thus had important military aspects as well as being an essential part of the work of Kuhnian 'normal science'. From the simple multiplication table of the schoolroom to complex topics such as lists of air densities as a function of altitude at a given time, the table, thanks to the printing press, is a basic tool of scientific culture. However, creating tables is a work of utter drudgery, involving 'intolerable labour and fatiguing monotony' as Babbage wrote in 1822.

Babbage planned and began to build a **difference engine**. A calculus of differences, the theory of which was to be comprehensively treated by Boole in 1860, created a method for converting multiplication and division into operations of addition and subtraction. A series of terms is determined wherein each term is subtracted from the term immediately preceding it (differences of the first order) and the results are similarly treated (differences of the second, third, etc. . . . . order). Such lists can be manipulated, by addition or subtraction, to produce multiplications or divisions. A mechanical device, not unlike the ratcheted wheels which record mileage in a car but more complicated, can automatically and without error calculate complex polynomials using such a calculus. Babbage himself, in working on this machine, was in a tradition already two centuries old in his time. Indeed, his exact proposal had been made, unbeknownst to him, by a Hessian Army officer, J.H. Muller, in 1786.[51]

Logarithms were the first modern mathematical aid, introduced by the Scottish mathematician and theologian, John Napier, in 1614. He too was driven by the drudgery of calculations. Three years later, he produced a simple logarithm-based device to perform multiplications. A set of rods with numbers inscribed on them, **Napier's Bones**, could be manipulated to give instant answers. These led to the development of the **slide rule**, by William Oughtred, another seventeenth-century divine. But already by this time, 1632, Wilhelm Schickard, a professor at Tübingen, had built an automatic machine for basic calculations.

'Rideres clare' . . . 'You will have to laugh when you see how it accumulates left carries of tens and hundreds by itself,' wrote Schickard to Kepler.[52]

As Schickard's plans were only rediscovered in 1935, the most widely known of these early calculators was the **Pascaline**. It was built by the 20-year-old Blaise Pascal possibly to relieve the labour of his father, the presiding judge of the tax court at Clermont-Ferrand and himself a respected mathematician. By the middle years of the seventeenth century Pascalines were well known. Leibniz improved the device by introducing cogged wheels milled in steps – Leibniz wheels – so that the processes of multiplication and division could be done more directly. In the Pascaline, multiplications and divisions were treated as a series of sequential additions or subtractions, as they were in all developments of the Pascaline, such as the engine built by Lord Stanhope in 1775. (Stanhope, apart from his calculator, built a simple slide-rule device to demonstrate, in analogue fashion, logical propositions, a step towards the Boolean concept of a calculus of logic.) All over Europe, throughout the eighteenth century and beyond, scientists worked on the calculator theme. In 1817, for instance, Adam Stern, government censor and inspector of Jewish schools in Warsaw, demonstrated to the Societatis Scientiarum Varsoviensis a machine that handled square roots in six digit units. In 1820, Charles de Colmar, a French insurance man, marketed a sophisticated version of the Leibniz-wheeled calculator, the **arithmometer**.

This was the background informing Babbage's thinking on the difference engine. It was also, perhaps, the key, beyond his own erratic temperament, to why he abandoned the project in the early 1830s. The difference engine was, in some ways, a step back from Leibniz's machine and its derivatives. Although it was much more powerful, it was also more limited. Leibniz's calculator could aid a human computer in any mathematical task. The difference engine needed far less human input, but produced results across a more limited mathematical field. Thus the universal engine was designed to have the broad capacities of the Leibniz calculator but to require as little human input as the difference engine.

If intelligently directed and saved from wasteful use, such a machine [the **universal engine**] might mark an era in the history of computation, as decided as the introduction of logarithms in the seventeenth century did in trigonometrical and astronomical arithmetic. Care might be required to guard against misuse, especially against the imposition of Sisyphean tasks upon it by influential sciolists. [our parentheses][53]

At the time this report was written Babbage had been dead seven years and his son, the Major General, was seeking funds to continue the work. In the event, he, like his father, failed to complete the machine.

There were others. For instance, a Dublin accountant, Percy Ludgate, worked on plans for an **analytic machine** from 1903 until his death, aged 39, in 1922. He left nothing behind except one paper published in 1909. In it he outlines the use of single punched papertape containing both data and instructions and the concept of the numerical storage address.[54]

Babbage's career incorporates many themes that were to reappear in the history of computing. He showed the way in being the recipient of government funding and in exploiting, for purely mathematical purposes, the military needs of the day. Many were to do exactly the same. More generally, in terms of our schema, he represents the very personification of the ideation transformation, since his entire life's work contains virtually nothing but ideas, rendered in the metal only in the most partial way. For these reasons it is, perhaps, not as inappropriate as it might at first sight seem that he, the builder of unfinished machines, should become the figure, with Alan Turing, around whom British popular national aspirations in computing history gather.

But to overprivilege Babbage and all who either designed or built calculators between 1840 and 1940 is seriously to skew the history of this technology and the particular achievement of the EDVAC team. Nobody, before 1944, understood the full potential of the stored variable programme and every machine, from the difference engine to ENIAC, is at best a *partial prototype* for the computer. It is to the history of such machines we now turn.

Phase two: technological performance – prototypes

Pehr Georg Scheutz builds an ENGINE, 1855

(i) The first statistical engineer

It was left to Georg Scheutz, Swedish lawyer, newspaper publisher and litterateur to build Babbage's **difference engine**. Scheutz's machine was based on an account he had read of Babbage's work in an *Edinburgh Review* of 1834. When finished, the Scheutz machine, which had 'four differences and fourteen places of figures', punched its results on to sheet lead or papiermâché from which printing stereotypes could be made. The machine was operable by 1834 and refined, to the point where duplicates could be made, by 1855. But this was 'after many years' indefatigable labour, and an almost ruinous expences, aided by grants from his Government, by the constant assistance of his son, and by the support of many enlightened members of the Swedish Academy.'[55]

Thus did Babbage, with rare generosity of spirit, introduce to his countrymen the work of the man who had succeeded where he had failed. Scheutz even made an engine for the British Registrar General's office and it worked, calculating actuarial tables, between 1859 and 1864. In our schemata it should be designated as the first of the *accepted prototypes* that the second stage of technological performance in computer history yields.

The Scheutz **engine** was built in advance of any real supervening necessity and, despite a few such applications of the actuarial type, the avoidance of drudgery was scarcely a motivating force in Victorian life and therefore no pressure existed to relieve the tedium of low-paid clerks or, even, economically marginal pure scientists. The calculator remained, more or less, at the technological level achieved by Pascal. It was not in any way a ubiquitous item. Only 1500 arithmometers, for instance, were sold by 1850. But by the 1860s and 1870s, forces of change were at work creating, in their train, a supervening necessity for a number of devices fundamental to twentieth-century life. In short, it was these decades that saw the emergence, in law, of the modern limited company, a development detailed below in connection with the telephone (see chapter 6, page 326). The new corporations necessitated the modern office and with it the telephone (1876), the geared hydraulic elevator (c.1885), the typewriter (which had been first patented in 1714 but finally only appeared in its modern shift-key form in 1878), and the modern mechanical calculator with the spring-loaded pin wheel – the **Baldwin** (1875). This new wheel was the first serious modification of the mechanical calculator since Leibniz and was successfully copied by the Swede Ohdner in 1878.[56]

Commercial desk-top calculators were in immediate production and the office equipment industry was born. The first key-driven calculator was demonstrated in 1887. Five years later, William Burroughs added an elegant roll-paper printing mechanism. A larger device of this period was built by Leon Bollee but his interest moved to racing cars and he founded Le Mans, leaving Otto Steiger to market the machine as the **Millionaire**. It had built-in (i.e. stored) multiplication tables and was manufactured continuously from 1899 until 1935.

The new motive power, electricity, was also used. In 1876, at the very same Centennial Exposition in Philadelphia where Bell made such a stir, an engineer, George Grant, showed an electrically driven piano-size difference engine. For the 1890 American census, Herman Hollerith designed a device of even greater significance both in its use of electricity and for the fortune it made his company, eventually to become IBM.

The decennial census was proving ever more difficult of completion.

By 1887, the Census Office (it became a Bureau in 1920) realised that it would still be processing the 1880 data even as the 1890 returns were being collected. In a public competition held to find a solution to this problem, Hollerith proposed an electro-mechanical apparatus. He resurrected the punched cards that Babbage had intended as the input for the analytic engine. (Babbage had, in turn, borrowed these from the French silk weaver Joseph-Marie Jacquard who, in 1804, pioneered them as loom-controlling devices whereby different cards could cause the loom to produce different patterns of cloth.) In the received history, Jacquard is an elegant and visionary figure whose peripheral insight was only to bear fruit a century and a half later. In fact he was working for the Lyons silk manufacturers to produce a machine which would eliminate operatives, especially skilled operatives. The workers burned the first Jacquard looms but he succeeded in his commission.[57]) Hollerith's cards contained the census data as a series of punched holes. The operator placed the card into a large reading device and pulled a lever. This moved a series of pins against the card. Where the pins encountered a hole, they passed through the card and into a bowl of mercury, thereby making an electrical circuit which activated the hands of a dial.

In the test which won him the contract, Hollerith's machine was about 10 times faster than its nearest rival. Six weeks after the census, the Office announced that the population stood at 62,622,250. Hollerith declared himself to be 'the world's first statistical engineer'. The census was completed in a quarter of the time its predecessor had taken and the Tabulating Machine Company was founded. Hollerith's enterprise became part of the Computing Tabulating and Recording Company (C-T-R), a holding organisation owned by Charles R. Flint whose fortune was built on what a friend described as:

> the more dramatic commodities – munitions, ships, explosives, speculative inventions. . . . He was a pioneer investigator of, and investor in, the automobile and the aeroplane. He had a direct hand in the earlier development of the submarine and the dynamite gun.[58]

In 1924 C-T-R became IBM. By then it had some rivals in the field of business machines. James Powers had been put in charge of the manufacturing division which the Census Bureau established after falling out with Hollerith, a notoriously difficult man. The company Powers founded upon leaving the bureau would eventually become Sperry Rand, IBM's great computer competitor in the years after the Second World War.

Although all these digital devices worked well for the purposes for which they were designed, they can be seen as calculators, *accepted prototypes* meeting imperfectly articulated needs; and this is true, too, of the line of analogue devices which marches step by step with the developments noted above.

(ii)   Substituting brass for brain

If one accepts the view that Stonehenge is an astronomical calculator, then it is an analogue one, representing heavenly phenomena by the placement of stones and holes in the ground. There are a variety of similar devices from an ancient Greek calendrical calculator built in bronze to the astrological and astronomical calculators of al-Kashi, chief of the observatory at Samarkand in the fifteenth century. Oughtred's slide rule which converts numbers into distances is also an analogue device. By the end of the nineteenth century analogue calculators were, if anything, more sophisticated than digital and this was to remain the case until the late 1930s.

Maxwell had built an analogue device for measuring the area of an irregular plane – an integrator – which he called a **planimeter**. By the early 1860s the Ulsterman (by birth) James Thomson (not to be confused with the Englishman J.J. Thomson) had made a sophisticated variant of this, a **disk-globe-and-cylinder integrator**, but he saw no use for it. Some years later James's brother, Lord Kelvin, did; and seeking, in his words, a design 'to substitute brass for brain in the great mechanical labour of calculating the elementary constituents of the whole tidal rise and fall', used it to make an elegant calculator to tabulate the times and heights of high tides at different points around the coast.[59]

In 1876, Lord Kelvin thought to extend the usefulness of this line of development and proposed a general purpose analogue device, a **differential** or harmonic **analyser**, which essentially required that a number of integrators be arranged in sequence so that the output of one inputed the next. Such an arrangement would allow complex linear differential equations to be calculated, such as 'vibrations of a non-uniform stretched cord, of a hanging chain, of water in a canal of non-uniform breadth and depth. . . .'[60] But since the integrator's output was measured by the small rotations of a shaft, there was insufficient torque to continue the process from integrator to integrator. The machine could not be built. Thus there was no progress from the publication of the Kelvin paper until the early 1930s.

Vannevar Bush, one of the most distinguished scientists of his generation, was Shannon's professor at MIT and an important enabling figure in computer development through his work directing the US Office of Scientific Research and Planning in the Second World War. He was interested in analogue devices from the late 1920s. The team he established at MIT first built a **continuous integraph**, using at its heart a meter of the sort that measures on-site consumption for the electric company. It could only cope with first-order differential equations which drastically limited its use, most problems facing physical and electrical engineers being second order.

In 1927, Bush produced a revision of this device which could solve a single second-order problem by putting the output of the meter into a Thomson integrator. In the same year an engineer with Bethlehem Steel, C.W. Niemann, built a torque amplifier, something like a ship's capstan in principle, which could so increase the movements of the integrator's shaft that another integrator could be thereby powered. When this device came to Bush's attention, he used it to build a **complex differential analyser** in which no less than six Thompson integrators were placed in sequence. It looked much like an enormous Meccano toy; and, indeed, Douglas Hartree, Lord Mountbatten's opponent in the earliest phase of the Artificial Intelligence debate, built a small replica of part of it out of Meccano. Hartree saw the original during a visit to America in 1933 and, on his return, raised the funds to build a full-scale duplicate for Manchester University. Another was built at Cambridge, UK. By 1940 there were 7 or 8 world wide, including one built for the Russians by General Electric, as well as the two involved in ballistics research at the Moore School and the Aberdeen Proving Ground, the American army's main testing ground. The Aberdeen machine took a commercial company two years to build and had 10 integrators. At the same time, the Moore school machine with 14 integrators was built by the Civil Works Administration as part of its programme to put skilled technicians, unemployed during the depression, back to work. (This same programme had 'a very small number of mathematicians and a fairly large number of people from the relief roles of the Depression . . . more than rusty in their arithmetic' calculating tables. Out of this effort was to come, after the War, the Applied Mathematics Laboratory of the National Bureau of Standards.)[61]

The analyser worked well, but it was slow and 'programming' it required days of effort with wrenches and hammers. It needed these tools to keep it going as well. Nevertheless, as with any *accepted* prototype, it did satisfy at least part of the appetite scientists had for advanced calculators and, in this, it matched the machines then being produced by IBM and others for business.

So well-accepted were these prototypes that business's requirements for automated office machinery were being met and the market for more advanced calculators was non-existent – to which situation the continued presence of Hollerith punched cards into the last decades of this century is eloquent testimony. IBM, to take the industry leader, had been making sorters, tabulators and punched card equipment since 1914. It introduced a printing tabulator in 1920, a horizontal sorter in 1925, an electrical calculator that could add, subtract and multiply (the first of the 600 series) in 1931 and an 80-column card as well as a subtracting accounting machine in 1938.[62] The IBM 601 was a crossfooting multiplier which used cards and was leased for payroll calculations and other tasks. It contained a plug-board which allowed it a degree of versatility – the arrangement was copied for ENIAC. These devices, not untypical of the most advanced business machines of the 1930s, were of a sophistication now forgotten. There was little reason to seek more efficient solutions for most big commercial computational tasks. Bush's differential analyser and electrically driven mechanical business calculators inputed by punched cards worked so well, it could be argued, that the development of purely electrical devices was inhibited and the *invention* of the computer was delayed. IBM had no interest in pursuing the matter. It was left to others to explore.

(iii) Sur l'automatique
Bush himself moved on to digital devices from 1936. This led to the development of **RAMP** (*Rapid Arithmetical Machine Project*) which was supported by the National Cash Register Company (NCR). A parallel project, funded by NCR and Kodak, was for a device automatically to retrieve and photographically copy data held on reels of 35mm film, but this work produced only *partial* prototypes because the team was broken up in 1942 as more pressing duties were assigned its members.[63]

The possibilities of building an electro-mechanical digital calculator along the general lines proposed by Babbage had been first outlined by the Spanish scientist Leonardo Torres Y Quevedo in the *Essais sur L'Automatique* published in 1915.

> An analytic machine, such as I am alluding to here, should execute any calculation, no matter how complicated, without help from anyone. It will be *given* a formula and one or more specific values for the independent variables, and it should then calculate and record all the values of the corresponding functions defined explicitly or implicitly by the formula. It should work in a manner similar to that of a human calculator, i.e. executing the indicated operations in the necessary sequence, taking the results of each as factors or arguments of the subsequent operation, until the final results are obtained. (italics in original)[64]

Torres y Quevedo suggested the way to do this was with switches 'i.e. *in the place of an index which runs over a graduated scale, we shall have a brush which moves over a line of contacts and makes contact with each of them successively.*' (italics in original)[65]

In 1920, Torres y Quevedo built a prototype to illustrate the feasibility of his suggestion, the first machine to have a typewriter as the input/output device. But his basic idea of a switch sounded like nothing so much as the then current mechanism used in telephone exchanges, the Strowger uniselector (see chapter 6, page 347). 'By 1925 the designers of switching systems had become most adept in devising complex logic circuits using general-purpose electromagnetic relays of telephone quality.'[66] Although Bell Labs began to think about the Torres y Quevedo line of inquiry in this same year, it was not until the late 1930s that any devices appeared in the metal. The Model K (for kitchen table) Bell computer was made of old bits of telephone exchange mounted on a breadboard, one weekend in 1937, by George Stibitz. Stibitz, a staff mathematician at the Labs, was convinced he could wire a simple logic circuit to produce binary additions because the ordinary telephone relay switch was a binary – on/off – device, and over the weekend that is exactly what he did. There was little immediate enthusiasm for the breadboard calculator but a pilot project was funded and the first **complex calculator** was built. The internal supervening necessity was the endless calculations necessary in the developing theory of filters and transmission lines. The complex calculator (a.k.a **Model 1**) was finished by 1940 and did the job the Labs required. Apart from building the first binary machine, at a meeting of the American Mathematical Society, Stibitz also performed the first remote control calculation by hooking up the teletype input keyboard in a lecture hall at Dartmouth and communicating via the telephone wire (of course) with the Model 1 in New York. In the audience were John Mauchly and Norbert Wiener.

Stibitz now looked to make a more complex device, but the cost, $20,000, 'frightened the Lab administration and no further computers were built for several years.'[67]

When the USA joined the war in 1941, Stibitz's expertise did find a proper outlet, in building a series of specialised electromechanical complex calculators. The **Model 2** had a changeable paper-tape programme and was self-checking. It was designed specifically to test anti-aircraft detectors. The **Model 3**, designed for anti-aircraft ballistic calculations, made, like its predecessor, for unattended operation, rang a bell in the staff sergeant's quarters if for any reason the computation was halted. By now Stibitz had left the Labs for the University of Vermont and the two copies of the next model, the 5, were built under the direction of S.B. Williams – one of these versions was placed at the Aberdeen testing grounds. The building of these machines was abandoned, in 1950, for a second and final time. The Labs had found a better way (see chapter 4, page 210).[68]

Stibitz's experience during these years parallels, albeit rather less dramatically, those of Konrad Zuse, another drawn to automata because of boredom.

[In 1934] I was a student in civil engineering in Berlin. Berlin is a nice town and there were many opportunities for a student to spend his time in an agreeable manner, for instance with the nice girls. But instead we had to perform big and awful calculations.[69]

In 1936 Zuse, having patented various elements including a stored programme concept, gave up his job at Henschel Flugzeugwerken in Berlin and began building electromechanical binary calculators out of relays, machines which were to occupy most of the living room of his parents' apartment. The **Z1** and **Z2** were test models. A friend, Helmut Schreyer, on seeing the Z1, suggested that vacuum tubes would be quicker than relay switches, but Zuse could not afford them and anyway considered them too unreliable. He had failed to interest the German office machine industry in his project except for some support from Kurt Pannke, a manufacturer of calculators.[70] In 1939, work on the Z2 was halted as Zuse was inducted into the Wehrmacht. According to Zuse's account Pannke wrote to his major explaining that he was engaged in developing a calculator which would be of use in the future to the aircraft industry. The major said, 'I don't understand that. The German aircraft is the best in the world. I don't see what to calculate further on.'[71] (Another version of this has Zuse being told by an intelligence officer that, unless the device was ready within 6 months, the war would be over and a triumphant Germany would have no need of it.[72]) Nevertheless, he was reassigned.

These, however, are typical of the anecdotes told by Germans of this generation seeking to distance themselves from their then government. It is likelier that a document of October 1939, written by Schreyer, which explains the potential of the Zuse machines and additionally argues for valves, played a part in Zuse's reassignment:

Since the calculations are done very quickly the machine is a useful aid for desk calculations which are required in laboratories. It can also be a valuable aid for setting up tables used by the artillery and on warships . . . in the calculation of weather charts . . . in the production of military equipment.[73]

Zuse explained to his American interrogator in 1946, the Deutschen Versuchsanhalt für Luftfahrt (the Aerodynamics Research Institute) was interested in the Z2, and so he was relieved and went to back to Henschel which was now building V2s. By 1941, still finishing the **Z3** in his spare time, he began building the special purpose **S1** to increase missile production by speeding computation. It worked so well that in the later part of 1943 Zuse in partnership with his Henschel colleague Gerhard Overhoff and supported by the Air Ministry, established his own small firm, Zuse Apparatbau.[74] The Z3 was, like the contemporary models in America, programmable, but by holes punched in

35mm film rather than papertape. It was a floating-point binary machine with a 64-digit store and was the first of its general purpose class to work, by December, 1941.[75] At the war's end Zuse was working on a **Z4**. As the Allies closed on Berlin he was given one truck on which he loaded the Z4 and headed south. By the surrender he was holed up in the village of Hinterstein in the Bavarian Alps close to the Austrian border, the Z4 hidden in a cellar. The Z3 is now in a Munich museum and, by 1949, the Z4 was installed in the Federal Technical University in Zurich.[76]

Howard Aitken, the builder of the third major line of electro-magnetic machines, came to computers, like many others in this history, while slogging through the calculations he needed for his thesis. By 1937 he had a proposal ready for mechanising the process. He showed it to the military but he also took it to a calculating machine company which decided the device was impractical, against the opinion of its chief engineer. The engineer then directed Aitken, via the IBM connection with Columbia University, to IBM itself. Using contacts at the Astronomical Computing Bureau at Columbia, which had IBM equipment, and the good offices of a business professor at Harvard, Aitken got to Thomas Watson Sr, the then president of IBM. He was clearly in the right place since in his 1937 prospectus 'Proposed Automatic Calculating Machine' he had determined to use IBM apparatus.[77]

IBM's Watson was a somewhat bizarre executive given to endless speechifying and having an absolute intolerance of criticism.[78] A master salesman, he had come into the company as Flint's man from National Cash Registers and, by the mid-1920s, he had more than justified Flint's trust. He built a gigantic empire and imposed a degree of conformity on his workforce unheard of outside of a totalitarian state; but he also established a history of support for projects such as Aitken's. In 1929 he had donated a full collection of machines to establish the Columbia University Statistical Bureau, which used them for educational testing. He then funded the development of an up-to-date version of the Scheutz engine. The University began to experiment with the machinery in performing astronomical calculations and after further gifts of equipment the operation, suitably enlarged, was renamed the Thomas J. Watson Astronomical Computing Bureau.[79]

Far from illuminating the corporate mind as to the potential of advanced calculators, all this simply reinforced IBM's by-now traditional indifference to them as a business proposition. These uses, in effect, obscured the potential for computers and the fancy applications the machines were put to in Columbia confirmed the view that profit lay in the simpler needs of commerce. When Aitken arrived this perception had changed slightly, for the imminence of war sharpened

Watson's appetite for new markets – he had been after military contracts for some years – and Aitken brought with him a potential naval connection.[80]

In 1939, Aitken was contracted to build his electromagnetical machine at IBM's Endicott Lab, the money to come from the US navy and a million-dollar gift from Watson. Aitken was still salaried through his Harvard professorship and was given the reserve rank of Naval Commander. His staff was navy, too. IBM furnished the space and the equipment, mainly, as with Stibitz and Zuse, relay switches, but it also provided a team of engineers led by Frank Hamilton.

Within 4 years the machine, the Mark 1, was working. It was transferred to Harvard, as a present from IBM. Watson insisted it be clothed in gleaming, aerodynamically moulded steel, an IBM vision of 1940s high tech. Aitken thought it should be left naked in the interests of science. Watson won, but Harvard virtually excluded him, and IBM, from the dedication ceremony. IBM called the machine the **ASCC** (*A*utomatic *S*equence *C*ontrolled *C*alculator); Aitken called it the **Harvard Mark I**. Watson did not get the expected honorary degree and IBM ignored the Mark I as a prototype for a business machine yet nevertheless set about building a bigger and better one; but the world got, in the publicity brouhaha, its first glimpse of a 'robot brain'.[81] Aitken's young naval help organised a routine which made the machine produce strings of meaningless numbers to the accompaniment of much clicking from the relays and many flashing lights. Visiting admirals were impressed.[82] Like the Z3 and the Model 2, the Mark 1 was externally programmable, using papertape. In its general scope and speed it was also of a piece with these other devices.

Given that IBM was not in the computer business but rather engaged in some high level public relations work, it is not surprising that 'Aitken did not always appreciate Thomas J. Watson Sr., and all the cooperation he got from IBM.'[83] (Aitken had seen the subservience demanded of the faculty in the Columbia facility.) Yet IBM could have been in the advanced electromechanical calculator business had it so desired, for Stibitz, Zuse and Aitken had *invented* nothing. Their machines were built out of readily available parts and this was to be the case with the all-electric calculators that came next. As with the delay in the development of television, these *accepted* prototypes, from the Pascaline to the Harvard Mark 1, speak directly to the lack of a real supervening necessity. Occasional tasks, like Bell Labs' need to crack filter theory, might produce a machine. Inspired amateurs, like Zuse, might do so too. And in certain corners of the military, needs were perceived and met. But basically the world worked well enough with these various prototypes. A universal engine was simply not needed. It was to take two wars, one hot and one cold, to change that perception.

(iv)   The use of high speed vacuum tubes
The Second World War acted as a major supervening necessity for a whole range of technologies, but its effect on the development of computers was not at all clear-cut. The war threw up no more obvious general need for a totally electronic computer than had the peace; and, although Aitken and Stibitz continued to work on electromechanical calculators, Zuse was initially drafted and Vannevar Bush's research team at MIT, developing digital devices on grants from NCR and Eastman, was disbanded for other war work. In the US, only the specific requirements of ballistics were to generate a supervening necessity for an advanced electronic calculator but, even then, the need was not of such an order as to give the project top priority.

The art of ballistics, until the mechanisation of warfare in this century, had been, one might say, a hit and miss affair. In a series of experiments between 1864 and 1880 the Reverend Francis Bashforth B.D. related air drag to velocity and produced the first firing tables to be consulted on the battlefield as a guide to the ranging of artillery pieces. These tables were refined by a French military commission working in Grave but, at the outset of the First World War, there was still a lot of guesswork involved. Many factors were simply not calculated. For instance, it was assumed that the air exerted a constant drag whatever the elevation of the projectile which, when guns became powerful enough to lob shells into the upper atmosphere, was demonstrably not the case.

Since, without firing tables, any weapon is much reduced in utility, the Americans took ballistics very seriously and established a tradition of research in the aftermath of the First World War. (Under the mathematician Veblen, one of these teams included the very young Norbert Wiener.) At the Aberdeen Proving Ground a succession of highly educated ballistics officers appeared during the 1920s and 1930s. Thus enlightened, the officials at Aberdeen made moves to acquire a Bush differential analyser as detailed above.

The ballistics work generated by the Second World War was enormous. Each firing table for every new weapon required the tabulation of dozens of factors across thousands of possible trajectories, any one of which represented half a day's work for a human computer with a desk calculator. By the summer of 1944 the computing branch of the Ballistics Research Laboratory was producing 15 tables a week, but the need was for 40.[84] This, the so-called 'firing table crisis', was the supervening necessity of the ENIAC project to the general history of which we now return. Professor John Brainerd, the Moore School's official ordnance liaison officer, wrote:

In the specific case of sidewise firing from airplanes, construction of directors for guns has been held up several months because it has been a physical impossibility to supply to the manufacturers the necessary ballistic data. The proposed electronic difference analyzer [the ENIAC] would, if successfully developed, not only eliminate such delays but would permit far more extensive ballistic calculations than are now possible with existing equipment. (our brackets)[85]

The extensiveness of the calculations needed was just being glimpsed. Norbert Wiener:

Even before the war, it had become clear that the speed of the airplane had rendered obsolete all classical methods of direction of fire, and that it was necessary to build into the control apparatus all the computations necessary. These were rendered much more difficult by the fact that, unlike all previously encountered targets, an airplane has a velocity which is a very appreciable part of the velocity of the missile used to bring it down. Accordingly, it is exceedingly important to shoot the missile, not at the target, but in such way that the missile and the target may come together in the future.[86]

Norbert Wiener spent his war working on the basic maths for a sighting mechanism that could predict the future whereabouts of a warplane in any given situation. The tasks were becoming too difficult to be mechanically computed.

It is also possible, though, that too much has been made of the 'firing tables crisis'. The Harvard mathematician Garret Birkoff often visited Aberdeen at this time and is of the opinion that the whole issue had been overblown. 'This is because it was not fixed position trench warfare, it was mostly pointblank fire, often at a moving tank.'[87] In support of Birkoff's view is the chronological fact that victory was secured before ENIAC came on stream, despite the 'crisis'. Yet, on the other hand, tanks were not the only target and there can be no doubt that the Ordnance Department gave a high priority to the creation of firing tables and was not adverse to mechanising their computation.

John Mauchly's original paper, 'The Use of High-Speed Vacuum Tube Devices for Calculating', was written in August of 1942. Like Schreyer in Berlin 34 months before, Mauchly presented the suggestion as an alternative, all-electric version of the calculators currently available or being built. Beyond speed of calculation, he suggested greater accuracy, enhanced checkability, programmability, easy fault-finding and labour saving would result from the use of valves but the idea had scarcely ignited imaginations and the paper itself was left to lie in a file for 8 months.[88]

Mauchly was another eager to avoid the slog of hand computations.

Indeed, in the 1930s, he had been driven away from one investigation, an analysis of meteorological data, because of the mechanical enormity of the task but he had built some experimental circuits, according to him, using valves, gas and vacuum tubes.[89] Others suggest that these early attempts were analogue, along Bush analyser lines, rather than digital.[90]

Immediately before coming to the Moore School in the summer of 1941, he had been in touch with John Atanasoff, an associate professor of physics at what was then the State College of Iowa. Until 1943, Atanasoff, it is generally agreed, was the only man to essay using electrical valves in a calculating device. Iowa College had a tradition of mechanised computation methods dating back to the work done in agricultural statistics during the early 1920s by Henry Wallace, before his political career, under Roosevelt, as secretary of agriculture and vice president.

In the last years of America's peace, Atanasoff, to whom the term 'analogue computer' can be credited, was building a binary digital machine having about 300 valves – a prototype of the electronic machine, the first such. In a 1940 paper, really a grant proposal, Atanasoff stated 'substantial progress has been made in the construction of the machine' and provided a series of photographs of it.[91] Its purpose was to solve linear algebraic equations and Atanasoff, in addition to pathbreaking valves in his logic circuits, used capacitors arranged in drums as the store and a built-in pulse by which the machine could internally time its own operations. The device, which Atanasoff designated in honour of himself and his assistant the **ABC** (the *A*tanasoff-*B*erry *C*omputer) was the size of a large desk. It was interesting enough for one of the photographs to appear in the local paper, dated January 1941. This story said the machine would be ready in about a year and by the following spring it was, at a cost of about $6500. As an entirely electronic and binary calculator, it was, on the one hand, extremely sophisticated; but, on the other, malfunctions on the electromechanical binary-card puncher prevented it from fulfilling its design purpose. Further, it was to do multiplication and division, like the seventeenth-century Pascaline, as a series of additions and subtractions; and, like Babbage's difference engine, it was a single-purpose machine. Nevertheless it was indisputably the first prototype electronic calculator to use valves.[92]

In December of 1940 Mauchly read a paper on his analogue analyser at a conference in Washington which Atanasoff attended. Mauchly was intrigued enough with the ABC to drive to Ames, in the summer of 1941, for a period variously described as between 3 days and a week.[93] Atanasoff subsequently claimed to have discussed in great detail with Mauchly plans for a general purpose calculator, obviously utilising valves. Mauchly claims that he advised Atanasoff against certain aspects of his design, notably the capacitors, but nevertheless subsequently invited him to join the Moore School team. Instead Atanasoff, who did not get round to patenting the ABC which was eventually broken up, went to the Naval Ordnance Lab on the outbreak of war and never returned to computing machinery.[94] A judge in 1974 was to hold that Mauchly got all his ideas from Atanasoff, yet, confusingly, found that Mauchly had nevertheless 'invented' ENIAC.[95]

After the Iowa trip, Mauchly went on his refresher course at the Moore School, obtained a faculty position and produced his paper of August 1942.

Herman Goldstine was charged with the responsibility of producing firing tables on the complex differential analyser housed in the Moore School's basement. This was exactly the device that Atanasoff had christened an analogue computer and that Mauchly thought to replace. An ex-student of Mauchly's, Joe Chaplin, had been hired to keep the machine running – no small job – and one day, with Goldstine standing over him while he tried to maintain it, he suggested that Mauchly might have a better way. Goldstine, who had been getting further and further behind in the work, was eager to explore Mauchly's proposal.[96]

Goldstine discovered that not only had Mauchly's original paper not been acted upon, it had been lost. It was now the spring of 1943. The proposal was reconstructed out of the secretary's old shorthand book and Goldstine used it to smooth the way at Aberdeen and with his own brass. Oswald Veblen, the chief scientist of the Ballistics Research Laboratory, fell off his chair during the presentation – presumably with excitement – and upon recovering simply said, 'Give Goldstine the money'; and so, by early April, after a bout of furious work, Goldstine and Mauchly presented detailed proposals and costings. The 'Report on an Electronic Diff.* (sic) Analyzer' was produced by 2 April 1943.[97] 'Diff.*' stood for both differential, i.e. the mechanical device in the basement, and the difference between that machine and the one proposed, i.e. that the latter would be all-electric. There was another difference. The machine was to be 60 times larger than the only valve calculator then in existence, the ABC, and was budgeted to cost 23 times as much.

On May 31st, a request for $150,000 having been approved, work began. In June the machine was redesignated 'Electronic Numerical Integrator and Computer' – ENIAC. Its cover, one might say, was blown. Mauchly no longer pretended he was simply building an electrical version of the mechanical analyser. This was going to be a new beast, although it was still, in essence, a calculator designed to work out ballistic firing tables and, in fact, to replace (or rather abet) the monster in the basement and a number of human computers doing the same work.

J. Presper Eckert Jr had grown up a mile away from Farnsworth's Philadelphia television laboratory, to which he was a constant boyhood visitor. He came to the Moore School in 1937 and, by his graduation in 1941, had shown himself to be a most promising electrical engineer. He was the graduate assistant on the defence training course that brought Mauchly to the School. They had spent time talking about calculators and the relationship continued after Mauchly was appointed to the faculty at the course's end. Mauchly states that, 'Then, and later, it was the knowledge that Eckert saw no technical obstacles to the realization of fast electronic computing that sustained my own confidence that it could be done.'[98] Mauchly immediately involved Eckert in the ENIAC project.

Eckert was prepared to defy conventional wisdom as to the reliability of valves since he understood that, if valves were left on constantly, the problem of failure was reduced. His hunch was that the failure rate would be low enough to allow Mauchly's plan to work. The importance of the decision to use valves for ENIAC cannot be overstated. (If Eckert were to be proved right – and he was – ENIAC would demonstrate that there was no technological reason for an electronic calculator not to have been built much earlier.) The big gamble in ENIAC's design, then, was in the number of valves required. The original plan called for 17,000 tubes operating at 100,000 pulses a second. It had (theoretically) 1.7 billion opportunities to break down every second it was switched on.[99]

Eckert and Mauchly began work on the ENIAC a week before the final contracts between the University and the Ballistics Research Laboratory were signed in June 1943. The earliest stages of the design were much influenced by Mauchly's understanding of the contemporary generation of IBM machines except, of course, in the matter of the valves where he had the example of Atanasoff; and his confidence was sustained by Eckert. Eckert's first task was to cut the odds by making sure the parts he was buying off the electronics industry's shelves met his specifications. Work began on building two accumulators and a cycling unit, the completion of which would trigger the balance of the Ballistics Research Laboratory funding. From the signing of the contract to the official hand-over, ENIAC was to take three years of laborious and painstaking effort.

In the first year the two accumulators, 'stores' in Babbage's terminology, were built and the project was finally fully funded. By August of 1944, Goldstine felt the machine would be operational the following January. It was at this point that the design was locked and the wiring began in earnest, freeing the team to concentrate on the ideas they had been having for the device that was to become EDVAC. Thus, this same August, Goldstine started the campaign for the more advanced design. We have seen above how Eckert and Mauchly were already thinking about storage earlier that year. It is a measure of the military commitment that was finally brought to bear on the project that the initial funding for EDVAC was forthcoming in September of 1944 despite the fact that ENIAC was now seasons behind schedule.

The reasons for ENIAC's slowness were manifold. The work itself was tricky but accomplishable. Getting the parts off the shelf, in the face of wartime shortages, was more difficult by far. It is likely that, had the Ballistics Research Laboratory set out having an ENIAC built in the late 1930s, it would have been finished rather more quickly than it was in the mid-1940s. Weeks were wasted, months even, waiting for electrical supplies. Sometimes the supplies were of inadequate quality. The racks were made by a kitchen cabinet manufacturer in New Jersey who was going out of business for lack of steel. ENIAC had just sufficient urgency to prevent this happening. Despite this low priority (or perhaps because most of the money had already been spent), when the war ended, the work still went forward.

By November 1945, ENIAC's debugging procedure started and, at von Neumann's suggestion, a team from Los Alamos came to Philadelphia with a problem for ENIAC to run. (Von Neumann had ensured that Los Alamos was kept abreast of progress at the Moore School.) Early in the New Year, the Moore School and the Army elaborately prepared to unveil ENIAC to the world. It had eventually cost $500,000 (but then so too had the latest Mark 5 Bell electro-mechanical device) and, although the war was over, ENIAC did what it was designed to do:

> [W]e chose a trajectory of a shell that took 30 seconds to go from the gun to the target. Remember that girls (sic) could compute this in three days, and the differential analyzer could do it in 30 minutes. The ENIAC calculated this 30-second trajectory in just 20 seconds, faster than the shell itself could fly.[100]

There was already a degree of awareness, from the 'Robot Brain' ceremonies for the unveiling of Aitken's Harvard Mark 1, that a new generation of super devices was emerging and on 15 February 1946, upon the public demonstration of ENIAC's powers, the *New York Times* quoted 'leaders' as 'heralding it as a tool with which to rebuild scientific affairs on a new footing'.[101]

(v) We might have lost the war

The brouhaha surrounding ENIAC stands in stark contrast to the treatment given by the British to their equivalent device, **Colossus**. Forty years after the end of the war Colossus still remains only partially declassified.

There is a contrast too between the gravity of the British and

American supervening necessities which lay behind the building of ENIAC and Colossus. While there can be some argument as to the seriousness of the firing table crisis, in the British case there is less room for debate. Advanced electronic calculators were needed for cryptanalysis and the work of the British teams which developed them in the secret war appears to have been essential to overall victory.

From the outset of hostilities in 1939 until the middle of 1941, the British Government's Code and Cypher School (GC & CS) struggled to break open the encoding of German messages. Alan Turing, who came to Bletchley Park, the war-time site of GC & CS, the day after war broke out, led one of the groups working on this problem. They had a certain head-start. They knew the Germans were using an encoding machine, the **Enigma**; they had a fair idea of the principles involved in it and its design. It was based on a commercial device developed by an American, Hebern, for use by banks and anybody else who needed to transmit secret messages. Inside were a series of 'hebern' wheels which substituted letters for the originals of the message. How it did so depended on the number of wheels involved and the way the machine was wired through its electronic rotors.

Thanks to the efforts of Polish intelligence, a lot of detail was known. The Poles became aware of encrypted German radio traffic in 1928 but failed to make much headway with its decipherment. From 1932 on, they kept a more serious watch.[102] With daily keys obtained from French Intelligence, Marian Rejewski, one of the three Polish cryptologists assigned to the Enigma, was able to decipher the messages, but the need was for a system which would allow for the keys as well as the text to be opened. Between 1933 and 1935, the Poles, using the stolen keys, the messages and a commercial Enigma, built their first machines, **cyclometers**, in an attempt to mechanise the search for the daily keys. From 1935 this work assumed a greater significance because, as more and more Nazi networks were created and the ordering and number of the drums within their Enigmas altered and increased more often, the flow of espionage material started to slip behind. Eventually the Poles produced a device called the **Bombas** (or Bombe in the Western European literature) which worked as Enigma's alter ego. Put crudely, it reversed Enigma. Encoded messages were fed into it and it worked through all possible combinations until it produced German. The Polish Bombe had been rendered obsolete when, in 1938, the Germans increased the rotors again. Inside the naval Enigma, for instance, there were now eight so that the machine had 336 rather than 60 possible rotor settings as with the 5-rotor version or 6 settings as with the original 3-rotor model. These increases geometrically complicated decipherment.

In another account, Richard Lewinski, a Polish engineer, found work over the German border, also in 1938. He was making pieces for what he knew was a type of signalling device. Lewinski was gifted with an extraordinarily precise memory to which he carefully committed the wiring details on which he was working. He then lost his job, following a routine security check which revealed him to be Jewish. On his return to Warsaw he was sucked into the great game for he had, in fact, memorised the plans of the new 8-rotor Enigma. British, French and Polish Intelligence shared and evaluated all they knew about Enigma, and Dilwyn Knox, one of GC & CS's top cryptographers, flew on 25 July 1939 to Warsaw to consult. Lewinski was shipped out to Paris and there built a mock-up of the machine.[103] Rejewski and his colleagues Henryk Zygalski and Jerzy Rozychi also found themselves in France after the outbreak of war.

Having a good idea of what Enigma was did not mean that the codes were open; only revealed were the enormous dimensions of the problem. Enigma's possible permutations in wiring patterns were of the order of 150 million. But, and here was – as ever – the cryptographer's bane and the cryptanalist's boon, the messages had to be in a language, in this case German, understandable by both. The Bletchley Park 'Golf Club and Chess Society' set about redesigning the Polish Bombe until it matched the developed Enigma. Since the code was set mechanically, it could be broken mechanically (especially if helped by brilliant human intuition) and broken it was, before the first year of the war was over.

Turing was in charge of Hut 8 at Bletchley Park, his official address Room 47, Foreign Office, Whitehall, London. Eventually he was to have a dozen mathematicians, four linguists and about a hundred other people, all women apparently, to do mechanical clerical work, and a number of 8 foot high bronze Bombes. British Intelligence had some breakthroughs on the German codes but it was Hut 8 that cracked the problem by building an improved Bombe. Using a 'probable word' system, the new Bombe had a diagonal board which allowed the increased complexity of Enigma's variable wiring patterns to be duplicated. By August 1940, this device was being used to speed the work of GC & CS. As a specialised electromechanical machine it was, if anything, a rather straightforward, if simple-minded, cousin to the Zuse and Bell machines and the Harvard Mark I then being built. In its particular design objectives it was not unlike the primitive electromechanical apparatus – a number sieve – created, in 1936 in his home in Pennsylvania, by D.H. Lehmer.

Number sieves solve sieve problems. Lehmer had built his first mechanical – photoelectric – apparatus for this purpose in 1932, following in the footsteps of his father who had been using card stencils to discover whether, for instance, any given number is a prime or a composite.[104] The son was still working on sieve problems, now in modern computer form, in the late 1970s. In public, Lehmer seems to be unnecessarily modest about this field for, when he was asked at one conference why, apart from intrinsic complexity, he was interested in such a topic, he replied, 'A reasonable man, like myself, wouldn't spend 12% of his time, maybe, worrying about building sieves, if there wasn't any real use for them. It's very esoteric, of course, and since I am practically the only man working in this field you can see how widespread the interest in it is.'[105] Professor Lehmer doth protest too much; beyond his contributions to algebraic number theory, he can also take pride in the relationship between his sieves and the Bombes, crucial weapons of the secret war.

There was also a British tradition of pure research calculators upon which to draw. In 1931, Dr C.E. Wynn-Williams had built an electronic binary counter to be used in nuclear research at the Cavendish Laboratory. By 1935, this apparatus had been developed to provide binary-decimal conversion and print-out. Dr Wynn-Williams was attached to that other centre of war-time academic expertise, the Telecommunications Research Establishment (TRE) at Malvern, working on **RADAR**.[106]

Hut 8, building on these traditions, had, however, not cracked the Naval Enigma, a more complicated machine than the one used by the other German forces. In February of 1941 a special raid into Norway brought the Naval version back to Turing. In response, the Bombes became more complicated, involving punch-cards and an army of clerks. By May of 1941, British Intelligence was able to read all U-Boat messages within a day. This was the situation until the Germans reprogrammed the Enigmas 8 months later.[107] There is therefore less of hyperbole than might seem to be the case in this remark by Jack Good, a member of Turing's team at Bletchley Park. Recalling Turing, he said,

> It was only after the war that we learned he was a homosexual. It was lucky the security people didn't know about it early on, because if they had known, he might not have obtained his clearance and we might have lost the war.[108]

The importance of this work can be seen in the subsequent event. Although unaware that Enigma was opened, the Germans introduced another refinement to it in January of 1942 and secrecy was again attained. Once more the U-Boat fleet was hidden from sight. Losses in the Atlantic began to climb precipitously. It is true that other factors can be cited apart from the intelligence blackout – the United States's need to divert ships to the Pacific would be one such and the American underestimate of the submarine threat to the Atlantic shores would be another. The ever-rising number of U-Boats and some failures on the Allied side in protecting the integrity of their own signalling codes, giving the Germans better foreknowledge of Allied convoy routes, also contributed. The result was very grave.

On 12 March 1942 Churchill signalled Harry Hopkins, Roosevelt's Secretary of Commerce:

> In January eighteen ships, totalling 221,000 dead-weight tons, were sunk or damaged; in February the number rose to thirty-four, totalling 364,941 dead-weight tons; in the first eleven days of March seven vessels, totalling 88,449 dead-weight tons, have been sunk. Yesterday alone 30,000 tons were reported sunk or damaged. . . . The situation is so serious that drastic action of some kind is necessary. . . . The only other alternatives are . . . to stop temporarily the sailing of tankers, which would gravely jeopardise our operational supplies.[109]

In December 1941, before the veil of secrecy fell, a U-Boat radio operator had sent a message using a hitherto neutral wheel in the Enigma. He had corrected it, but the error, when recalled, revealed to the British and the Americans that the problem was the introduction of this new extra wheel. The U-Boat Enigmas produced 1680 possible rotor orders and the starting positions, crucial for any cryptanalysis, were now of the order of 9×10^{23}.[110] More Bombes were needed and indeed a whole new class of calculator was needed. All these solutions were put in hand. Turing and Gordon Welchman built a better Bombe; the Americans built a vast number of Bombes and by the autumn, Max Newman had created the first of a series of prototypes for the machine that would be designated Colossus.

Max Newman had come to Bletchley Park as a civilian volunteer in the summer of 1942, the height of the second Enigma blackout. (The GC & CS commandeered the very best of academic minds, Newman being but one more among the many involved. Wynn-Williams, for instance, was at Bletchley Park by the winter of 1941.) Newman and his team, in F Hut, known as the Newmanry, created a series of machines whose closest antecedents were to be found in the work just then being abandoned by Vannevar Bush at MIT. Three machines were built, the **Peter Robinson**, the **Robinson and Cleaver** and the original, named with some justice, the **Heath Robinson**; for these were temperamental in the extreme. Designed for Boolean operations and built (in part) by Wynn-Williams, the Heath Robinson used relays, photoelectric papertape readers and a teleprinter output as well as between 40 and 80 valves. It could have processed 2000 characters a second were it not for the limitations of the punch-paper input mechanism. The paper kept breaking, the relays produced electrical interference and on occasion the whole apparatus threatened to burst into flame.

The work of the Robinsons was absolutely necessary since the

Germans were now using extremely sophisticated encoding devices known as **Geheimschreiberen**, secret writers.[111] The work created a crucial learning curve. Heath Robinson and its variants were built in part by Post Office engineers working in the GPO's Research Station at Dollis Hill, London. The machines were constructed out of available components in fairly short order, the first being on stream, in so far as it ever was, by April of 1943. The latest German encoders having been captured the previous October, the results of all this effort, in the Newmanry and elsewhere, paid off. Once more the U-Boat fleet could be accurately tracked via its messages.

The final example of the significance of intelligence in the Battle of the Atlantic can be seen in the events of March 1943. This month was to be the worst of the war in the North Atlantic for the Allies; nearly three-quarters of a million dead-weight tons were lost, simply because the Germans broke the merchant marine code. The Allies changed the code and the tide turned in their favour. The Germans could never believe their machines were unlocked and so never altered them again for the duration of the war. Instead they searched for spies around the U-Boat bases in France. They should have consulted Konrad Zuse on the matter.

T.H. Flowers of the GPO was the main engineer involved in the work of the Newmanry. He had been in charge of the switching group at Dollis Hill and had spent the 1930s grappling with the problems involved in substituting electrical for mechanical parts in telephone exchanges, exactly the move now required in automating calculations. He had seen the Manchester University analogue differential analyser in 1937 and had been in contact with an X-ray crystallographer who was interested, à la Stibitz, in building a specialised differential calculator from telephone exchange relays. At the outset of the war he was working on an electromagnetic digital device for ranging anti-aircraft guns and, as the one member of this group cleared to receive information about radar developments, he was uniquely aware that the machine in hand was going to be rendered obsolete before it could be built.

In all of this, one thing distinguished Flowers – he shared Eckert's faith in valves, unlike Wynn-Williams who preferred electromagnetic relays, as did most contemporary engineers. Flowers's solution to the requirements of Bletchley Park Hut F, an assignment for which Turing apparently suggested him, was to avoid one of the major problems of the Robinsons, the physical synchronisation of the two tape inputs, one containing instructions – the patterns the machine was to look for – and the other, the data – the German messages. He proposed doing this by cutting out one paper input and storing the patterns internally, on valves. This would mean the machine would have a huge number of valves, 1500 in its first version.

Bletchley Park was not convinced; so, with Newman's backing and supported by a general highest-priority instruction from Churchill for whatever GC & CS needed, the work was done entirely at Dollis Hill. The Colossus project, as a result, was given top priority. When the team asked for electric typewriters, these were immediately flown over from the States. When another thousand valves were demanded from the Ministry of Supply, they were forthcoming, albeit with the comment, 'What the bloody hell are you doing with these things, shooting them at the Jerries?'[112] In 11 months, by December of 1943, it was finished.

Donald Michie and Jack Good, members of the Newmanry, the latter having moved there from Hut 8, found that they could manipulate the machine while it was running, acting, as it were, something like a human version of the stored programme, albeit a limited one designed simply to seek out, at each stage of choice, the solution closest to German. Immediately work was begun on a Mark II to automate this function in the machine. The Mark II, with 2400 valves and an operating speed five times as fast, i.e. 25,000 characters per second by a combination of parallel operations and short term memory, was working by 2 June 1944. It had taken a mere 3 months to build. Five copies followed.[113] The current whereabouts of all the Colossi is unrevealed.

The original Colossus, by recognising and counting, was able to produce the best match of a given piece of pattern with the text. The new Colossus, by automating the process of varying the piece of the pattern, was able to work out which was the best one to try. This meant that it performed simple acts of decision which went much further than the 'yes' or 'no' of a Bombe. The result of one counting process would determine what the Colossus would do next. The Bombe was merely supplied with a 'menu'; the Colossus was provided with a set of instructions.[114]

Nobody on the British side knew that at the same time precisely the same solution, serried ranks of valves, was being proposed in the Moore School in Philadelphia. Indeed Eckert's faith in valves perforce outstripped Flowers's since ENIAC was a digital machine and it needed to have 18,000 of them. As with ENIAC, so with Colossus; the valve decision was central.

ENIAC and Colossus were the first all-electronic calculators to be built. In the development of the computer, therefore, they stand as the last, most powerful of the *accepted prototypes*, lacking only in stored programme capacity of the computer. This lack, though, was not at all clearcut and the degree by which these two machines missed being computers is a matter of debate, especially the Colossus Mark II.

It has been suggested above that a number-cruncher metamorphosed into a computer when it became a general purpose, programme-controlled, digital device having a high-speed store. It is still, because the British Government restricts information, unsettled how close the device Colossus Mark II came to fulfilling all these requirements – but it certainly came close. 32 years after the end of the war the following information about it was declassified: the apparatus was entirely electronic, fully automatic, programmed by a punch papertape system (read photoelectronically) and it contained conditional branching logic facilities. The machine, like the ABC, generated its own pulse. It could count, do binary arithmetic and perform Boolean logical operations. It used 2400 valves and worked at a rate of 25,000 characters per second, clattering out its results on an electric typewriter. Almost certainly nothing but cryptanalytic problems were ever put to it so in that sense it is, technically, a special purpose machine. In the opinion of some who worked with it, 'The use of Colossus to do other things unconnected with the purpose for which it was built, although possible, would have been strained and artificial.'[115] Others disagree: 'It was general just because it dealt with binary symbols but it wasn't designed to be an ordinary number cruncher.'[116] The more becomes known, the closer Colossus gets to meeting the requirements of a true computer. 'There were in fact many ways in which Colossus could be used, exploiting the flexibility offered by its variable instruction table.'[117]

Although there is nowhere in the record anything comparable to the ENIAC team's conceptualisation of the electronically stored programme and extensive memory (and the British computer scientists, unlike their television counterparts, have never hidden their direct reliance on those concepts), nevertheless there is now some evidence that the Mark II Colossus was, if only by a hair's breadth, a little more sophisticated than ENIAC; it could even be 'almost' set up to perform numerical multiplication.[118]

It was on the brink of being a computer.

The second transformation: supervening necessity

Nicholas Metropolis needs a COMPUTER, 1946

The war was the supervening necessity for the advanced electronic calculator and ENIAC and Colossus were the last *accepted* prototypes for the computer. We have seen how those building ENIAC had conceived of the stored programme concept before ENIAC was finished. At war's end, all these teams were in danger of being returned to other projects. Neither the firing-table crisis nor cryptanalytic needs was sufficient to maintain them; and general scientific or business requirements were no more visible in 1945/6 than they had been throughout the 1930s. However, the idea of the stored programme machine – the true computer – was not lost because a new supervening necessity came into play during the first years of the peace.

ENIAC was finally working by November of 1945. As indicated above, Nicholas Metropolis and some colleagues from Los Alamos had already the previous summer, at John von Neumann's suggestion, come to Philadelphia to explore how the machine might aid the computation of some problems in thermonuclear ignition. In November, Metropolis returned and ran the calculation. The results appeared on punch cards containing no 'indication of what they were or what they were intended to represent'.[119] Los Alamos was so early on the machine, its problems were used to debug the device.[120]

> The study of the implosion problem gave one of the great impulses to the development of fast computers. There were many others of the same sort of equal importance to Los Alamos. . . . And because it had to solve these problems Los Alamos, consciously or not, made a great and fundamental contribution to the development of computing.[121]

One war was over, but another, with a nuclear arms race at its heart, was starting. There was a hiccup in computer development as the implications of this were imperfectly realised; after all, the Russians would not explode an A-Bomb for four years. The first hydrogen bombs were exploded in November 1952 (US) and August 1953 (USSR). Aside from the academy, elsewhere in America, even at Los Alamos, computer plans were being abandoned. Mauchly and Eckert, having left the Moore School and gone into business for themselves, nearly went bankrupt trying to sell computers (see below page 162). The development of the path-breaking real-time machine that MIT was building for the Office of Naval Research, to study aircraft stability design, was on the point of closure when it was realised it would serve to meet the perceived threat of Soviet air attack.

The perceived threat, Russian nuclear capability, cleared the hiccups. IBM's proto-computer, the **Card-Programmed electronic Calculator** (CPC), built in essence out of available IBM bits (a **603 calculating punch** and a **405 alphabetical accounting machine**), was commissioned by Northrup as part of guided missile development and was delivered in six weeks.[122] Thomas Watson was, at last, in the defence business. He telegraphed President Truman offering IBM's full services to the government. The first IBM computer proper, the **701**, a development of the CPC, was originally designated as a **defence calculator**; and the first 701 customer was, again, Los Alamos. And of the next seventeen 701s, 9 went to aircraft companies, 2 to the navy, one to the Livermore Atomic Research Laboratory, one to the Weather Bureau and one to the National Security Agency.[123]

Even within the purest bastions of the academy, at the Institute for Advanced Studies (IAS) at Princeton, where von Neumann was building a machine specifically for the purposes of advancing computer science itself, the Cold War intruded. Giving evidence in the matter of Robert J. Oppenheimer, von Neumann testified:

Von Neumann: We did plan and develop and build and get in operation and subsequently operate a very fast computer which during the period of its development was in the very fast class. . . .
Q: When was it finally built?
Von Neumann: It was built between 1946 and 1952.
Q: And when was it complete and ready for use?
Von Neumann: It was complete in 1951 and it was in a condition where you could really get production out of it in 1952.
Q: And was it used in the hydrogen bomb program?
Von Neumann: Yes. As far as the Institute is concerned . . . this computer came into operation in 1952 after which the first large problem that was done on it, and which was quite large and took even under these conditions half a year, was for the thermonuclear program. Previous to that I had spent a lot of time on calculations on other computers for the thermonuclear program.[124]

This is not to say that non-military uses for computers were not essayed in the early days after the war. Everybody, from the pari-mutuel betting company in America to Joseph Lyons tearooms in the UK looked into its potential. Few, though, were prepared to write cheques. Against this, it is certainly the case that within meteorological circles as well as the Census Bureau the device was quickly regarded as a *sine qua non*; so too within the scientific academy. But the fact is that all of these latter requirements antedate the war and they were not then, in the 1930s, of sufficient import to get computers built, despite the readily available parts. The new factor, maintaining the impetus created by the Second World War, was neither old scientific needs nor the spotty interest of business. It was the Cold War.

Perhaps the ultimate proof of this can be seen in the British experience. The British rapidly picked up on the thinking done by the ENIAC team and, still propelled forward on the wave of expertise created during the hostilities, built the first undeniable electronic stored-programme computer actually to compute. Yet this lead was lost by the early 1950s, for why would a post-imperial Britain need to be in the forefront of such developments?

There is an appositeness in the fact that the International Research

Conference on the history of computing, the published record of which has been much used in these pages, should have convened, in 1976, at the Los Alamos Scientific Laboratory in New Mexico rather than at IBM in Endicott, New York or the Institute for Advanced Studies at Princeton, or the Moore School in Philadelphia, or the Manchester University Laboratory in Britain. The computer survived the Second World War *in utero* to be born as a child of the nuclear age.

Phase three: technological performance – invention

Williams and Kilburn invent the COMPUTER, Manchester, UK, 21 June 1948

(i)   The Hartree Constant

At war's end, the computer stood ready to be *invented*. The concept was to hand and much of the hardware had been effectively demonstrated by ENIAC and Colossus Mark II. It was, however, to take a number of years before a stored programme machine was to work. The immediate post-war period, from 1946 until the first years of the 1950s was subject to an iron rule called in America the 'von Neumann Constant'. In Britain the same law applied, only under the title the 'Hartree Constant'.

> Some people have mentioned it as being part of von Neumann's machine, but as I remember Johnny developed it as a universal constant – that constant number of months from now until everyone's machine is expected to be completed. After the Eniac, many computers had been started, but at that time [1948] none had reached the promised land.[125]

This constant, a species of the 'law' of the suppression of radical potential, was the result of a number of factors. Although ENIAC had created an enormous stir and the 'First Report' on EDVAC clearly outlined what needed to be done to produce a large-scale computer in the metal, to *invent* it, there were many obstacles – none of them, though, attributable to technology.

The only technological question mark over the nascent computer was the form of its memory, crucial since this was the device's distinguishing characteristic. Although nobody had yet used them for this purpose, Eckert's mercury lines were to hand and were found, eventually, to work as planned; so these cannot have been a source of delay. Alternatives to the delay lines also appeared at this same time, before any computer was operational, so it is obvious that the development of these was not a major retarding factor either. Rather, some projects languished because the purposes for which they were planned disappeared. Some were thwarted by the obstruction of science bureaucrats or the shortsightedness of potential commercial buyers. In short, the supervening necessity of the Cold War had not fully operated to make the machines indispensable and the whole trajectory of development rocked uneasily between this lack of need and the forward impetus generated by the war-time work.

The result was, in contrast to the speed of Colossus but following the model of ENIAC, a leisurely pace. Even within Los Alamos, work proceeded without computers or, indeed, plans for computers. The computer was the child of a war that had yet to break out in earnest. Nevertheless, although this lack of a supervening necessity slowed the development of the computer, it did not halt it and the impetus given by war was particularly evident among the British.

The British did not know of the EDVAC plans until 1945. Then a flurry of visits – Douglas Hartree and T.H. Flowers among others – apprised them of the latest developments, making them aware of the stored programme concept and the mercury delay line solution. Among those stimulated was another ex-TRE scientist, F.C. Williams, whose alternative to the mercury lines, the 'Williams electrostatic memory', was to be the most important single British contribution during the years of the 'Hartree/von Neumann Constant' – the incunabula period (as it might be termed).

Williams patented his data storage device, 'in which numbers are represented by distributions of electric charge on a surface inside a vacuum tube', in this case a cathode ray tube, in December 1947.[126] This solution to the memory store problem used a suitably modulated beam of electrons to 'write' the information on the surface in the following way:

> [T]he 'mud puddle' analogy is useful. Imagine a flat area of thick mud (analogous to the phosphor coating of a cathode-ray tube) and above it, a source of drops of water (analogous to the electron beam). If water is dropped at a particular location a crater is created (analogous to a potential well in the phosphor) and a binary '0' is stored. If the water source is moved slightly (technically known as a 'twitch') and more water is dropped, the original crater is partially refilled, and a binary '1' is stored. Sensing of the stored information is achieved by returning the water source to the original location. When water is dropped, more mud will be displaced for the stored '1' following the twitch than for the stored '0' without the twitch.[127]

The system was far from perfect but it did use a readily available and quite cheap component. It was many times quicker to access than a delay line. Eventually any word could be reached in about

30 millionths of a second as opposed to 0.5 thousands with the mercury lines; which meant that, although the phosphors were constantly decaying and therefore in constant need of being recharged if the data was not to be lost, the 'Williams Memory' was still a lot faster than the delay lines. It was to be a popular solution, computer operators even learning to read some aspects of the machine's mind, as it were, from the patterns displayed on the CRTs.

Williams accepted the chair of electrical engineering at Manchester. The university had also appointed Max Newman to the chair of pure mathematics. Newman, having spent a semester with von Neumann at Princeton, obtained the support of the Royal Society to establish a 'Calculating Machine Laboratory'.[128]

In order to demonstrate the efficacy of his memory system, Williams and another colleague from TRE, Tom Kilburn, set about building a computer.[129] In contradistinction to von Neumann's monster plan, their design philosophy was to make the machine as small as possible, but this was not because they had escaped the influence of 'The First Report'. Rather it was so that a machine, which was not really thought of as a computer because it did not match American specifications as to size and power, could be built quickly and electrostatic storage effectively demonstrated. With this ambition, progress was swift. They circulated a paper, 'A Storage System for Use with Binary Digital Computing Machinery' which was written by Kilburn and dated 1 December 1947.[130] Six months later, on 21 June 1948 the Baby Mark I ran a 52-minute programme. It was the first fully electronic stored programme device to work anywhere in the world – in fact, the *invented* computer. And it was also to be the first and last baby machine for 30 years (except for two built for the similar purpose of demonstrating a potential component – one, the magnetic core memory in the early 1950s (see page 159 below) and the other integrated circuits, in the mid-1960s (see chapter 4, page 214)). But the Baby's creators declared, 'The machine is purely experimental and is on too small a scale to be of mathematical value', and proceeded to subject the apparatus to 'intense engineering development' over the course of the next twelve months so that it was big enough to warrant the name computer.[131]

Although in British fashion not quite so gigantic as the machines the Americans were proposing, the Manchester University Mark I or the *Manchester Automatic Digital Machine* (the **MADM**) proper, which was working experimentally by April of 1949, was what everybody expected a computer to be – a room full of valves. In attracting some publicity that summer, the MADM diverted attention away from the unveiling of EDSAC in Cambridge University which was, by contrast, fully operable. (The MADM only became available for 'regular' problems in the autumn.)[132]

As with ENIAC and the ABC, so with the MADM and the Baby Mark I. Von Neumann's 'First Report' made it clear that computers were not to be 'babies'. The Manchester tradition therefore took another form – that of 'large machine design and fruitful cooperation with industry'.[133] The giant ENIAC, which was not a computer but a prototype, is inscribed as the first computer in popular consciousness while the Baby Mark I, which is the *invention*, is seen, if at all, as a prototype.

In the literature that acknowledges the significance of the stored programme – that is, the literature that refuses ENIAC the title of first electronic computer – the Cambridge EDSAC (the MADM's rival) is accorded that honour. EDSAC, though, is another machine designed under the influence of the EDVAC team.

The pioneers who had worked on ENIAC and designed EDVAC, in contrast to their increasingly strained internal relations, presented to the outside, albeit minuscule, world of fellow computer scientists a face open and friendly in the extreme. A summer school was held at the University of Pennsylvania in 1946 to share thinking and among the participants were a number of British scientists including M.V. Wilkes.

In the 1930s, Wilkes had worked on the Cambridge differential analyser and had, like Williams, spent the war at the TRE in Malvern. He was appointed director of the Cambridge University Mathematical Laboratory after the war and he resolved, upon his return from America that summer, to build a computer along the lines suggested in the 'First Report' – in fact, an Edvacide. He obtained support from the DSIR, the University Grants Committee and J. Lyons & Co and, by May of 1949, 30 months after he had begun, and shortly before the MADM was fully operational, EDSAC worked. Wilkes' designation describes the device, *Electronic Delay Storage Automatic Calculator*, that is, an electronic calculator with the capacity to store data and programming instructions in its delay lines – a computer. EDSAC was to be the first Edvacide, born, as it were, before its parent and, indeed, with its 32 mercury lines, it was the first full-scale electronic computer to incorporate Eckert's memory system and be used for proper mathematical work. Although the delay line memory was not, finally, the option preferred as the computer developed, Wilkes did produce elegant new systems for control and arithmetic hardware design using closed sub-routines, micro-programming, which made it 'much easier to find one's way about'.[134]

These three British devices (Baby Mark I, MADM, EDSAC) are incontrovertibly the first electronic computers, but they were preceded by one other machine for which stored programming claims are made.

IBM's response to Harvard's war-time rebuff was to behave like nothing so much as a great ship cast adrift on an electromechanical sea. In complete defiance of ENIAC, the 'First Report' and all they portended, IBM built, tested and put into operation, between 1945 and 1948, the last giant electromechanical machine, said to be 6000 times faster than the Harvard Mark I (ASCC). The *Selective Sequence Electronic Calculator* (**SSEC**), had 12,500 vacuum tubes and 21,400 electromagnetic relays. It was dedicated at IBM, New York in January 1948, with all the pomp and circumstance from which Watson had been excluded at Harvard. Although in most ways an obsolete machine, a part electronic and part electromechanical hybrid, curiously it did have something very closely resembling a stored programme.

All instructions are given in numerical form. Each unit of the machine and each operation it can perform have been assigned a number, and the presence of this number in the instructions calls into action the associated unit or operation. This procedure permits the use of all the numerical facilities of the machine, not only for problem data *but for handling operating instructions as well.* (our italics)[135]

In fact, in the electronic part of the SSEC (as opposed to its electromechanical part), there was an 8-word store and it can be said that this was the first machine (assuming Colossus Mark II to be *hors concours*) to have a stored programme capacity. It was, in fact, despite its outmoded electromechanical elements, a computer. IBM's Director of Pure Science, Walter Eckert (no relation of Presper), who had for years headed the Watson Lab at Columbia, first used it to calculate the coordinates of the moon; but it also received the distinguishing mark of every American computing incunabula – a Los Alamos problem, in this case to do with hydrodynamics.[136] Despite all this, some claim that its measure of stored programmability is insufficient to justify it as a computer and the SSEC certainly suffered the fate of a dinosaur. Without successors, it was dismantled in 1952. The future lay with the Manchester University Baby Mark I.

All this activity in England can be traced to a quite brief period during which Britain, in victory still a great imperial power, continued independent nuclear development. She needed the computers because she had been cut off, momentarily, from her main source of information and primary research, the USA, by an act of Congress. In 1946, Winston Churchill enunciated the concept of the Iron Curtain. That same year, the Americans stopped sharing atomic information with Britain. The British decision to proceed without this information was made in the first months of 1947. Thus the British created for themselves the same supervening necessity that the Americans had. Ferranti, based in Manchester, was given $100,000 and, in one of the vaguest of government contracts, was required 'to construct an electronic calculating machine to the instructions of Professor F.C. Williams'.[137] This was marketed in February of 1951. To go with the machine, the university produced a 'Programmers' Handbook', written by Alan Turing who, between the end of the war and this point, had suffered as much as anybody from the 'Hartree Constant'.

Turing had been recruited from Bletchley Park by John Womersley to the National Physical Laboratory at Teddington. Womersley had been running the NPL's Mathematical Division since its establishment in September, 1944. He had, under the highest possible security clearance, visited the Moore School before the end of hostilities, being the first foreigner to see ENIAC and he had been given a copy of von Neumann's 'First Report'. It was this document that he used to entice Turing. Turing took the view that the NPL's invitation was a sort of repayment for his war-time efforts. Now, in return, the British government would fund a universal Turing machine.[138]

Official British thinking, in the first year after the war and before the Americans ceased to supply nuclear information, was that all the nation's computing needs could be done if not on one machine then certainly in one centre and that the NPL was to be the place. This meshed with war-time planning when, in advance of the electronic machines, it was hoped that the Mathematical Division would be able to concentrate all military computational needs under one roof. The scientific bureaucrats having made this decision, it was difficult for others to fund their projects. Turing's device at the NPL was to be the computer, the only one. Newman, writing to von Neumann in 1946, said, 'Once the NPL project was started, it became questionable whether a further unit was wanted.'[139]

The NPL had an imperialist view of computing. Wilkes was involved, at these early stages, as a potential builder of the Turing blueprint yet he responded somewhat gingerly to these overtures. He wrote to Womersley that, 'I am beginning to do a little work on electronic calculating machines and that I am anxious to co-operate with you.'[140] However the anxieties did not translate into cooperation, and the 'little work' became a rival project which yielded the EDSAC.

Similar attempts were made to interest Williams in the project and some abortive meetings were held. Newman only got his funding by arguing (after the fashion of von Neumann) that, in contrast to the NPL plan, the Manchester machine would be used for theoretical problems in pure mathematics, and, with considerable prescience, he mentioned 'testing out the four-colour theorem' which would only be proved, with the help of enormous computing power, thirty years later.[141] Wilkes, Williams and Newman were aided in their floating of rival projects by the fact that the official and unique computer, although first off the mark by many months, never really got under way and, by the time that it did, the Cold War had overturned the official plan which had the NPL alone in the field.

Before the end of 1945, Turing had produced on paper plans for a stored programme computer to be designated ACE (*Automatic Computing Engine* – 'Engine' in honour of Babbage). He also acknowledged his debt to his pre-war teacher von Neumann by stating that his design was to be understood in terms of the 'First Report', although his architecture was very different from that of EDVAC; and, curiously, he made no mention of 'On Computable Numbers'. Douglas Hartree then being the member of the NPL executive most responsible for such schemes, the project was agreed on the 19th of March 1946, a month after ENIAC had been unveiled.[142]

Turing's proposals for ACE envisaged a very ambitious machine involving 200 mercury delay lines and a capacity some six times greater than the biggest of its planned rivals – 6400 32-bit words. Many more refined blueprints of the ACE were developed in the course of the next two years but getting the thing into the metal proved ever more difficult. Originally it was decided that the building of the machine should be contracted out. Wilkes was approached, as we have seen, but elaborate discussions led nowhere. (Wilkes was not one of Turing's admirers.) Williams was courted. Turing's attempt to get the GPO to put Flowers and sufficient resources on to ACE failed. It was eventually decided that the design was to be executed in house by another division of the NPL. The stage was set for a classic British confrontation between theory and practice. H.A. Thomas, head of a new Electronics Section of the Radio Division, was charged with developing industrial devices to help rebuild Britain. ACE, which became his responsibility, apparently did not fall within this basic requirement, as he saw it. He obstructed the machine. 'It was a curious sequence of events, in which the NPL administration had done everything possible to avoid "building a brain".'[143] It was as if the NPL, well before anybody else in Britain, understood that the nation no longer had serious military computational needs and there was therefore no reason for ACE.

Within the team various schemes were developed to facilitate ACE. Hartree had recruited the American Harry Huskey, a member of the ENIAC team, to spend a year's sabbatical at Teddington. Huskey, who accomplished little during his time in England, proposed that they build the smallest possible machine that would illustrate Turing's architecture; but this did not happen either.

Turing, an increasingly frustrated victim of the 'Hartree Constant', took a sabbatical in Cambridge (where he barely visited Wilkes) and then accepted Newman's offer to go to Manchester. The Royal Society agreed that his salary could come from its grant and he was made deputy director of the lab where he arrived in September of 1948, too late to influence the shape of the MADM but in time to write its manual. This, Turing's last move, was to prove fatal because it was in Manchester, during the aftermath of an investigation for burglary at his house, that the police began proceedings against him for homosexuality. His suicide followed shortly thereafter. There is an opinion that had he returned to Cambridge, the worse excesses of official homophobia would have been spared him.[144]

At the NPL, Thomas having moved on, progress was at last being made. There was now some pressure since Edsac and the MADM were working – so, in 1949, plans were drawn up for a **Pilot Ace**, following Huskey's proposal for a less ambitious device. Pilot Ace was very much smaller than the machine Turing had designed three years previously, having only a 300-word store, but it was built and ran for the first time, in a preliminary way, in May of 1950. It worked well enough to do years of service at the NPL, one advantage of its limited physical size being that it could be moved, unlike its competitors. Pilot Ace was also the effective mould for a commercial version built by English Electric and marketed as **Deuce**.

Pilot Ace, despite this history of prevarication and delay and although it had been rendered obsolete by Wilkes's development of programming techniques and Williams's more advanced memory store, was nevertheless the fourth electronic machine to work anywhere in the world.

Actually it is unclear if it was the fourth or fifth machine because in the spring of 1950 the first American computer was coming on line. 'A computer, at least in those days, was not something you put together all at once and turned on to find out whether it worked or not.'[145] Its builders had begun to find out if **SEAC** worked during April of this same year.

(ii) The von Neumann Constant
In America the role of the NPL was being played, with a good deal more despatch, by the National Bureau of Standards (NBS). This institution had grown out of the New Deal programme mentioned above for the hand computation of tables done by mathematicians with the help of the unemployed. The NBS, according to American official thinking which was along the same lines as the British, was to be the organisation coordinating the entire computational needs of every branch of the government. Consequently the NBS, in conjunction with its most obvious governmental client, the Census Bureau, simply commissioned outsiders, in this case the company formed by Eckert and Mauchly, to provide it with a computer.

The 'von Neumann Constant' operated to such effect in Eckert and Mauchly's enterprise that no machine appeared (see below, page 165). Under pressure from the Air Force (whose needs were more significant than the Census Bureau's), the NBS therefore revised its strategy and began to build an 'interim' computer in-house, using the basic design outlined in the 'First Report', to plug the gap. With such elements as a magnetic wire cartridge from a commercial dictating machine, as well as Williams tubes and magnetic tape, it became fully operational in May 1950 – the fourth (or fifth) computer in the world to work, the first in the USA. It was designated SEAC (*S*tandard *E*astern *A*utomatic *C*omputer). Of course, 'It was not many weeks before Nick Metropolis . . . showed up with a problem from Los Alamos.'[146]

SEAC was not only the first stored programme machine to be operational in the US, in fact the American *invention*, but it was also the first computer to use solid state components – diodes – for its logic circuits, reserving valves for amplifying functions. Eschewing the latest type because of the uncertainty of their operation, SEAC used instead 1N34 'whisker' diodes, 10,000 of them, as well as 750 valves. It was, then, a 'cat's whisker' computer but, despite this, it remained operational well into the age of the transistor, finally being closed down in 1964.

For all the innovation of its design, in its architecture SEAC remained true to the 'First Report' and was, like EDSAC, another Edvacide born before its own parents.

During the summer of 1945, the EDVAC team began to fall apart. Von Neumann had decided to build his own machine for purely scientific purposes, although, given his connections with both IBM and Los Alamos, there was clearly some overlap with the world. To accomplish this scheme at Princeton, where the Institute of Advanced Studies (IAS) had set its face against anything but theorising, von Neumann took advantage of his status as one of the men of the atomic hour and orchestrated a campaign of offers from MIT, Harvard and Chicago. He got his way, his funding and his machine, which he assured Princeton would not be used to solve any problems in any field of study; rather – as is the case with thousands of contemporary underused microcomputers in schools and homes – it was the machine itself which was to be the scholarly object. With $100,000 from the Institute, $100,000 from RCA, which had an option on a possible memory system as is described below (page 158), $100,000 from Princeton, $100,000 from the army, $1,000,000 from the navy and taking Goldstine as well as a number of others from the Moore School with him, von Neumann was in business – the business of pure science, but with the atomic connection looming very large. By 1946, the Atomic Energy Commission was also supporting the project.

Given the IAS's hesitancy about exploiting developments in the metal, the price of all this government support was that the team's thinking and designs be distributed as quickly as possible to a number of other parties more willing to dirty their hands. These included not only Los Alamos but the national laboratories at Oak Ridge and

Argonne as well as the Rand Corporation and the University of Illinois.

Von Neumann opted for a 40-bit word. During the first period, up to June of 1947, a building to house the work as well as test equipment and possible memory systems were all being constructed. Only after these were completed did serious thought start on the design of the machine itself. It was to take two years to build, eventually having a store of 1024 words of 40 bits and using 2600 tubes. By January of 1950, it was ready for testing, which was to take over a year to complete. Needless to say, among the earliest users, in the spring of 1951 when the computer finally became available, were scientists from Los Alamos. Contrary to von Neumann's promise, they were less interested in the machine itself and more in what it could do for them. They ran one nuclear problem for 60 days without cessation, 24 hours a day. The IAS computer was in use until 1960 when it was given to the Smithsonian.[147]

Von Neumann had been eager to engage Eckert as his chief engineer but had failed to secure his services. Perhaps because of this, given the centrality of Eckert's idea of the mercury delay tubes for the store, the IAS team was eager to find another method for data storage. In the summer of 1947 the team was experimenting with a magnetic wire recording system which, in its physical design, relied heavily on bicycle wheels. But the possibility of an electrostatic solution was well understood from the beginning; iconoscopes had been mentioned by Burks, Goldstine and von Neumann in an early Princeton paper. Indeed von Neumann had involved RCA in the project to develop a cathode ray tube for memory purposes, to be marketed as the **Selectron**.[148] The magnetic wire devices were a stop-gap while waiting for Jan Rajchman at the RCA lab.

RCA had been interested in firing-aid calculators in the early part of the war but when Rajchman discovered that the Moore School had obtained the contract to build ENIAC, he passed his thinking on and RCA's interest ceased. At von Neumann's behest, Rajchman designed the Selectron, a digital, as opposed to the normal analogue, cathode ray tube. It took four years to bring this to the point of manufacture during which time Williams had demonstrated how to use readily available ordinary CRTs.[149] In June of 1948, the IAS team had the Manchester paper in hand – 'A Storage System' had been brought to them by Douglas Hartree on a visit. It seemed to offer the most promising solution.[150] Julian Bigelow, who had taken the job von Neumann first wanted Eckert to have, was dispatched to England. His first encounter with Williams and his memory system, then in prototype form, was vivid:

I can remember him explaining it to me, when there was a flash and a puff of smoke and everything went dead, but Williams was unperturbed, turned off the power, and with a handy soldering iron, replaced a few dangling wires and resistors so that everything was working again in a few minutes.[151]

By January of 1950 a full scale 'Williams Memory' was built into the IAS computer. But this was not the only option; as has been seen, wire recording devices were being actively used in some machines and it was not that computer engineers were unaware of Bing Crosby's interest in tape. The Dirks, Doctor G. and Engineer G., had registered a number of German patents for the use of magnetic tape in calculators in 1943.[152] Eckert's 'Disclosure' document of 1945 discussed magnetic recording techniques as one of a number of possible data stores and devoted time during this period to designing superior methods of accessing. Turing had examined a captured Magnetophon before leaving Bletchley but, envisaging the extent to which the tape in 'On Computable Numbers' would have to move, he rejected it as a medium. Others, given the available tape movement mechanisms, agreed and avoided tape not just because its access time was slow but also because of its variable quality.[153]

We had the same problems with magnetic tape as everybody else. The tape had blemishes, bits of dust or other flaws that would push the tape away from the head as it went by and cause a dropout.[154]

The full-scale MADM used a variant on tape, a **magnetic drum store** as a back-up to its 'Williams Memory'. An English scientist, A.D. Booth, who had spent the war at the University of Birmingham studying the crystal structure of explosives, built in 1947 a revolving drum with the data encoded magnetically around it on tracks. Drums were made of brass or bronze and either nickel plated or sprayed with iron oxide and, although still slow, were more accessible and reliable than linear tape. Eventually Jay Forrester of MIT – who is usually credited with this development – would produce, in the final stages in cooperation with IBM, a core memory drum which by the mid-1950s became the industry standard. (Forrester was working on the most significant of these early machines, **Whirlwind** (see below page 161).) For the memory he first used a special magnetic material marketed as Deltamax and then ceramic ferrite which he made into a sort of chain mail. He had one of his team, Kenneth Olsen, direct the building of a small computer to test this magnetic core memory system – the second abandoned baby computer. The core memory system was working in the main Whirlwind machine by the summer of 1953.[155]
(Booth was another British pioneer with a typical computer career. After the war he moved from explosives to rubber and worked in the

lab of the British Rubber Producers Association. He returned to Birkbeck College in the University of London where he built a series of machines culminating with the **APEX**s, *A*ll *P*urpose *E*lectronic *C*omputers. He used a grant from the Rockefeller Foundation to buy the tubes for the prototype of these last. Apex was, like Pilot Ace and MADM, marketed – as the **HEC**, Hollerith Electronic Computer. It was manufactured by the British Tabulating Company which John Womersley, Turing's old boss, had joined from the NPL. Booth went to Canada in the early 1960s.[156])

All these memory devices were available for incorporation into the machines – the Iasides – built as a result of the IAS team's circulation of its plans.

At Los Alamos, the need for advanced calculators had been apparent long before von Neumann mentioned the existence of ENIAC. Dana Mitchell, a Columbia physicist seconded to the laboratory, knew of the use to which IBM commercial equipment had been put in the Watson Astronomical Calculations Lab back in New York and a similar range of IBM machines was installed at Los Alamos.[157] In this situation, however, they were inadequate and the usefulness of a fully electronic device was nowhere better appreciated. Los Alamos was, therefore, a prime recipient of the IAS plans. However, even here, there was a hiatus. Nicholas Metropolis had left for the University of Chicago in 1946. He was called back two years later and began to build his Iaside. It was running by March of 1952, by which time magnetic tape as well as a 10,000-word magnetic drum was being used. Metropolis, irked by the acronymic christening then fashionable for these machines, called his MANIAC (*M*athematical *A*nalyzer *N*umerical *I*ntegrator *a*nd *C*ounter) in a futile attempt to kill the fad.[158]

The distribution of information insisted on by the US government meant that other Iasides were produced within the same time frame. **AVIDAC**, *A*rgonnes *V*ersion of the *I*nstitute's *D*igital *A*utomatic *C*omputer and **ORACLE**, *O*ak *R*idge *A*utomatic *C*omputer and *L*ogical *E*ngine were in operation by 1953. On the last, 2 inch magnetic tape was used as a back-up to the 'Williams memory'.[159] It was the Rand version, the **JOHNNIAC** (named for von Neumann), which used for its memory the only RCA Selectrons ever sold.[160] The University of Illinois was part of the IAS network so that it could build two Iasides for the Ballistics Research Laboratory. Although the Illinois team relied on the preliminary design specification of von Neumann's group, the resultant machines, **ORDVAC** and **ILLIAC**, contained much that was contributed locally, notably the configuring of the Williams tubes to provide for the random access of data.[161] Both were handed over to the military in 1952.

Not all the American incunabula of this period can be traced to IAS or EDVAC. At MIT a special machine of great long-term significance was built. It had begun as an analogue real-time calculator to build a sort of flight simulator – an airplane stability analyser – for the Special Devices Center of the navy. It was to be a cockpit and a computer and it would 'fly like an airplane not yet built'.[162] The computer part of the device went digital in 1945 when its designers, led by Jay Forrester, became aware of ENIAC and the other planned digital machines. By 1947, the year Forrester began developing the core memory, the plan was ready. It was working about 4 years later – the first device to use a visual display system and (by 1953) the first to use a magnetic random access core memory drum system.

Whirlwind, as the machine was known, is the best illustration of the hiccup caused by the disappearance of one supervening necessity – the hot war – and the emergence of another – the cold, as is mentioned above. The Special Devices Center had been disbanded. The Office of Naval Research (ONR), now the funding agency, was not interested in an aircraft stability analyser and even less in spending around $1 million a year for it – some 5 per cent of its total budget. MIT was just beginning its successful search for industrial funding and had established an office for this purpose. Forrester thought perhaps that by redirecting Whirlwind toward automated tool applications, the Industrial Liaison Office could find some backers. The administration was certainly unwilling to lose its one general purpose computing development without a fight.[163] In 1950, when the ONR was about to pull the plug, Whirlwind re-acquired a purpose. It did so without help from industrial liaison.

About that time the USSR had developed atomic weapons and intercontinental aircraft, and considerable attention was being given to air defence. A major threat came from low-flying aircraft. At low altitude, radar range is very short, and information from many radars had to be netted to cover large areas. George Valley at MIT was concerned about radar coverage. He met Jay, and it turned out that what he needed was a [real time] digital computer and what we needed was air defence. So Air Force financial support appeared in the nick of time.[164]

In August 1949, the Russians exploded an atomic bomb and George Valley, a member of the Airforce Scientific Advisory Board, raised the spectre with Vannevar Bush of a Soviet nuclear attack over the North Pole. Forrester joined this committee in March 1950.[165] In June the Korean War began. By November instead of being abandoned Whirlwind had acquired airforce backing and 175 personnel. The number was to grow as it became more and more indispensable for its new task. Eventually it metamorphosed into the central control for the Cape Cod air defence system, the initial machine in the entire

Semi*automatic* Ground *Environment* (**SAGE**) Air Defence system.[166] Whirlwind in its developed form was the basis of an IBM series later in the 1950s and, although it was the biggest of the early machines, it was, in its overall architecture and design as well as in use, the closest of them to the present generation of mini and microcomputers.

The contrasting histories of these various projects is the proof of the supervening necessity thesis as it applies to the computer. Von Neumann, being closest to the heart of the Cold War action, suffered no interruption of his developmental work. Whirlwind, being slightly further away, was nearly abandoned until it could be made to serve Cold War purposes.

1952 was the last year of the computer's period of *invention*, the age of the incunabula, of one-off machines. In November the USA exploded an H-Bomb. Already in the UK the world's first two commercial computers were in production. In January of 1953 IBM was to ship its first production-line computer, an Iaside designated the 701, to Los Alamos. More significant for the supposed revolution than these almost secret applications was Eckert and Mauchly's commercially produced **UNIVAC** which had been used, with an enormous flurry of publicity, to predict the outcome of the 1952 presidential election on television. In the public mind, the computer age was in full swing. In fact, the better part of nine years had already passed since Clippinger and the Eniac team had thought of the idea and there were barely two dozen machines worldwide.

The incunabula period closed off many options. The machines were all massive and unbelievably expensive, designed for extremely complex work at the heart of the Cold War, each, as one of the IAS's builders remarked, 'an exceedingly complicated thing'.[167] The need for secrecy alone meant nobody had any real interest in the possibilities of accessible, smaller and cheaper devices. The 'law' of the suppression of radical potential was as much at work in the US as it was in Britain.

The third transformation: the 'law' of the suppression of radical potential

Bryan Field needs a MACHINE to make book, spring 1946

(i) Adventures in the skin trade
The operation of the 'law' of the suppression of radical potential balanced the Cold War supervening necessity to ensure that the computer 'revolution' would continue at the unrevolutionary pace evidenced from the start. During the incunabula period three factors contributed to the 'law'. The first was the indifference of commerce.

[V]irtually every highly civilized country realized the great need for these instruments. This caused a great awareness of the scientific importance of computers to the university and government community, and the earliest and boldest developments are in general in the university rather than in the industrial world. Indeed, at this stage of the field's development most industrialists viewed computers mainly as tools for the small numbers of university or government scientists, and the chief applications were thought to be highly scientific in nature. It was only later that the commercial implications of the computer began to be appreciated.[168]

Von Neumann's decision to be in the business of science was therefore far sounder than Eckert and Mauchly's desire to be in the business of business. Although UNIVAC, as their commercial machine was called, worked at about the same time as the IAS computer, its developmental path was very much rockier. And even though Univac was designed for commerce, it was actually as much a government device as any of the Iasides. How Eckert and Mauchly got from the University of Pennsylvania to the studios of CBS in time to predict the 1952 election outcome reveals, better than received opinions as to the nature of technological change, the realities of the innovative process.

Even as Eniac was unveiled to a startled world, Eckert was finding life at the Moore School increasingly difficult. A new policy requiring that all patent rights be vested with the school was particularly galling. Eckert was considering the von Neumann offer but decided to throw in his lot with Mauchly. They resigned on 31 March 1946 to form the Electronic Control Company (ECC), and began to seek backers in the financial community. They were unsuccessful. Only the Census Bureau was interested in their plans but was uncertain about passing the large sums of money necessary for computer development to the fledgling private firm.

The most sustained interest shown in their plans came from Bryan Field, a racetrack manager in Delaware, eager to dislodge American Totalisator which had a monopoly on pari-mutuel betting equipment. After reading of ENIAC in the *New York Times*, he presented himself to the tyro entrepreneurs as the one businessperson eager to put at least racetrack affairs 'on a new footing'; but Eckert and Mauchly – although Howard Aitken had told them, somewhat mysteriously, that there was a lot of money to be made at racetracks – were unwilling to build the specialised machine Field needed and so this supervening necessity was left unfulfilled.[169]

The Census Bureau, with its history of effective mechanisation dating back to the Hollerith breakthrough of 1889, was as eager to exploit the new technology as was the military. All these requirements were being coordinated by the National Bureau of Standards

(NBS). The NBS was encouraging and in May of 1946 told Mauchly and Eckert that it and the Census Bureau were sufficiently impressed with their proposal for a general computing device, in fact the EDVAC, that it would let contracts for it in the immediate future.

At this point in 1946, the Moore School, which still intended to build EDVAC itself, was about to hold its seminal summer course, von Neumann was staffing the Electronic Computer Project at the IAS and Turing's ACE had just been approved as a working scheme by the NPL. Worldwide, three computers were funded and underway, and another – the Baby Mark I – was being put together without fullscale official sanction.

Apart from ECC, there was just one other commercial organisation in existence dedicated to computer work – Electronic Research Associates, founded by ex-naval cryptanalysts who had been attached to 'seesaw', or rather, CSAW, Communications Supplementary Activities, Washington – an American equivalent of Bletchley Park. One of them, Ralph Meader, had headed the Naval Computing Machinery Laboratory which operated out of an NCR factory in Dayton, Ohio. NCR was uninterested in continuing to work for CSAW once the war had ended and refused to invest in Meader's new venture. ERA's other founders, Howard Engstrom and William Norris, had obtained a tentative offer of support from their senior officers in Washington who otherwise, in a peace-time navy, had no way of keeping the engineering skills Engstrom and Norris represented together.[170] Funds were fortuitously found from the Northwestern Aeronautical Corporation, owned by a John Parker, who was desperate for projects to replace his war-time glider-making activities.

Both ERA and ECC were in difficulties. A major one was their procurement records which, as new organisations, obviously they had not established. Government agencies were unwilling to be the first to throw money at them. ERA did a lot better than ECC because its parent, Northwestern Aeronautical, had an excellent procurement record, albeit for gliders rather than computing equipment. By becoming a direct subsidiary of the glider firm, the problem was eradicated. ERA was happy to have the navy as its 'Delaware Racecourse' and build whatever specific bits of equipment were needed, such as advanced drum storage systems for specialised cryptanalytic devices. Building a general purpose computer was not the company's raison d'être, unlike ECC. The commission for a computer – **Atlas** it was to be called (named for a comic strip character) – did not come from the navy until the summer of 1947; it was ERA's thirteenth job. In its second year, ERA received more than $3 million from the navy, still its only customer, and employed over 400 people.[171] Eckert and Mauchly did not find their 'Northwestern Aeronautical' to help with the procurement record problem; they did not find any backers at all. At NBS, their proposal was submitted for review. It went to George Stibitz, Bell Labs' electromechanical computer expert. Stibitz had been a member of the National Defence Research Committee during the war and, as such, he knew all about ENIAC and was not impressed. As a Bell man he felt switches (which Bell made and which he had put into complex calculators) were just as good as valves (which Bell did not make). He wrote, 'I see no reason for supposing that [the Bell Labs machine] is less broad in scope than ENIAC'. He added, 'I am very sure that the development time for the electronic equipment would be four to six times as long as that of relay equipment'; which, while true, somewhat missed the point – similarly, circa 1810 say, the development time for a horse-drawn wagon was quicker than that of a steam locomotive. In 1946, Stibitz concluded that as far as the new Eckert-Mauchly proposal was concerned, 'There are so many things undecided that I do not think a contract should be let for the whole job.'[172] NBS also consulted Aitken who told them: 'There will never be enough problems, enough work for more than one or two of these computers . . . stop this foolishness with Eckert and Mauchly.'[173]

Thus armed with the best available opinion, NBS did not let the contract for the entire computer, giving instead a development grant of $75,000 to have models of the delay-line system built. This began a pattern that was to bring the firm to the brink of disaster, for the money was insufficient to maintain the establishment necessary to do the work for which it was provided. Further prospects had to be found but these either distracted Mauchly or, in some ways even worse, came through in the same partial way, causing more work and more expense.

In the metal, during the incunabula period and thanks to the 'First Report', the only fairly open options lay in developing methods for data storage. Eckert had written of magnetic systems in the 'Disclosure' paper of 1943. They knew of Bing Crosby's German-style tape and they now had an input device based on magnetic recording techniques. This at least was saleable. It was, for instance, just what the Prudential Company needed to cope with the new government regulations that were causing an upheaval in the American insurance industry. Its expert in the area, Edmund Berkely, had been on Aitken's team at Harvard, but the company as a whole was not convinced by Mauchly's pitch that it needed a fullscale machine.[174] Although he pressed for a commitment to buy a whole computer, still referred to as the EDVAC 2, the Prudential determined it only wanted the magnetic tape device.

The same pattern was repeated with the A.C. Nielsen company, a consumer research organisation now famous for its assessments of the broadcasting industry's audience. Eckert and Mauchly knew the younger Nielsen because he had overseen the erection of the ENIAC building at the Aberdeen Proving Grounds. Mauchly set about selling Nielsens a computer for $100,000, but again there were questions as to the fundamental value of the device and the basic financial stability of ECC. These led to the same sort of result, a monthly stipend to aid research.[175]

These incidents can all be explained because for industry, as in the 1930s, there was still no supervening necessity for a large computer. Mauchly had bargained that the Nielsen company would be, like the Delaware Racetrack, eager to avoid paying rental fees amounting to many thousands a month, in this instance for punched-card machinery. It was, on the contrary, exactly the existence of these machines, offered in a more-or-less open market – for IBM, unlike American Totaliser, had some competitors – that choked off enthusiasm for EDVAC 2.

In the spring of 1947, the Moore School then announced that it was building Edvac (which it did, in conjunction with the Ballistics Research Lab by 1951).[176] ECC had to come up with a new name for its version of the machine which, the *New York Times* wrote, would be 'capable of making the Army's world famous ENIAC look like a dunce'. UNIVAC (*Univ*ersal *A*utomatic *C*omputer) was chosen.[177]

ECC was building a model memory for NBS, a magnetic reader for the Pru and, on a monthly stipend, a general system for A.C. Nielsen; and it had no firm commitment from any of these to buy. Eckert and Mauchly had planned UNIVAC as a decimal machine to attract commercial interest, but the market, such as it was, was governmental and scientific, preferring binary devices. When the Cold War supervening necessity finally intervened in the affairs of ECC, the machine required was not UNIVAC but another device both binary and specialised. In October 1947, two months after ERA had obtained its contract for Atlas, Northrop Aircraft Corporation contracted ECC, in return for $100,000, to build a binary computer for its guided missile project.[178]

Bankruptcy thus averted, the company now faced the major problem of overcoming the 'von Neumann Constant' and building, with dispatch, BINAC. Northrop, driven by this real supervening necessity, meant real business. It put its own man into ECC who immediately proposed that the company incorporate as a way of finding much-needed investors. In December of 1947, Year 3 of the computer revolution, the ECC gave way to the Eckert-Mauchly Computer Corporation.

Henry Straus had not been unaware of Bryan Field's approach to Eckert and Mauchly the previous year and was most interested in it. He owned American Totalisator, having, in the late twenties, designed and patented the equipment upon which it relied. For Straus, now a wealthy 'prominent sportsman', investing in the Eckert-Mauchly

Computer Corporation represented a sure bet on each horse in a two horse race; either the company would fail, and a potential competitor would be removed, or else it would succeed and he would still be in a monopoly position in the betting industry, as well as making further fortune from computers. He put in $400,000 and got 40% of Eckert-Mauchly and seats on its board.[179]

BINAC was not immune from the 'von Neumann Constant'. Contracted to be completed by May of 1948, it was not actually delivered until September of 1949; but, despite the thrice-extended deadlines, it did not suffer too badly. Eckert and Mauchly had built the world's sixth computer, specialised though it was. However, the $178,000 overspend had to be absorbed and it and the other projects had so slowed work on UNIVAC that the NBS had set about building its own general purpose machine, completing SEAC (as described above) before BINAC was delivered to Northrop.[180]

Early in 1948, through its National Applied Mathematics Laboratory division, which had been set up at the behest of the navy, NBS began another computer on the west coast, designated therefore **SWAC** (*S*tandards *W*estern *A*utomatic *C*omputer). Like the UNIVAC it was built to fill in for, it was an Edvacide. Harry Huskey, the ENIAC team member whom Hartree had involved in the ACE project at NPL, began work on SWAC in January of 1949, eventually providing it with 37 cathode ray tubes of 'Williams memory' to store 256 37-bit words. Huskey worked, by the standards of the day, with real speed and the computer's assembly was completed 20 months later. By mid-1953 it was computing 53 hours a week. Like its sister SEAC, it also used, in addition to 2,600 valves, a number of crystal diodes – 3,700.[181] The computer was eventually passed on to the University of California, Los Angeles, and retired in 1967.

NBS, who had also contracted the Raytheon Corporation – a valve manufacturer which had become the US computer industry's third firm – to build a machine for the air force, were still waiting for this device and UNIVAC. Eckert-Mauchly was now devoting its full time to UNIVAC's development but, a month after BINAC was completed, Straus was killed in a light aircraft crash. Without him, the American Totalisator Company panicked and plunged the computer firm into a fresh round of uncertainty by seeking to bail out. The company once more veered towards bankruptcy. Bendix Aviation, International Telephone and Telegraph, Hazeltine Electronics, Westinghouse, Federal Telephone, Burroughs Adding Machine, Mergenthaler Lino-type and Hughes Aircraft were added to the list of those corporations who saw no future in owning Eckert-Mauchly.

Remington Rand (RemRand) was an integrated office equipment company which had grown by absorbing a whole slough of specialised firms making things, anything, for the office. It was IBM's major competitor and co-defendant in a Justice Department case against monopoly practices in the punch-card trade in the mid-1930s. Among the firms RemRand acquired long before was the Tabulating Company which Hollerith's rival Powers had established.

James Rand, who had succeeded his father as RemRand's chief executive, was not only conditioned by this pattern of growth via acquisition; he was also virtually alone among American business-persons with interests in areas congruent to computing to show any awareness of the supposed new age. In 1946, he had established a small research laboratory and had secured the services of General Leslie Groves, late of Los Alamos, to run it. Rand kept American Totalisator dangling until he bought it out in March of 1950. A year later the first UNIVAC was at last delivered to the Census Bureau. It had taken the better part of six years to build and when it was shipped it bore not the name Eckert-Mauchly, but Remington Rand. However, it was to be the first of five.

As with the Air Force rescue of Whirlwind, Rand's takeover of Eckert-Mauchly came at the same propitious moment, between the first Soviet A-Bomb test and the North Korean strike across the 38th parallel. The Cold War hotted up and with it the computers' supervening necessity.

A paradox of this situation, documented by Augarten, is that the same Cold War that secured a future for Mauchly's machines prevented the man himself from working on them. The same month the first UNIVAC was shipped, he was denied clearance on the grounds that he had been a member of the American Association of Scientific Workers (a supposed red-front organisation), had signed a petition for the civilian control of nuclear power and, even, that his wife had drowned while taking a midnight swim with him. It is a measure of the non-existence of the market he had been trying to exploit that this banning, because of the dominance of governmental work, neces-sitated that he remove himself from the UNIVAC division of RemRand until he was 'cleared' six years later.[182]

With four more government UNIVACs to build and another four orders on the books, Rand set about consolidating his position against his seemingly slumbering business opposition. ERA was still under-capitalised, although being closer to the heart of the military-industrial complex it had performed far better than Eckert-Mauchly. Neverthe-less, 'under the frugal defence funding of the Truman administration before the Korean War', employment at ERA started to drop back from its 1948 high.[183] John Parker, the glider maker, was still the chief executive and he, like Mauchly and Eckert, began to look for a buyer. NCR refused (for a second time) as did Raytheon, Honeywell, Burroughs and IBM, but James Rand bought it, concentrating a fair percentage of America's computer talent under one corporate umbrella. Indeed, because security prevented Rand from making an assessment of ERA's potential, the sale price was arrived at by multiplying its 340 engineers by a factor of $5000. ERA's Atlas, the thirteenth task assigned it by the navy, was put into commercial production as the **1101** (13 in binary notation). Machines were sold to the Office of Naval Research (which had nearly killed off Whirlwind), the Civil Aeronautics Administration, various airforce bases and one to the John Plain Company of Chicago, the first to be used as an electronic inventory system in a commercial setting.

The apex of Rand's dominance was reached on the night of 4 November 1952 when UNIVAC was programmed to guess the result of the Stevenson/Eisenhower election. Walter Cronkite introduced it thus:

> It's going to predict the outcome of the election, hour by hour, based on returns at the same time periods on election nights in 1944 and 1948. Scientists, whom we used to call long hairs, have been working on correlating the facts for the past two or three months. . . . Actually, we're not depending too much on this machine. It may be just a sideshow . . . and then again it may turn out to be of great value to some people.[184]

It was probably best not to depend on the machine in the studio which, given the tradition of the fake that television is heir to, was only a mockup. The deal that got the mock UNIVAC to CBS, and a real one at the other end of a line in Philadelphia, grew from a simple request by the television network to Remington for the free loan of around 100 typewriters and adding machines which would, in the gently corrupting way these things are arranged, appear on camera during election night coverage; free commercial for free machines. Free computer for free commercial sounded even better. With only 3 million votes (7 per cent) counted UNIVAC predicted an Eisenhower landslide, 438 electoral college votes to 93 for Stevenson. Nobody in Philadelphia believed it, so the programmers fixed the machine to give a more acceptable result. However, in the event, Eisenhower secured 442 votes to Stevenson's 89. When the night's story got out, Ed Murrow said, 'The trouble with machines is people.'[185] But it was a brilliant publicity coup. One IBM senior computer executive recalled that: 'The name UNIVAC appeared to be on its way to becoming the "Frigidaire" of the data processing industry.'[186]

The publicity, like that for Harvard's 'robot brain' and the army's ENIAC, contributed to public awareness. And, slowly, more machines were being sold. Parker, now chief salesman for RemRand, placed UNIVACs in General Electric, US Steel, Metropolitan Life and Westinghouse (which had refused to buy Eckert-Mauchly a few years previously) – 46 non-governmental sales in all.[187] But RemRand's dominance was not to last. Computers were a rental and maintainance affair, the tradition in tabulating equipment industry, so the more Parker sold, the more capital Rand needed to stay in business. And RemRand now had a competitor, twice its size and more than three times as profitable. In the very month Parker went to New York to start selling computers, IBM shipped its first production model to Los Alamos.

IBM had spent the 1930s encouraging advanced calculation in the academy. It had spent the war building an electromechanical device for Harvard and the navy and in the aftermath of the war it had built for itself a hybrid, the SSEC, the world's first and last electromechanical computer. Significantly, computing, within IBM, came to be referred to as the 'magnetic-tape area', highlighting the crucial stored programme capability of the new machines. But the company had a long distance to travel before it could get itself into the business of 'magnetic-tape'. The operation of the 'law' of the suppression of radical potential is as perfectly exemplified by IBM's changing attitude to the computer as by Eckert and Mauchly's adventures in the skin trade.

The company remained well aware of computing science (as evidenced by the SSEC's measure of stored programme capacity), yet it still failed to connect these activities to the mainstream of its business. The SSEC, at $300 an hour, was booked up a solid six months in advance, mainly by the armed forces, throughout the late 1940s – but this made no difference.[188] It only served to reinforce the belief of Watson and IBM that complex calculators had more to do with government and the academy than the world of commerce.

The point was further confirmed in 1948 by a device IBM lashed up in six weeks, for Northrop's guided missile programme. The original Card-Programmed Electronic Calculator (CPC) used standard IBM machines – a type 603 calculating punch, no longer in production, in connection with a 405 alphabetical accounting machine. In the summer of 1949 a revised CPC was marketed with an electromechanical store capable of 16 10-digit words.

> Tracking a guided missile on a test range now is the only way to make sure of its performance. At one Department of Defence facility this is done by planting batteries of cameras or phototheodolites along a 100 mile course. During its flight, the missile position is recorded by each camera at 100 frames per second, together with the camera training angles. Formerly these thousands of pictures from each of many cameras were turned over to a crew of [human] computers, to determine just what happened. It took two weeks to make calculations for a single flight. Now this is

done on the International Business Machines (IBM)
Card-Programmed Electronic Calculator in about 8 hours and the
tests can proceed. [our brackets][189]

By 1952, IBM had installed 250 CPCs, in universities, defence
establishments (for neutron shielding calculation, jet engine thermodynamics, helicopter vibration analysis) and, in civilian life, for such
purposes as utility billings, all at $1500 a month rental. At Northrup an
engineer named Reiss 'developed a technique for using the CPC to
program some of its own operations' – which is to say, pushed it from
calculator towards computer.[190] Still the decision to build a computer
was not taken.

IBM did not, despite Watson's PR efforts in the 1930s (or perhaps
because of them), look to the academy for talent. It was essentially a
company wedded to its late-nineteenth-century technology, as refined,
which continued to be enormously profitable. Before the war, it was
not remotely a high tech operation, as were, say, RCA and AT&T. It
had never been threatened by radio, and so it had made no real
investment in research and development and had hired no science
graduates as a matter of policy. Watson, whatever else he was, was
one of the century's great salesmen. He, and the company he built,
tended to allow the market to establish needs and then ask the
engineers to fulfil those needs as quickly and as cheaply as possible,
using as many interchangeable parts as could be accommodated. The
resultant patents were as often as not jointly filed by Watson and the
engineer.[191] It was this sophisticated responsiveness to the market that
held IBM back. IBM machines rented for hundreds of dollars a month
but, given the company's pricing policy, a computer along EDVAC lines
would have to cost thousands. Although the CPC fetched $1500 a
month or more, received opinion within the company was that the
market would not stand for EDVAC, a view confirmed by the shaky
progress of the firms in the fledgling business. Watson had decreed 'No
reasonable interaction possible between Eckert-Mauchly and IBM.'[192]
It was only when the Census Bureau ordered its third RemRand
UNIVAC that IBM saw the light. According to one IBM man the whole
episode 'frightened . . . the old man, who was convinced he had lost
his grip'.[193] 'We went into an absolute panic', said Thomas Watson Jr,
the executive chosen by his father to meet the threat. 'Here was our
traditional competitor, whom we had always been able to handle quite
well, and now, before we knew it, it had five of these beasts installed
and we had none.'[194]

Not for the last time in its history as a computer manufacturer, IBM
began to play catch-up. Internally, PR work for large computers was
being done. The success of the CPC was played down, although it was
possible to see the future in terms of cheaper (and more fully

programmable) CPCs rather than larger, more expensive devices. Ralph Palmer, the Director of Engineering at Poughkeepsie, was taken to meet enthusiasts for large computers; and, most importantly, the Pentagon had been visited and a positive response had been obtained. It must be remembered that Watson had offered the services of the company to Harry Truman in the defence effort occasioned by the Korean War, although his eagerness to help the American government did not extend to allowing IBM apparatus to be attached to other people's equipment – the IAS team was so prevented.[195] By Christmas of 1950, Year 6 of the 'revolution', Watson Jr called the meeting at which these various strands came together and the decision to build a 'defence calculator' along Iasides lines was finally taken.

In January 1951, the engineering calculation that the proposed machine ought to cost $5500 a month was 'rounded off' to $8000 by the financial vice-president. In 1952, von Neumann was contracted to IBM as a consultant for 30 days a year – and the IAS team was allowed to use IBM card handlers. The fully tested specifications of the defence calculator were released to production. In March 1952, the machine, now designated the 701, was demonstrated for the first time, without its drum memory. A month later, at a dedication addressed by the Watsons, Sr and Jr, and by Oppenheimer, the demonstration problem was of neutron scattering, prepared by Los Alamos but of a suitably unclassified nature for such a public occasion. In May the price went up to a maximum $17,600 per month, but, nevertheless seventeen 701s were leased in the years 1953/4, the vast majority by US military agencies and the aircraft industry.

The 701 represents, even more than do the other commercial machines (Ferranti Mark I, the English Electric Deuce, HEC, UNIVAC, ERA's 1101 series, etc.) the end of the incunabula period. The 701 was of modular design in that this designation properly belongs only to the Electronic Analytic Control Unit. The memory, along Williams lines with 72 CRTs, was named a 706 Electrostatic Storage Unit; 731 was the Magnetic Drum Reader and Recorder, Power Supply and Distribution Box; and so on. 'It was already intended that there would be a series of improved machines not only in the type 701 central processor itself but in the memory and the peripherals.'[196]

The general pace of developments allowed IBM to find its way into the new computing age with no disruption of its business. In 1952, while building the 701 giant for governmental work, the computer lobby within IBM was looking to replace the CPC, ignoring its potential as a stored-programme device. They came up with what was called a medium size computer having an external magnetic drum store, the 650. Against opposition even more strenuous than that mounted within the company to prevent the 701, the first 650 was installed in December 1954 and cost between $3000 and $4000 a month. Within a year 120 were placed and by the time the model was phased out fifteen years later, 1500 had been used.[197]

The success of the 650, like that of the CPC before it, did not push IBM or the computer industry in general in the direction of smaller machines. On the contrary, computers became larger and larger. A more commercial version of the 701 was developed and announced in 1953 as the **702** *E*lectronic *D*ata *P*rocessing *M*achine (**EDPM**), although of the first fourteen of these, four went into the military-industrial complex just the same. However, one was sold to the Bank of America and one to the Prudential. The **703** was a one-off built for the National Security Agency (which already had a 701 and a 702). The **704** was announced in 1954 and eventually incorporated a ferrite core memory system developed in connection with the Forrester team at MIT (itself – as we have seen – an integral part of the effort to build a computerised air defence system). Since one went to France and another to the British Atomic Research Establishment at Aldermaston, it was with the 704 that IBM became a multinational computer manufacturer. About one hundred 704s and its successor **705**s were placed.

The MIT connection was of great importance to IBM. Whirlwind was the first realtime computer, the first with magnetic core memory, the first to multiprocess, the first to network, the first to have interactive monitors. IBM was building Whirlwinds for SAGE and acquiring a great deal of expertise. In 1954 it began, for instance, a 10-year development process that eventually provided American Airlines with a cut-down real-time computer for airline reservations.[198] This flurry of activity continued until, simply, the competition was outrun and IBM was left dominating the field. RemRand, anyway late with UNIVAC **II**, could not match the capital IBM had available for the building and placement of large numbers of machines. But IBM's success throughout the 1950s was still very heavily reliant on the defence needs of America. The computer industry was now a reality, but its supervening necessity was still the Cold War.

In other countries, such as Britain, the lack of such a large ready defence market curtailed commercial development. Contrast with IBM the experience of Ferrantis who managed to sell nine Mark 1s between 1951 and 1957. Only one went to a weapons establishment, Aldermaston, the only such establishment in Britain. Two went to the aero industry and the Ministry of Supply took two – of which one went discreetly to GCHQ, the successor to GC&CS, where Turing had continued as an advisor. Two more went to academic institutions, one in Rome and one, in the University of Toronto, to help design the St Lawrence Seaway; and one went to an industrial lab, Royal Dutch Shell's in Amsterdam.[199] In all by 1960, the British computer industry had sold only 300 machines and was five years behind. It was not because of the quality of personnel because many war-time computing experts continued to hold commanding positions. For example not untypical was W. Gordon Welchman, late of Bletchley Park. He worked on Whirlwind and offered the world's first university digital computing course at MIT before returning to Ferrantis.[200] Ferrantis still tried to compete with the Americans building, for instance, a British **Atlas** to challenge IBM's 7030 (**Stretch**), its first million-instruction-per-second (I MIPS) computer, in the early 1960s.

> But in Britain, apart from the scientists and a small progressive
> minority of industrialists, the new electronic machines were treated
> with suspicion and the Government was either ignorant or
> indifferent to their implications.[201]

It is here suggested that the British responded correctly by ignoring machines not appropriate to their new reduced post-war position in the world. Perhaps because of this fundamental, if unacknowledged, truth Harold Wilson's attempt to revive computer pre-eminence by fiat in 1965 struck a somewhat comic chord, and the rhetoric of the 'white heat of technology' drew forth little significant response.

Between its ideation in 1944 and the mid-1950s, the computer had increased enormously in capacity and speed and it had become a regularly produced industrial item – but it was, within no stretch of the term, diffused much beyond the government, attendant client-firms of the military, limited parts of the academy and only the very largest of civilian businesses. The media-fed confusion as to its supposedly innate capacities, coupled with its increasing scope, produced the fertile soil in which the idea of a 'revolution' was planted. But the increasing scope, although impressive, involved no breakthrough to new techno-logical areas – any more than the machine itself had done; and it was achieved against a faltering sense of demand, at least outside of the needs created by the supervening necessity of the Cold War. As is demonstrated above, this was the enabling factor in the development of practically every machine, incunabula as well as first production models. This is not to demean its significance, for the market it created was no small thing and companies within the business, especially IBM, grew vigorously. Yet, despite these successes and the endless background hype, the result was that, in failing to make its case to the world at large, the computer's announced capacity for radical change was effectively contained by the comparative uninterest of business. By 1955, Year 11 of the 'revolution', there were still just 250 computers in the world.

This argument is strongly sustained by the historical record but the account offered above leaves aside another more vexed and hypo-thetical issue – given that diffusion was effectively contained by the failure of commerce to perceive computing possibilities, was it also *necessarily* contained? Could the machines have been less complicated, cheaper, more accessible than in fact they were, earlier than they were? Did the machines need to be so inaccessible, so 'user-unfriendly'?

(ii) A great feast of languages

The annals of the Computation Laboratory of Harvard University for 1948 contain the following remark made at a symposium the previous year: 'One of the most important problems now facing us is to achieve a reduction in preparation and set-up time proportionate to that already achieved in actual computation time.'[202] The coming of electronic machines did not ease the problem. In 1953 it was possible to state:

> The difficulty of programming has become the main difficulty in the use of machines. Aitken has expressed the opinion that the solution of this difficulty may be sought by building a coding machine. . . . However it has been remarked that there is no need to build a special machine for coding, since the computer itself, being general purpose, should be used.[203]

The debate about programming in the early fifties speaks to a crucial constraint in computing development. Specifically the slow emergence of programming languages was definitely an obstacle, and an unnecessary one, to computer diffusion. The debate begun in these years still echoes in the perception, created by the present computer industry and those who uncritically accept its pronouncements as prescriptions for structuring society, that 'computer literacy' is a vital need. It was reflected then, as it is now, in contrasting design philosophies which have tended, up to the very recent past, always to favour the machine rather than its user – a continuing element in suppressing the device's radical potential.

In its first appearance, the computer literacy argument was between those who favoured using some of the expanding capacity of the machines to produce decimal numeration and others, the purists, who felt that, if one could not read binary rows of noughts and ones, one should get out of computing. Of course, this latter group was engaged in the time-honoured pursuit of protecting its mysterium, the secrets of its guild.

> This feeling is noted in an article in the 1954 ONR (Office of Naval Research) symposium: 'Many "professional" machine users strongly opposed the use of decimal numbers . . . to this group, the process of machine instruction was one that could not be turned over to the uninitiated.' This attitude cooled the impetus for sophisticated

programming aids. The priesthood wanted and got simple mechanical aids for the clerical drudgery which burdened them, but they regarded with hostility and derision more ambitious plans to make programming accessible to larger populations. . . . Thus they were unalterably opposed to those mad revolutionaries who wanted to make programming so easy that anyone could do it.[204]

Were the mad revolutionaries possessed of anything more than dreams? Indeed they were and had been for some years. The German electromechanical calculator engineer, Konrad Zuse, was contracted after the war by IBM and eventually set up in business in West Germany by Rand. But, during his enforced idleness on the top of the mountain at war's end, unable to work in the metal, he could only think. What he thought of was the **Plankalkul**, the programme calculus, the first high-level computer programming language. It antedates the first electronic computer by three years.

Before laying this project aside, Zuse had completed an extensive manuscript containing programs far more complex than anything written before. Among other things there were algorithms for sorting; for testing the connectivity of a graph represented as a list of edges; for integer arithmetic (including square roots) in binary notation; and for floating-point arithmetic. He even developed algorithms to test whether or not a given logical formula is syntactically well formed, and whether or not such a formula contains redundant parentheses – assuming six levels of precedence between the operators. To top things off, he also included 49 pages of algorithms for playing chess.[205]

The high-level language is the key to the computer. As Zuse wrote:

The first principle of the Plankalkul is: data processing begins with the bit. . . . Any arbitrary structure may be described in terms of bit strings; and by introducing the idea of levels we can have a systematic code for any structure, however complicated. . . .[206]

Zuse's scheme was scarcely known, even in the German-speaking world, and others developed the concept of the programme language without reference to him. But his is the first annunciation of an idea which was as necessary to the computer's diffusion as was von Neumann's 'First Report' to its *invention*. That it was done so early is evidence that the programming bottleneck of the mid-1950s was artificial. Collectively the fledgling industry was not really interested in discovering how to sell as many computers as possible and devoted almost no attention to programming as a result. The programmers presented themselves as an integral part of the 'exceedingly complicated thing' that was a computer and, for many – the majority – engaged in security tasks, the very incomprehensible nature of computer operations was a distinct advantage. Nicholas Metropolis and his Los Alamos colleagues could run highly classified problems in virtually open circumstances on every computer in the country as it came on stream. The continuing interface of computing and the Cold War encouraged the persistence of the mysterium.

Of course, in the incunabula period few if any general languages could be written because each machine was different; but the incunabula period was prolonged by the lack of commercial opportunity for computers, in part a function of the lack of languages. Nevertheless, the unread Zuse apart, some progress was made. Wilkes had developed 'assembly routines' to combine numbers of sub-routines and one of Turing's last acts as a computer scientist was to produce for the Ferranti machines a manual explaining how programmes were to be written.[207] There were early attempts at 'compilers', although none was so broad in scope as Zuse's. Mauchly suggested a **Short Code** which was implemented in the design of the first BINAC but it was not apparently popular with UNIVAC users.[208] Arthur Burks, a member of the original IAS team who had moved to the University of Michigan, explored the possibilities of going from 'ordinary business English' descriptions of a computational problem to the 'internal program language' description of a machine-language programme appropriate to that problem. However, the result was still too complex in its symbolism for general business use.[209]

For the MADM, which was distinguished by a 'particularly abstruse' machine language, Alick Glennie, of the Royal Armaments Research Establishment, developed the **Autocode**. Glennie was working on the British bomb and his language was designed for highly skilled professional users, which is to say it was still very much machine oriented but nevertheless it was 'an important step forward'.[210]

Most significant of these efforts was that of J. Halcolme Laning Jr of MIT who, with Niel Zierler, after two years' work, ran an **algebraic compiler** on Whirlwind in 1954:

> The system is mechanised by a compilation of closed subroutines
> entered from blocks of words, each block representing one equation.
> The sequence of equations is stored on the drum, and each is called
> in separately every time it is used. The compiled routine is then
> performed interpretively.[211]

It used instructions like 'Stop' and 'Print' but it slowed Whirlwind down ten-fold. Yet each new computer was so much faster than its predecessor that, even in the short term, this was no real obstacle to the development of programme languages. And, moreover, the slowdown, in commercial settings, would still leave the machines infinitely faster than any mechanical, electromechanical or human alternative. The real problem continued to be objections from programmers. Programming was a skill just born and already seemingly threatened. It is an unremarked irony that the first to fear computerised automation were among those whose very work had been necessitated by the computer.

> At that time most programmers wrote symbolic machine instructions exclusively (some even used absolute octal or decimal machine instructions). Almost to a man [sic], they firmly believed that any mechanical coding method would fail to apply that versatile ingenuity which each programmer felt he [sic] possessed and constantly needed in his [sic] work.[212]

This obstructionist tendency was not universal. Among the most senior of programmers was Grace Hooper, a Vassar mathematics professor who had been on Aitken's naval Harvard Mark I team and who was a strong believer in automatic programming. She was working for RemRand in 1951 when she developed an 'automatic programming' system, a compiler, for UNIVAC which went well beyond Mauchly's Short code.[213] As a member of the naval reserve – she was to achieve the rank of Captain in 1973 – she set up the ONR conference at which Laning and Zierler made their experiment known. John Backus, then assembling a group within IBM to explore the issue, had gone to the conference with a paper which queried, 'Whether such an elaborate automatic programming system is possible or feasible has yet to be determined.'[214] He came away knowing that it was both possible and feasible and, after visiting MIT, set about developing what would become **Fortran I** (*For*mula *Tran*slating System) against a background where, 'our plans and efforts were regarded with a mixture of indifference and scorn'.[215] The IBM 704 thus owed not only its core memory to MIT but also the initial idea for its programme language. Although the most significant, this was not the only result of Ms Hooper's initiative in calling the conference. Boeing developed **Bacaic** for the 701 and other algebraic compilers appeared.[216]

IBM (as usual, scarcely the innovators) took the idea and ran with it. Backus became confident enough to suggest that 'each future IBM calculator should have a system similar to Fortran accompanying it.'[217] Now, instead of days of hands-on experience, an average programmer after one hour learning Fortran notation could understand a programme written in it without further help. It was still to be some years, till the spring of 1957, before the system finally worked. In 1958, though, a survey of 704 users revealed that only half the time did half of the users take advantage of the language. It was therefore more than a decade after the Baby Mark I and EDSAC and fifteen years after the Plankalkul that the principle of compiling instructions into languages to ease programme writing came into general use and with it perhaps the most essential tool necessary to the diffusion of the computer. A computer that used everyday icons as the basis of its instruction mode would not appear – again to the derision of the serious computing enthusiast – till the early 1980s.

Computer history has its share of obstructionists like H.A. Thomas, the man charged with building Turing's ACE, who could not see any industrial applications for it and therefore hindered it as much as possible; or George Stibitz, pushing for the electromechanical solution after its day had passed; or the entire management of IBM for the better part of a quarter of a century ignoring the potential of complex calculators of all sorts. This first generation of programmers was in the same category, working as an important agent of the 'law' of the suppression of radical potential. But what must be stressed is that the programmers were in thrall to something other and greater than what is commonly, and most erroneously, called Ludditism.[218] However easily one can read the programmers' attitude as simply defensive of their jobs, it would still be wrong to draw from that the conclusion that such defensiveness was all that was at work.

The programmers are scarcely believable as representatives of the innate conservatism which supposedly governs all human response to change. They were at the innovative cutting edge of a civilisation which, for over four centuries, has anyway in every aspect rebutted this putative natural obstinacy. Innovations, properly grounded in supervening necessities, are not thwarted by threatened workforces. Workforces in such circumstances are always shaken out, for King Lud's writ seems never to run very far. Our industrial history strongly suggests that the objections of the programmers would have been overwhelmed had it only been a matter of their jobs. Rather, for a number of years, the protection of their jobs was synchronous with the failure of commercial vision within and without the computer industry and both fed the broader need to contain the computer. The result was that the gradual process of development continued into the computer's second decade even after the working of the Cold War supervening necessity.

The human factors of management shortsightedness and labour protectionism were not the base issues in this slow process. Rather these were expressions of a more fundamental need to contain the disruptive potential of the computer. We have here argued that a greater attention to the development of higher computer languages was not only necessary to the wider diffusion of the machines but also was possible far earlier. To the indifference of commerce must therefore be added the failure of language-creation as the second element in constraining the computer's diffusion.

Both these elements contributed to the design philosophy, conditioned by war-time approaches, which stressed ever larger and more expensive machines. The military-industrial complex that was the prime market for these devices fed upon advanced technologies of the very largest scale. Thus neither buyers nor sellers of computers had much interest in breaking the circle that had been created. The question remains, though, were more user-friendly, smaller devices possible during this period?

We have already seen how the ABC was transformed into ENIAC and the Baby Mark I became MADM. The suggestion that the IBM CPC be made programmable was ignored within the company in favour of the development of the 701 giant and even the 'mid-sized' runaway hit 650 was not constantly replaced by a succession of machines as was the 701. But the real answer to the question can be found in the history of the inter-relationship between the computer and the semiconductor. Our contention that the 'law' of the suppression of radical potential operates in computer history depends finally on an argument denying the supposed 'synergy' of these two technologies.

Transistors were *invented* late in 1947, so every machine came on stream in the semiconductor age yet, contrary to received opinion, transistors did not become the norm in computers. Beyond gigantism, beyond general obstructionism, the mysterium of programming and applications failures, the slow, hesitant use of solid state technology especially before the mid-1960s stands as the most eloquent testimony to the 'law' of the suppression of radical potential. Solid-state electronics constitute a third element, apart from size and language provision, in the operation of the 'law' as it applies to the computer. To understand how, we must now turn to the history of the transistor.

DIGRESSION – 'THE MOST REMARKABLE TECHNOLOGY'*

The received opinion of the relationship between microelectronics and computers is as follows:

> It all began with the development thirty years ago of the transistor: a small, low-power amplifier that replaced the large power-hungry vacuum tube. The advent almost simultaneously of the stored program digital computer provided a large potential market for the transistor. The synergy between a new component and a new application generated an explosive growth in both.[1]

Thus Robert Noyce, one of the fathers of the integrated circuit and a founder both of Fairchild Semiconductor and Intel – a man who should know.

The received history is that computers ate transistors ravenously. The historical truth is – and herein lies the proof of the third element of constraint – that they did no such thing. The transistor was not crucial to the development and growth of the computer. Indeed, it will be our argument that the transistor was not, of itself, a significant *invention* but rather the signpost to one. It took 21 years to get from the transistor to the invention – the microprocessor; 21 years in which the computer industry played very little part, except towards the very end, and was certainly not a major consumer of solid state devices. IBM, for example, never marketed a commercially successful fully transistorised computer, and valve-based computers were still being shipped to customers a decade after the transistor was commercially available. The two industries did not come together in a synergetic way until the 1970s.

To make this case, we must now digress and deal with the entire history and development of the microprocessor.

---

* Sir Ieuan Maddock, sometime Chief Government Scientist.

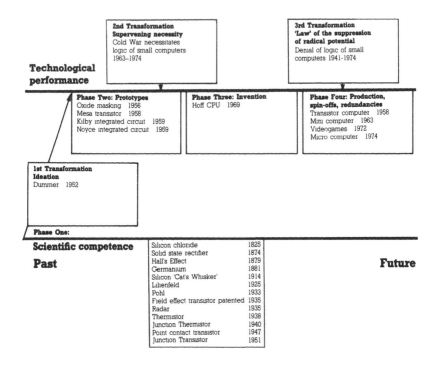

Figure 6 Central Processing Units

Phase one: scientific competence

Professor Braun invents the CAT'S WHISKER 1874

> Goodyear was not a very impressive mind or talent. But he
> developed an obsessive idea that it would be nice to do something
> with rubber. He therefore did a great many different things with
> rubber, from freezing it to burying it in the ground, until one day an
> accident provided him with the happy thought, and he popped the
> substance on the stove – hence vulcanized rubber. The case of
> Goodyear is a little too close to turning monkeys loose on the
> typewriter to produce Hamlet as a statistical probability, to be of
> use, but it does suggest the difference between then and now. . . .
> The difference between then and now is, to a considerable extent,
> the difference between the way in which Mr. Goodyear went about
> things in getting vulcanized rubber and the way in which the Bell
> Laboratories got to the transistor.[2]

The transistor is the twentieth-century device *par excellence*, its history, supposedly, the most refined of all examples thus far of the highly directed orderliness of contemporary technical innovation. This is the common view. A rather more informed vision uses the example of the transistor to argue the somewhat contrary point that its development is the most telling justification for undirected basic scientific research in the industrial setting.

The transistor, even more than the computer, the laser or television, seems to have burst upon society. It is without infancy, without past history, born fully formed in Bell Labs on 23 December 1947. The internal Bell 'Memorandum For Record' dated 27 December 1949, commences:

The history of the transistor begins with the decision to study intensively the properties of silicon and germanium. This decision was made in 1946 as a result of series of conferences intended to establish a plan for semiconductor research. . . .[3]

The impression of instantaneity is created at the cost of an amnesia profounder than that affecting our memories of the development of the other devices herein explained. The transistor had not been publicly promised for decades as was television; nor had it been celebrated by the media as was each new 'robot brain', nor was it predicted in science fiction as a 'death ray'. The world was unprepared and, anyway, the transistor was promoted only as an electrical amplifier, a curious sort of valve.

The history of the transistor illustrates a basic way in which media and their technologies become misunderstood. Simply, the device's antecedents have been struck, as it were, from the public record. Yet not only did it have antecedents, but an application of the same species of electrical phenomena was among the most widely diffused of all such innovations in the early part of the century; for at the heart of the Cat's Whisker crystal radio was a solid-state device, of the same genus as the transistor.

Crystal detectors, all relying on the semiconductivity of various metals, were a staple of early wireless telephony and subsequently of radio enthusiasts eager to build their own receivers. Semiconductors, as the history of selenium in chapter 2 reveals, were understood in the nineteenth century as being substances whose conductivity varies according to either thermal (Faraday's experiment of 1833 with silver sulphide) or light conditions (Willoughby Smith's experiment with selenium previously described).

The scientific competence for the transistor and its successor devices thus accumulated over a similar timespan to that of the other technologies here described. Among the most important developments in the slow growth of the understanding of semiconductors is the Hall effect, first demonstrated by E.H. Hall in 1879, which showed that, when a wire carrying electricity is placed in a magnetic field, a further electromotive force is produced perpendicular to the current and the field, creating a flow across the wire. In some materials the effect is reversed, which indicates that the current is being carried by charged particles within these materials. The possibilities of using the Hall effect for amplification were first outlined in a series of patents by the physicist Julius Lilienfeld in 1925. Lilienfeld's proposal was for a quite complex sandwich of semiconductor and conductor layers which, whether it would have worked or not, is nevertheless in vague terms the ideation transformation for the transistor.

At Göttingen in the 1920s, R.W. Pohl was making a study of the photoelectric properties of solids in the course of which he elaborated a picture of the behaviour of the electrons within crystals. In 1933, supporting Lilienfeld's line of investigation, he suggested that vacuum tubes in radios would give way to small crystals wherein the movement of the electrons could be controlled.[4] In 1935, Heil patented another solution to the problem, a field-effect solid-state amplifier which, although an elegant exploitation of the Hall effect, refused to work.[5] Anyway, in a world of valves, there was no supervening necessity for such amplifiers and semiconductivity for other purposes was already being used effectively.

Proto-solid-state devices were widely diffused. Frederick Braun, at the University of Marburg in 1874, had demonstrated that a metal wire when placed in contact with lead sulphide produced a rectifier – a device that would only conduct electricity in one direction. With this was *invented* a tuned circuit which could be adjusted to pick up the radio waves of a single frequency.[6] Although rendered obsolete by the De Forest (and Armstrong) valve, the crystal rectifier took a long time to disappear, despite the fact that only one person at a time could listen on the headphones. BBC engineering, for instance, was giving advice on receiving a new transmitter via crystal radios as late as 1929.[7]

Crystal sets could be and were indeed made at home, but also enjoyed some commercial exploitation. Many substances were patented in this connection – zincite, chalcopyrite, 'cerusite', molybdenum, tellurium and graphite. One crystal detector, produced by Greenleaf Pickard and marketed before the First World War as the **Perikon** consisted of 'a brass point resting on the surface (either rough or polished) of a piece of silicon'.[8]

Silica is common in nature, being the principal ingredient of sand. In crystal form it is quartz and, when coloured, produces such gems as amethyst and jasper. The same Berzelius who had isolated selenium prepared, in 1825, a silicon chloride; silicon steel was *invented* in 1889 and silicon itself was being produced commercially by the

Carborundum Company of New York before 1911. The existence of the substance called germanium had been predicted by D. Mendeleff in 1871. He named this putative relative of zinc and lead, ekasilicon. It was discovered in 1886 in Freiburg by C. Winkler and, although very rare in nature, it can be obtained by smelting certain zinc ores.

There is a supposed hiatus in the use of solid-state devices between the coming of the valve and further developments in semiconductors in the late thirties but the gap is more apparent than real. Crystals had barely disappeared from every scientifically inclined child's repertoire of radio tricks and experiments in the 1930s, only to reappear immediately in the thinking, experiments and devices of physicists and electricians (many of them with a Cat's Whisker radio in their background). Valves, although a staple device, were nonetheless bulky, comparatively unreliable and generated much heat – all of which made Pohl's opinion as to the viability of crystals most attractive – and needs were being generated, supervening necessities, for applications beyond the capacities of valves.

In telephone exchanges, for instance, great gains would obviously be made if unreliable valves and physical switching systems could be replaced. Steps were being taken in this direction with the development of 'crossbar' central switching equipment. Mervin Kelly, the then director of research at Bell Labs, believed that the all-electronic exchange was worth pursuing. Needless to say he was not alone. T.H. Flowers, the man who built Colossus for Bletchley Park, was, it will be recalled, interested in solid-state telephone switching problems; at Bell Labs, though, the inquiry was to be more focussed. The labs were already investigating silicon, but as a detector of short-wave signals. By the end of the decade this interest was important enough to justify a team of chemists, including R.S. Ohl, being given the task of producing purified silicon.[9]

Kelly, who had been head of the vacuum tube department at the Lab, instructed William Shockley, a recently hired PhD in that same department, to work on the electronic exchange problem in 1936. (Shockley was to produce the mercury delay line which would give Eckert the wherewithal to plan a computer memory.) He involved William Brattain, a colleague investigating **copper-oxide rectifiers**. These, which were being manufactured commercially, were used in electrical apparatus where diode valves would neither work nor fit, primarily in situations with low currents and/or tight spaces. The investigation of these rectifiers produced a theoretical literature in the late 1930s which explained metal semiconductor contacts. By the eve of the Second World War, Shockley and Brattain had determined that semiconductors, about which in general little was still understood, might indeed yield a device which would amplify electrical signals without generating the heat of valves, as Pohl had suggested. Pohl had by this time created a *partial prototype* for the transistor by demonstrating how a crystal of potassium bromide could be made to imitate the action of a valve and amplify a current. The paper on this was published in 1938, but since the crystal had to be heated to obtain this effect, the point was rather lost. Such devices would eventually be called **thermistors**.[10]

This then was the quite extensive tradition in which Shockley and Brattain began work. Far from being a situation unique to the twentieth century because the inquiry was theory-led, on the contrary, it was just like those nineteenth-century investigations where practical work proceeded parallel with or even in advance of total theoretical understanding. Now, contrary to received opinion (and all too like Mr Goodyear – who was a far better theoretician than he is usually given credit for), Brattain and Shockley began to do something nice, as it were, with semiconductors – specifically, following Heil, with copper oxides. Brattain is clear that this was somewhat unoriginal: 'Anybody in the art was aware of the analogy between a copper oxide rectifier and a diode vacuum tube and many people had the idea of how do we put in a grid, a third electrode, to make an amplifier'.[11] They had the ideas of Lilienfeld and Heil and the example of Pohl before them but after only two months, with results still rather poor, they stopped, diverted by the impending war to work on underwater radar for submarine detection.[12]

The war halted Bell Labs' interest in semiconductor amplifiers but in other ways it speeded progress in the area. Fessenden, the radio pioneer, used waves for depth sounding in 1904 and scientists in the mid-1920s determined the height of the ionosphere with reflected radio signals. The similar reflection of radio waves from aircraft had been observed in 1931 and three years later the British began to develop a serious interest in the possibility of using this for detection. By 1935, Robert Watson-Watt, superintendent of the Radio Division of the NPL, was working on a practical system. Five RADAR (*Ra*dio *D*etection *A*nd *R*anging) stations were built.[13] To detect very high frequency signals used in radar, though, requires low capacitance in the receiver, lower than could be provided by vacuum tubes. In 1939 Randall and Boot, at Birmingham University, produced an unconventional triode, the **magnetron**, which was five times as powerful as a valve and reached frequencies four times as high. It was the basis of airborne radar equipment. (The work on radar impacted on television because better CRTs with more sensitive phosphors were developed during the war.) Cat's Whisker detectors were also used and refined, the wire eventually being of tungsten and the metal, silicon.

The importance of semiconductors in microwave applications generated a most extensive research blitz to uncover their secrets. The general need to improve radar, the rareness of germanium and the comparative ineffectiveness of common silicon (when compared to germanium) all contributed to the supervening necessity of this work. During the war, as many as forty American laboratories were organised, at the behest of the US government, by MIT. One lab was at Purdue and it was here discovered that semiconductivity does not depend, as had been thought possible, upon impurities in the materials.

The very choice of silicon and germanium as the objects of study in post-war semiconductor investigations is clear evidence that semiconductor research reveals no new mode of human endeavour. These metals were used and investigated heavily during the war because they were known to work – there was no time to worry about whether or not others might be more effective. The impetus thereby created carried on into the post-war period, effectively delimiting the range of inquiry and preventing an equally serious investigation of other metals even of the same group, IV in the periodic chart.

But who knows what materials we might be using today if these two had not established such a substantial lead during the war, a lead that could have been challenged only at enormous cost and which was never seriously rivalled by the dozens of compound semiconductors proposed at various times.[14]

This trajectory continued into the present although, in the early 1980s, despite the supremacy of silicon, other substances, notably gallium arsenide, are coming into specialised use, for instance in high-speed logic circuits. The notion that the work at Bell was an entirely free investigation in basic science must accommodate the fact that it did not start from scratch. The course of the investigation was, in fact, a prisoner of a well-established tradition, outlined above, which had closed off many options before they were opened and which was adhered to in the interests of producing effective, i.e. marketable, devices as quickly as possible.

Although much had been learned, there were a number of unexplained phenomena. For instance, in 1940 Ohl had shown Brattain a silicon slice doctored so that a particularly sharp boundary existed between n-type silicon (where the majority of electron carriers are negative) and p-type (where they are positive). A light shone at this junction produced an unexpected flow of electricity. Ohl went on to build an unreliable semiconductor amplifier – a thermistor, like Pohl's of 1938, which worked when heated and which he used to power a radio.[15]

Two Purdue graduate students, Seymour Benzer and Ralph Bray, continued the war-time work and investigated the spreading resistance resulting when a point contact wire was placed on germanium. The resistance measured was lower than theory predicted. Through MIT's co-ordination, Bell Labs knew of the work at Purdue, this included, and Shockley visited the university early in the peace. Bray was interested in the resistance caused when the current flowed forward, Benzer in the reverse.[16] They publicised their results at a meeting early in 1948 when Brattain was in the audience. He spoke with Bray after his paper.

> Bray finally said, 'You know, I think if we would put down another point on the germanium surface and measure the potential around this point, that we might find out what was going on.' And I couldn't resist saying, 'Yes Bray, I think that would be probably a good experiment.'[17]
>
> As Bray said later, 'What's perfectly clear is that if I had put my electrode close to Benzer's electrode . . . we would have gotten transistor action.'[18]

This was, in fact, the experiment that Brattain himself had done the previous November, the experiment that for the first time demonstrated the transistor (*trans*fer of electricity across a re*sistor*) effect. It was still a Bell secret at this time. As has been said, 'The amazing thing . . . is why Purdue didn't invent the transistor.'[19]

One difference, a crucial one it is claimed, between Bell and Purdue was John Bardeen, a senior theoretical physicist who had joined Shockley and Brattain in the semiconductor group at the Labs.

In the summer of 1945, Kelly had authorised Shockley's group to research semiconductors with a view to 'obtaining new knowledge that can be used in the development of completely new and improved components and apparatus elements of communications systems.' But this was not quite as broad as it might seem. The Labs' management had a view of which aspect of semiconductor research might prove most profitable:

> From the point of view of the communications art, it appeared that the most important development likely to arise from semiconductor research, and quite possibly from any branch of solid state research, would be a useful semiconductor *amplifier*. This consideration influenced the emphasis of the work in various parts of the solid state area. (our italics)[20]

Shockley therefore returned the group to the investigation of Heil-type field-effect amplifiers despite the poor results obtained in 1939. The months went by. Then Bardeen, on 19 March 1946, committed to his notebook a theory to account for the failure of Shockley's field-effect experiments. (At the Bell Labs, as everywhere in the scientific world, all notebooks are filed and entries time-dated. This is a standard practice dating back to the beginning of modern scientific culture:

> It was the need for safety in approaching land . . . that made quantitative observation a habit among seagoing people. . . . The plotting of the ship's course and the transcription of topographic data to maps instituted the higher bookkeeping of science. And finally, the keeping of the ship's log, the prompt record of observed events, set the meticulous pattern of the laboratory notebook.[21]

Such a tradition of meticulousness has the added advantage of allowing disclosures, a crucial element in establishing patent rights, to be properly witnessed – normally under the signed rubric, 'read and understood'.) The problem, Bardeen hypothesised, was that electrons were being trapped on the surface of the semiconductors. The experiments were not working because they depended upon the electrons moving as freely on the surface as they did with the body of the substance. But, in reality, this was not what was happening. The attempt to build an amplifier was abandoned for the time being so that surface states could be investigated.

Shockley's analysis of the group's notebooks reveals that it was 13 months before the next significant step occurred, an insight into the nature of electron and hole densities. (Holes are unfilled vacancies or absences in a valence band – a band of energies in a solid wherein electrons are fixed to atoms and cannot move freely.) This April 1947 entry of Shockley's uses the n (for electron densities) and p (for hole densities) terminology which was just then coming into currency.[22] The summer and autumn of 1947 passed.

Then, on 17 November 1947, at the suggestion of a chemist on the team, Robert Gibney, Brattain applied voltage to the semiconductor while it was immersed in an electrolyte. A strong Hall-effect-like field was created, for Brattain had liberated the trapped surface electrons with the help of Gibney's electrolyte. The best substance to facilitate the effect was found to be 'Gu', which was extracted from 'electrolyte capacitors by using a vice, a hammer and a nail'.[23] (Scientifically, 'Gu' is glycol borate.)

A new series of experiments started. With some difficulty, the electrodes were placed close to each other on the germanium surface and it was found that a small positive charge to one greatly increased its capacity to carry current. Bardeen then hypothesised that by even more closely spacing the two electrodes an amplifying effect could be achieved. Brattain – who has said he had 'an intuitive feel for what you could do in semiconductors, not a theoretical understanding' – began to mess about with the crystal, the wires and the 'Gu'.[24] He found on December 16th that two wires 2 thousandths of an inch apart produced a 40-fold amplification – the transistor effect. On December 23rd the team demonstrated the effect to some Bell Lab colleagues. By the following July, Bell unveiled what was described in the rather low-key press release as 'a device called the transistor, which has several applications in radio where a vacuum tube is ordinarily employed'.[25]

But the demonstration of the transistor effect had not really produced a device to, as it were, rebuild electronic affairs on a new footing. The **point contact transistor** was in its general design a throwback to the Cat's Whisker. Although manufactured in the early 1950s, it was not the foundation upon which microelectronics were to rest.

> [I]t never really emerged from the laboratory stage mainly because its construction limited very severely the currents which it could handle. The real importance of the point-contact transistor was that it demonstrated the possibilities of the practical application of electron flow in crystals.[26]

In other words, the point contact transistor functioned somewhat like the daguerreotype in the development of photography – it was not the technique which prevailed but rather an effective demonstration of the basic practicality of the technology.

While Brattain and Bardeen continued experimenting during the month from November 17th to December 16th, Shockley, having abandoned field effects, was nevertheless convinced that a better device must be possible, a belief buttressed by the poor performance of the Brattain/Bardeen solution. The unexpected effect Ohl had shown Brattain nearly a decade before with the mixed n and p-type silicon had been much in his thoughts throughout these years, as evidenced by his April 1947 notebook entries. Inspired by his colleague's breakthrough, he evolved the concept of the **junction transistor**, a transistor built like a sandwich. This idea, after all, was in gross general terms along the lines that Lilienfeld had suggested 23 years earlier; but, by January of 1948, Shockley had formulated the precise way in which a sandwich transistor could be made to work – by placing a doped n-type semiconductor between two slices of p-type. He reached this conclusion, though, while still 'trying to invent new experiments to improve the scientific aspects of the research on the p-type surface layer of the point-contact transistor'.[27] 'Improve the scientific aspects' is the key phrase since the Brattain/Bradeen device was secret and its theoretical underpinnings still shaky enough to threaten the patenting process. Filled with a 'competitive urge to make some transistor invention on my own', Shockley finally conceived, by an 'accident for which my unsuccessful inventive efforts had prepared my mind', the junction transistor 'having wandered close to this idea on a number of occasions'.[28] This was the shape of the future for, if such a junction transistor could be made, it would depend on entirely internal operations of electrons and would have nothing of the Cat's Whisker about it. While considerable efforts began to refine the point contact transistor, Shockley worried about how junction transistors might be made at all.

The Bell technique, as it evolved, required adding doping substances to drops of melted germanium, in three separate stages, to make the p-type/n-type/p-type sandwich within the growing crystal. Then the crystal was sliced and the leads attached.

The junction transistor, like its point contact cousin, was another albeit more fruitful element in the ground of scientific competence for the microprocessor, for what was to prove significant in microelectronics was not devices replacing valves or other discrete components but ones which functioned as entire circuits. Bell, however, was not seeking such devices. It merely wanted a better amplifier. As a result 'Constraints were imposed upon the use and development of the new technology which were, in a sense, self inflicted.'[29] But this is no mystery. It simply means developments in microelectronics fit our schema as well as does the history of television and the computer.

It took till 1951 for Shockley to produce a comparatively reliable transistor of the junction type. By 1952 Bell's manufacturing arm, Western Electric, was making just 300 of them a month; it was making 28 times as many point contact devices.[30]

To what extent can the development of either transistor be described as being theory-led? It is clear that Bardeen's theoretical approach pointed Brattain in the right direction – just as it is clear that Shockley's knowledge of theory and the work of his predecessors prevented any real way forward for the first years after the war. The effect of learning and theory on the team would seem to balance itself out as far as the point contact transistor is concerned. And Bell kept that innovation under wraps for months while feverishly trying to tie the theory of the effect down more securely so that the patent position could be better assured.[31] (Shockley's contribution was deemed to be too far from the work to warrantise a patent award and the point contact transistor is essentially credited to Bardeen and Brattain alone.)

Shockley's subsequent procedure in developing the junction transistor makes the more impressive case for theory since he articulated its design principles fully two years before it was successfully produced. He himself privileges this reading by highlighting theoretical considerations in his own 1976 account of his work, stressing for instance, the influence of Fermi and other theoretical physicists on his thinking while omitting any mention of the experimental tradition.[32] But his theory for the junction transistor was rejected by a leading scientific journal as being insufficiently rigorous, and he can be quite explicit as to the gap between the practical plan he had thought of and its theoretical basis.[33]

> The record . . . shows that I was aware of the general considerations needed to develop the theory of the junction transistor. But I did not formulate several key concepts until *after the junction transistor was invented*. (our italics)[34]

that is to say, after 23 January 1948, the date Shockley gives for this event. In our terminology the notebook entry of this date was the specific point of *ideation* for his particular solution to the problem (not the *invention*); but the account quoted above suggests that his was a practical proposal in advance of at least some of its own theoretical underpinnings. The difficulties of building it were neither eased nor hindered by theoretical understanding. 'Nobody at that time could make the junction transistor Shockley had devised.'[35]

At a practical level, a method of manufacture had slowly to be determined and the junction transistor *invented*. The patent was applied for on 26 June 1948. In April of the following year Sparks and Mikulyak made a junction device by dropping *p*-type germanium on a strip of hot *n*-type, and slowly, by empiric testing, in the course of the next twelve months, the double doping system was teased out by Sparks and Gordon Teal, who had been creating thin silicon films in another part of the Lab.

Thus neither this 'teasing out' nor the development of the point contact transistor required a full theoretical understanding of the properties of semiconductors. Bray or Brattain, given the fact that germanium was so much to hand, could have stumbled across the effect without understanding it at all. Such ignorance would have been no bar to its exploitation. Gibney's contribution of the 'Gu' and his suggestion on 17 November 1947, which marks the beginning of the breakthrough, is similarly inspired rather than argued.

After all, the mysterious semiconductor had been in use for a variety of purposes for three-quarters of a century without anybody having much idea as to why these metals behave the way they do. Theory was of importance for Brattain, of course, but it would be stretching the point to claim it was absolutely crucial to the discovery of the transistor. The transistor effect was demonstrated in a totally unexpected way, after months – if not years – of work looking for an amplifier, while investigating something else – surface states, and using, for no sound theoretical reason, an electrolyte.

It is equally difficult to maintain that the environment of the industrial lab, beyond providing the extremely expensive germanium, created conditions significantly different from any other sort of laboratory. And, in their day-to-day labours, Brattain and Bardeen also seem to offer no exempla of some new method of innovation. Beyond Brattain's intuitive approach, Shockley himself refers to the month of November/December 1947 as 'the magic month', echoing, as a Bell historian has pointed out, Thomas Watson's description of an exactly similar period of feverish activity on the part of Alexander Graham Bell as 'that exciting month of June 1875' when 'the first vocal sound ever transmitted by electricity' was heard.[36]

This is not to deny that Shockley is in a slightly different case from Bardeen and Brattain. All the messing about with wires and germanium in the world would not have produced the grown crystal that was needed to make his junction transistor. Yet, using Lilienfeld's basic concept of a sandwich and Ohl's device as leads, it is possible that the *p-n-p* sandwich idea could have occurred independently of a theoretical understanding of why it might work. But the execution of the idea, again in contrast to the work of Brattain and Bardeen, does seem to demand a lab. There was talk of making point contact transistors in the home, just like Cat's Whisker radios. There was no such talk about junction transistors. However, despite these significant differences in the development of the point contact and the junction transistor, it is still clear that the claims made for the methods involved in both instances are overblown.

In short, when one reads in the most scholarly available account that, 'The transistor is one of the supreme examples of an invention truly based on science', one must be permitted to ask, what can this possibly mean?[37] The argument that the point contact transistor represents some perfect example of a new style of innovation is seriously flawed. It is an argument that can be better attached to Shockley's work – but that is not what is normally meant by talk of 'how Bell got to the transistor'. 'Theirs was not a blind empirical fumbling' is typical of such descriptions and refers to Brattain's demonstration of the transistor effect, not Shockley's painstaking building of the junction transistor.[38] But what was the team doing if not empirically fumbling, in however opened-eye a fashion?

The events of the winter of 1947 – the junction transistor story – are inscribed in general consciousness as the prime example of modern innovation which, if they are indeed that, means there is precious little new about 'modern innovation'. And even Shockley's work, although more appropriate as evidence, can be better understood by referencing more time-honoured notions of creativity – blind empiric fumbles with globules of melting germanium, say. One needs to account for the period from 1945 until 1951 during which Shockley struggled with the problem and one is hard pressed to believe that it reveals a day-to-day activity 'truly based on science', except in the most obvious and banal way.

The transistor has become a model because distinctions are drawn between its development and the development of other devices. The above simply argues that these distinctions are imperfectly grounded in the record. And to this must be added the fact that the junction transistor, far from being of itself a crucial device, was a passing stage in the growth of semiconductor scientific competence and is now as obsolete as the valve.

Like the Cat's Whisker, like the point contact transistor, Shockley's transistor was a signpost to a major innovation, albeit the first to indicate accurately the direction to be followed. It took 18 years from the introduction of the junction transistor to the *invention* of the **4004** chip, announced on 15 November 1969. Perhaps because we are so close to these years and certainly because they are systematically misrepresented by information revolutionists, the speed and impact of the invention is misunderstood. Obscured by changes of product within the electronics industry, this technological history commonly conflates and telescopes the diffusion of the transistor in the decade to 1960, the decline of the valve to the late 1960s, the development of the **integrated circuit** (IC) in the two decades from 1952 and the *invention* of the **microprocessor** in the 1970s. The resultant historical stew is one of the information revolution's main dishes.

It is here proposed to treat each of these ingredients as either marking advances in the development of the ground of scientific competence or as partial prototypes for the microprocessor. The result is that, in contrast to television, in the case of the microprocessor many devices were successively in the field before being vanquished by the *invention*. The transistor v. valve battle is still unresolved by the time development of the IC begins; and ICs are still gaining acceptance when **LSIs** (large scale integrated circuits) are introduced, which in turn are just about being diffused at the time of the *invention* of the microprocessor.

All this confusion did not, however, cause constant upheavals in the electronics industry, nor does it evidence a supposed speed-up in the rate of innovation. On the contrary, the overlaps are the result of the persistence of the previous device, the containment of the radical potential each had to destroy its predecessor. The overall result is, rather, that society gained forty years of preparation wherein to contain the small-scale computer and its microprocessing heart.

This process begins with Bell Labs itself where the operation of the 'law' of the suppression of radical potential was immediately apparent in 1948. 'Everybody started putting probes on pieces of semiconductor and studying processes like "forming" which was a bit of black magic to make transistors work better.'[39] So much for rational progress; but more to the point, the management was scarcely encouraging. Gordon Teal had each proposal he made for germanium research turned down between February and August of 1948, and, given his background, it is unlikely that every last one of his ideas was inadequate. He took his expertise to Texas Instruments and was responsible, in 1954, for TI's silicon transistor, the first to be commercially produced. William Pfann, whose zone refining technique was crucial to the building of the junction transistor, was prevented from working in this area. James Early had to struggle to have his idea for the superior design of the junction transistor accepted.[40] In fact, after the initial success, progress in making junction transistors took place outside of Bell altogether.

AT&T, as was seen with television (and as will be further detailed below in chapter 7), had evolved a complex corporate stance when it came to new technologies. It sought to balance its needs for technological protection (which required if not dominance then at least a significant patent presence in each new field) with its worries that an ever-watchful Justice Department would pounce upon it for techno-logically based monopolistic practices. In the case of the transistor, this stance accounts for Bell's initial actions to secure the patents and its subsequent willingness to share them with all comers. From 1952 on AT&T began to divulge not only the basic science of the transistor but also the know-how gained in its manufacture. Of course, this was not done as a free gesture. $25,000 a seat was charged potential licensees to attend the first Bell Labs symposium on transistors in that year. The money was to be deducted from licence fees should the firm go into production. The licensing system was leaky. As a basic protection against accusations of monopoly, AT&T had established extensive cross-licensing agreements with RCA. Companies happy to forgo explanations of Bell's know-how could pick up transistor information from RCA without paying the entrance fee.[41]

Bell, apart from buying a measure of protection from the Justice Department, was thus able to share the developmental load, specifically the problems of mass-producing the junction transistor. General Electric and Raytheon produced alloy semiconductors using idium dots on either side of the germanium which were easier to manufacture than the Shockley model, and Philco demonstrated the first etching technique.

Transistors had been developed to replace valves, but valves did not need replacing, at least not badly enough to cause an upheaval. The transistor was more expensive and less reliable than the valve, and was to remain so until the early 1960s.

For that decade or so, from '53 to '63, we had no choice but to go with vacuum tubes because they did a better job, and up until that time they were cheaper. You could get a perfectly good vacuum tube for about 75 cents.[42]

As we have seen, the Bell Labs remit to the Shockley team was translated from a general inquiry into semiconductors into a specific search for an amplifier, a valve. The further research and development of the device was in like case narrow rather than broad, and very often in the hands of technologists dedicated to valves who thought of the transistor as 'the enemy'.[43] Valves were themselves still a dynamic technology capable of doing practically everything a transistor could do, including playing the central role in the manufacture of computers. (Raytheon, the US computer industry's third firm, was in that business specifically because it was established in valves.) Valves were cheaper and more reliable than transistors and were becoming ever cheaper, ever more reliable and smaller. In the US military, where the greatest aficionados of new technology were to be found, little practical notice was taken of the transistor. Project Tinkertoy, an electronic miniaturisation programme, was started in 1950 and, ignoring transistors, was valve based.[44]

Even AT&T did not begin a wholesale rush to convert, although transistors were placed in long-line switching situations (card translators which worked in combination with cathode ray tube stores along 'Williams Memory' lines) and some experimental telephones by 1953.[45] Transistors were bound up with the computerisation of the central and public branch telephone exchanges and their utilisation within the Bell System marched with the introduction of ever more complex memories by the computer industry; but they marched slowly. The first fully 'electronic central office' did not appear until the end of the 1950s and the experiment ran for two years, until 1962.[46] The transistor was still unreliable enough for AT&T to hesitate about putting it into submarine cables in 1966 and a decade after that it was possible to state: 'the fully electronic public telephone exchange is still very much a research proposition'.[47]

The supervening necessity for electronic exchanges, which Bell Labs Director Kelly had articulated in 1936, was, forty years later, still being constrained by the inertial weight of established, expensive and, above all, profitable plant. When this plant finally came to be replaced in recent years the transistor was as much a thing of the past as the valve. The thunderous general silence that had greeted Bell's initial announcement in 1948 was prescient.

Only hearing aid manufacturers leapt on to the transistor. The first transistorised hearing aids were made in late 1952. AT&T, as a gesture to the interest Bell himself had in the deaf, waived the licence fees for this application in 1954 and Raytheon established itself as the primary manufacturer.

It was Texas Instruments, with Teal's cheaper silicon transistor, that began to create a more general market. Texas Instruments was the typical outsider company, having been, up to 1950, an oil industry service firm providing exploration surveys. Moving from an interest in electronics as applied to its own business, no sooner did it have the silicon transistor to hand than it set about putting it into a consumer item – a portable radio. (In the very first internal demonstration of the transistor to AT&T executives, '[t]o emphasize the Transistor's possible role in radio reception', the Lab showed 'a popular make of portable receiver whose vacuum tubes had been entirely replaced by Transistors. Reception was excellent.'[48] TI's production model was marketed in October of 1954. A battery-driven model was introduced the following year by Regency of Indianapolis. The synecdoche which made 'trannie' the British slang for such a radio is not a measure of the ubiquity of transistors, but rather the reverse. 'Trannie' came to mean portable radio because that was the only widely diffused device to contain transistors.

Transistors were of further value, especially to the military, in airborne or high temperature situations. Teal's silicon was more effective than germanium for such uses and, following Sputnik, the Signal Corps doubled its R&D grants for transistor research but only to a paltry annual million dollars.[49] The transistor industry did not benefit, to the same extent as others, from the enormous increase in military spending during these years. Contrast the nascent computer industry which was virtually a military subsidiary. The military subvention, though, was crucial in furthering the transistor's refinement. Texas Instruments sold more transistors to the armed services than they did to civilian radio manufacturers. Nevertheless the whole market was still small. In the years before Sputnik, 1954-56, transistors were worth $55 million, valves $1000 million and the entire electronic component industry $6500 million.[50] In 1957, before the full effects of Sputnik were felt, for every transistor, 13 valves were sold.[51]

Sputnik was launched in October of that year and worked to create a more effective supervening necessity for transistors (see chapter 5, page 235 below). The publicity attached to the Nobel Prize won by Shockley, Brattain and Bardeen in 1956 also helped. By 1959, sales overtook those of valves for the first time.[52] The value of military business to the transistor manufacturer between 1955 and 1960 climbed from a third of the market to nearly half. The space programme alone in 1936 used $33 million of them.[53] For the newcomers, this transistor trade was of major significance. Texas Instruments shares increased in value 38-fold between 1952 and 1959.[54] Yet for the old valve firms the transistor was not the disaster it is always presented as being. All eight major US valve companies were making transistors by 1953 and it was not until the second half of the 1960s that valves were finally phased out. It took television manufacturers, to take one crucial example, this long – fifteen years – to move to solid-state electronics, by which time the transistor, as is described below, had given way to the integrated circuit.

Of the eight major valve firms, only one (RCA) was still in the business twenty years after the semiconductor industry began and this is commonly read as clear evidence of the industrial equivalent of massacre.[55] But all the firms involved had histories of flexibility in their approaches to new technologies. They had adopted and abandoned a number in the period since the First World War and they had survived. One would need to know less than nothing of American life to assume that because such entities as General Electric and Westinghouse were no longer making transistors in 1975, the new technology had destroyed them. Further, by 1975, the list of semiconductors' top ten manufacturers, although dominated by newcomers, also contained such well-established electronics giants as Motorola and Philips who had not been there twenty years before.

Another view of this history is possible: the slow exploitation of the transistor, including paltry R&D support, gives the valve industry a decade to respond, followed by a further decade of semiconductor diffusion in which to run down. In all, the transistor, far from causing a massacre, is a classic example of what results from the balance between a slow working supervening necessity and the suppression of a device's radical potential, to wit – a fairly well stabilised change-over period. Again, here is a chicken-and-egg situation. If transistors had been more urgently needed they would have been developed faster.

The first transformation: ideation

G.W.A. Dummer glimpses the INTEGRATED CIRCUIT, 1952

In 1952, G.W.A. Dummer, of the Royal Signal and Radar Establishment, Malvern (RRE), delivered a paper at a symposium in Washington:

> At this stage I would like to take a peep into the future. With the advent of the transistor and the work in semiconductors generally, it seems now possible to envisage electronic equipment in a solid block with no connecting wires. The block may consist of layers of insulating, conducting, rectifying and amplifying materials, the electrical functions being connected directly by cutting out areas of the various layers.[56]

This represented the specific *ideation transformation* that was to lead to the microprocessor, the *invention*, because Dummer's concept revealed the true potential of semiconductors as the replacements, not of valves, but of entire circuits. He thought of it as commercial transistors were coming on to the market, some years in advance of the development of techniques which were to make it viable.

The British government was uninterested in *inventing* micro-electronics and Dummer 'never received the backing that his degree of inspiration would have justified.'[57] Dummer himself understood the lack of supervening, specifically military, necessity: 'In the United Kingdom and Europe a war-weariness prevented the exploitation of electronic applications and it was probably not realized how important electronics was to become.'[58]

The situation was to be different in the United States. In September of 1957 at an RRE conference, Plessey displayed a non-working model of what an integrated circuit might look like, Dummer having secured the firm a small research contract the previous April; but even Plessey treated the job as a 'laboratory curiosity'. Americans at the gathering were far more impressed than was the Radar Establishment which did not renew the Plessey agreement. Thus the IC idea crossed the Atlantic again, this time to bear fruit.

The received view is to see the IC and its successors as spin-offs from the transistor, not, as is here suggested, to see the transistor and IC as precursors – the former (because its existence prompts the *ideation transformation*) as an element arising from the *ground of scientific competence* leading to the microprocessor, the latter as a *prototype* of that device.

One major reason why the development of the transistor has been privileged over the microprocessor *invention* is that the micro-processor's emergence is viewed as being 'purely technological', requiring no advances in science – a bias reflecting the prejudice which elevates the transistor as the science-based artifact par excellence. It is true that, unlike the transistor, the microprocessor marks no theoretical breakthrough, because it is only a further exploitation of the same physics. Jack Kilby of Texas Instruments, who was the first to patent such a general purpose chip, states:

> In contrast to the invention of the transistor, this was an invention with relatively few scientific implications. . . . Certainly in those years, by and large, you could say it [the IC] contributed very little to scientific thought. [our brackets][59]

But what Kilby says of the IC can be said of any communications technology device, virtually by the nature of the case, i.e. the production of such devices is a *performance* of a previously established *scientific competence*. That the history of the microprocessor should be skewed in this way is not without significance. The distortion contributes to making 'the revolution' more purely scientific and therefore, supposedly, more modern.

Phase two: technological performance – prototypes

Jack Kilby builds an INTEGRATED CIRCUIT, 1958

Jack Kilby had attended the 1952 Bell symposium on behalf of Centralab, a large mid-Western radio and television component firm. In 1958 he joined Texas Instruments, where the person who hired him, Willis Adcock, remembered conversations with Dummer. Kilby (a man who in his own words, 'gets a picture and then proceeds doggedly to implement that picture') arrived just as the firm shut down for its annual holiday.

> Since I had just started and had no vacation time, I was left pretty much in a deserted plant. . . . I began to cast around for alternates – and the [integrated circuit] concept really occurred during that two week vacation period. [brackets in original][60]

Kilby began by building, by September of 1958, an oscillator device. In this he was following Harwick Johnson of RCA who had constructed, in 1953, the first simple circuit for a **phase shift oscillator** by exploiting the p and n junctions on a single germanium chip.[61] But then Kilby went further. In essence he mounted more than one junction transistor together in a solid slab and wired input and output terminals to each miniaturised section. This was exactly what Dummer had been groping for six years earlier. However like the point contact transistor, it was cumbersome – a minute block upon which the entire circuit had to be created by hand, and the more complicated the circuit the more vexed the wiring. Nevertheless, by October it worked – 'a body of semiconductor material . . . wherein all the components of the electronic circuit are completely integrated', in the words of Kilby's patent application of February 6th 1959.[62] Within weeks Kurt Lehovec, the research director of a Massachusetts firm, designed an IC in which the components were separated by the p-n junctions themselves.[63] Now there were valves, discrete transistors of two types and ICs available.

In 1959, the first year in which transistor sales overtook those of the valves in America, a new production technique was introduced which was both to make the discrete transistor very cheap and, shortly thereafter, obsolescent. The breakthrough to full-scale planar methods, which laid the foundation for the economically viable production of IC and LSI chips, resulted from seeking improvements in junction transistor manufacturing. The technique was developed by Fairchild Semiconductor, the TI rival founded by refugees from the organisation Shockley himself had formed after winning the Nobel Prize and leaving Bell.

Shockley had gone to his home town Palo Alto where the University of Stanford was building on the tradition of electrical research begun by De Forest. Encouraged by the university, a number of electronics firms were already located in what was to be nicknamed Silicon Valley, among them, for instance, the company founded by two Stanford graduates, William Hewlett and David Packard.

Shockley's seven dissidents had disagreed with the basic thrust of his Semiconductor Laboratories. Robert Noyce, who worked at the labs, has explained that:

Shockley was concentrating on four-layer germanium diodes at a time when there seemed to be a swing in the semiconductor community from germanium to silicon. . . . The people traditionally in the business are so committed to the previous course of action that they don't really get on with the new thing.[64]

Noyce, an Iowan with a PhD from MIT, was not, however, among the original seven dissidents. It was after they had involved Fairchild Camera and Instrument through the good offices of a New York investment banker, that he was invited to run Fairchild's new Palo Alto enterprise. The pattern of corporate scalping in the semi-conductor business, aided by the minimal disruption to families caused by job changing within the valley, thus began. (Shockleys was eventually acquired by ITT which closed the Palo Alto plant. Shockley, in recent years, achieved notoriety as a eugenicist with the idea of a sperm bank for Nobel laureates such as himself as a central plank of his thought – a sort of junction transistor for improving the race.)[65]

Fairchild was well placed to exploit the 'new thing'. At that time the cutting edge of design involved having the emitter protrude above the silicon base which acted as the collector, the so-called **mesa transistor**. These were none too robust, being difficult to wire and the 'mesa' attracted contaminants. In 1958, Jean Hoerni of Fairchild refined processes of diffusion and oxide masking (which another Bell symposium, in 1956, had introduced to the industry) by utilising advances in photolithography, specifically photoresist techniques. He bypassed the troublesome mesa altogether.[66]

In Hoerni's method, the silicon wafer had to be oxidised, which covered it with a layer of insulation. It was then coated with a polymer – a photoresist – which was sensitive to light so that, after the pattern of the transistor had been photographed on to it, it could be washed away from the exposed parts, allowing the base silicon to be etched. The doping substances – to create n and p-type material – could then be diffused on to the exposed portions of the silicon. The advantages of the method are that many copies of the same transistor can be placed on the one wafer which can then be sliced up.[67]

By 1959 Fairchild was manufacturing silicon transistors by the planar method and within three years the price of its product fell by a factor of ten. The result was simply a market glut.[68] The real significance of planar techniques was not cheaper transistors but ICs, as Robert Noyce began to appreciate in the early months of 1959.

Kilby and Lehovec ICs posed even more of a production nightmare than had the point contact transistor, and, because of this flaw, Kilby's claim to the integrated circuit has been seriously disputed by Noyce. Noyce saw that the planar process could not only yield many copies of the same transistor but could also allow different conduction levels of complex patternings to be placed on the one wafer – an IC which would avoid impossibly expensive hand wiring. By that summer, Noyce had produced a viable IC by the planar process.[69] Since it is this technique rather than the one developed by Kilby which is in current use, the argument as to precedence continues in the courts. TI filed for patent interference in 1963 but the court found, in certain aspects, for Noyce in 1969. The companies, despite these differences, collectively licensed all comers to exploit ICs.[70]

The planar process was already causing a transistor glut and the industry was clearly unable to accommodate so fundamental a development as the IC. Like the transistor, then, the IC was initially left confronting a very basic problem – what use was it? The answer was the same in both cases; by December of 1963, the first IC hearing aid was marketed. Otherwise there was little immediate demand. The US military were intent on imitating the Bourbons and, having learned nothing from Tinkertoy and the transistor, were now involved in the so-called 'micromodular plan' which used transistors but ignored ICs. The army continued to fund the scheme into 1963, spending $26 million on it. Thus to the slow decade of transistor diffusion can now be added a second decade of slow IC diffusion. But the military, still reeling from the implications of Sputnik, were evolving weapons systems, such as the **Minuteman** intercontinental ballistic missile, which would depend on ICs.

Military applications, because of their limited scale, did not transform the world of electronics. The demand was not great enough and its pace, thanks to procurement procedures, was too slow. By 1963, ICs accounted for only 10 per cent of the circuits sold in the US. The valve, never mind the transistor, was still very much in evidence in the early 1960s. For instance, it took an act of Congress, the All-Channel Law, which required that 1964 television sets be able to receive both UHF and VHF to get valves out of the sets.

There was an attitude on the part of the television makers that if the integrated circuit was such a hot new product, why hadn't it come from research at RCA, or GE, or even CBS? Fairchild, and Texas Instruments, were out West and just not that well known yet.[71]

Fairchild built a solid state TV simply to show the manufacturers that such a thing would work and would, indeed, have advantages for UHF.[72]

Once more supporting technology comes into play. By 1967, finally, Heil's patent of 1935 was realised and a cheap *m*etal *o*xided *s*emiconductor (**MOS**) was made into a field effect transistor. This was done at General Microelectronics, a company founded by dissidents from Fairchild – Fairchildren, as they were now being called. The MOS enabled the emerging IC, because it permits a multitude of high-density circuits to be etched on the chip, to become *l*arge *s*cale (LSI) and then *v*ery *l*arge *s*cale (VLSI). There was still no rush though, and General Micro's pioneering of MOSs did not translate into instant commercial success. The firm was bought out by Philco-Ford in 1966. It was not until the widespread popularity of the pocket calculator in the early 1970s that MOS/IC technology became widely diffused.[73]

The first semiconductor calculator had been introduced, with discrete components, i.e. transistors, in 1963 by a British firm. It was the size of a cash register. Texas Instruments brought out one with an MOS chip in 1967 but without much response.[74] But within four years the ancient abacus and, at around 325 years, the comparatively newly created slide rule ceased to be the most ubiquitous of arithmetical aids. The pocket electronic calculator, aided by Japanese production methods, arrived. The IC calculator was initially a device, like the Pascaline or the Baldwin, for the office. However, unlike every other mechanical aid to calculation ancient or modern, including the contemporary computer, it required no skill to use and was battery powered. Accessible enough to take the labour out of even everyday arithmetic, it attracted far more than those who customarily worked extensively with figures. As sales skyrocketed, the price fell – from some $200 in 1970 to around $15 for the same model by 1975.[75] Elsewhere, within this time frame ICs made their way into a host of devices, all of which had been well established with either conventional electronics or even older technologies.

Digital watches were the other early wonder of the semiconductor age, but here again there is a history which goes back to the first quartz timepiece of 1928. Greenwich Observatory had one in 1939 and by 1968 they were small enough to be battery powered on the wrist – at around $1000 per piece.[76]

It might be thought that the subsequent flood of cheap electronic watches and the attendant demise of the dominant Swiss mechanical watch industry is one example where the 'law' of the suppression of radical potential did not work. After all, between 1956 and 1980, the number of Swiss watch firms dropped from 2332 to less than 900 and the number of jobs was halved to 26,228.[77] The 'law' is not to be used to argue that changes never occur – only that upheavals are contained when they threaten the central arenas of the economy. So, indeed, here is a major shakeout based on a technological innovation, but its success crucially depends on the fact that the trade was an export one and there was therefore no constraint internally within any one market – except Switzerland, of course. Also, the innovation eventually, which is to say after a decade, began to provide a device markedly better than the traditional one – as the electronic watch developed multiple alarms, internal lights, stopwatch facilities, etc. etc., all of which were found to be of value by users.

Nevertheless the Swiss were not wrong-footed with electronics in the sense that they were unprepared. They had established a collective research laboratory, the Centre Electronique Horloger (CEH) in Neuchâtel and by 1967 working models were being made there of sufficient quality to be entered in the traditional competitions for accuracy.[78] The productivity of the Swiss industry in conventional manufacture increased significantly during the 1970s, the crucial first decade of the electronic challenge. In 1978, despite the fact that Japan was tripling its production, Switzerland, although declining sharply, still had the biggest slice of the world market, 63 million of 265 million pieces. After all, the Swiss had never been in the business of cheap watches and had survived the attack of Timex – the first best-selling cheap watch – throughout the 1950s. Also, it was some time before the self-evident superior attractiveness of the electronic watch was clear. Initially they were just a new sort of cheap watch and, in the early years, were screened by the public. If it didn't work, and there was a nearly 50-50 chance that it wouldn't, it just got shipped back and replaced. The manufacturers found it cheaper to do business that way. As late as 1976, J.C. Penney, America's third biggest retailer, was still returning 4 out of every 10 watches it got from one supplier. Finally, the investment needed for electronics on a mass scale was more than the small Swiss enterprises could afford, and they were not prepared to regroup.

> In short, the Swiss watch industry had everything it needed to enter the new world. The only thing lacking was entrepreneurship; the manufacturers of watches were not interested. Those years of comfortable, sheltered monopoly rents had cost the industry what once had been its most precious characteristic – its *Neuerungsfreundigkeit*, its joy in innovation.[79]

Neuerungsfreundigkeit was the driving force in Silicon Valley, but it did not help the semiconductor industry share in the débâcle of the traditional watch market. Many American semiconductor firms joined in the rush to replace the Swiss, constantly lowered their prices and flooded what was a seasonal market. The 'law' of suppression, which did not operate to protect the external Swiss, did work to help the American watch manufacturers resist domestic challenges. By 1977, all but the three largest microelectronics firms had bailed out of digital watch manufacture. For instance, Intel, the company Noyce had created when he himself joined the Fairchildren, had lost a fortune on its watches and sold out its interest in the field to Timex.[80] It was the Japanese, understanding marketing as well as electronics, rather than the Silicon Valley folk, who took up the slack created by the Swiss decline.

The second transformation – supervening necessity

Intel markets a IK RAM CHIP, 1968

Computers were to act as the supervening necessity for the micro-processor, the *invention* in solid-state electronics. As the railways had necessitated the telegraph a century and a half earlier (see chapter 6, page 298 below), so now in the late 1960s, the further development of a fourth generation of computers necessitated the microprocessor. The technology of the microprocessor, like the technology of the telegraph previously, was to hand. Indeed, so much was it to hand that the common perception is that it was already being fully used – but that was not the case. Computer manufacturers had moved slowly to ICs which, as off-the-shelf components, allowed them to design their machines as they liked. Computers were still expensive – even those third-generation devices designated as 'minis' – and the market for them remained, comparatively, small. But the logic of the computer – a logic that had allowed outsiders to the industry's establishment to build and market those same minis – could no longer be denied in the third decade of computer development. As Robert Noyce put it:

> The future is obviously in *decentralising* computer power. . . . As it turns out many of the problems addressed are not really big problems at all but lots of little, simple problems. The difference between a computer and a microprocessor is like the difference between a jet airplane and an automobile. The airplane carries a hundred times as many people and it carries them ten times as fast, but ninety per cent of the transportation in this country is done by automobile, because it's the simplest way to get from here to there, and the same is true of the microprocessor.[81]

The computer industry, from its inception, had been blind to this possibility. Now, as its eyes were opening, it was hostile to it.

> Computer engineers often looked at integrated circuits as a way of putting them out of business. . . . This was the era of big powerful computers, and little computers just weren't that interesting.[82]

Far from being in the symbiotic relationship that hindsight suggests, the two branches of the electronics business were at a distance from each other. The *invention* of the microprocessor did not initially bridge this gap for, on the contrary, it implied that semiconductor engineers, not computer engineers, were to be the real designers of computers. The microprocessor was not a component; it was, in design terms, the whole heart of the machine. After its introduction, for instance, semiconductor firms had difficulty in hiring programmers to exploit it. Just as valve people felt the transistor was 'the enemy' so computer people responded to the microprocessor in much the same way. Because of this background, the computer's operation as a supervening necessity to the microprocessor occurred in a series of tangential events and parallel tendencies. It was to be a fait accompli.

The key element concerned the repositioning of some major semiconductor firms in the late 1960s. Noyce's new company, Intel, which he had founded with Gordon Moore and Andrew Grove, was created to exploit a large market opportunity which was emerging – only at this late date, it should be noted – in computer memory chips. Moving from the general Fairchild business of electronic components, Intel was to be specialised. The hostility of the computer industry was minimised for, after all, the memory store was only a peripheral. Intel already had a rival, AMS, and others were known to be interested in helping to replace Jay Forrester's 20-year-old drum which was still the standard data storage device. For the semiconductor manufacturers, having just successfully broken into the television market with ICs, it was clearly time seriously to sell some state of the art electronics to the computer manufacturers.

The computer had reached, in the three generations of incunabula, mainframe and mini, the end of its development as a species of large capital goods. Massive diffusion required smaller computers and smaller computers required microelectronics of a greater sophistication than the IC or even the LSIC. The computer industry in the late 1960s, having grown rich on its existing markets, did not conceive of its situation in these terms. Rather, elements of the semiconductor industry, looking to break into computers, created the wherewithal which allowed others to fulfil the computer's micro promise. The drive towards the personal microcomputer begins with this repositioning of Intel. The development of specialised memory chips can be thought of as the exposed portion of the logic that makes computers and microelectronics such a perfect fit. It was a way into the central business of *inventing* an integrated circuit that could perform logical operations, a central processing unit – a CPU.

Phase three: technological performance – invention

Ted Hoff invents the MICROPROCESSOR, 1969

Just as Bardeen and Brattain found the transistor effect while looking at surface states, or, better perhaps, as Eckert and Mauchly were forced to build memories and BINAC while wanting to make UNIVAC, so the microprocessor emerges from Intel, not as part of its specific work on computer memory chips but out of a side deal.

In 1969, Busicom, a Japanese manufacturer, approached Intel with an ambitious plan to make an IC specifically for a series of commercial desk-top calculators, then the industry's hottest prospect, which could, through the installation of different ROMs, perform specialised functions. The preliminary plan called for twelve chips containing from ten to fifteen times as many integrated transistors as were then currently being used in such devices. Marcian (Ted) Hoff, one of the firm's brightest Stanford graduates and a man interested in possible applications of solid-state technology to logic circuits, was deputed to examine the Japanese designs. Hoff simply suggested that the calculator's arithmetic and logic circuitry should be put on the same chip. It would be a blank slate, as it were, capable of taking specific instructions from specially written ROMs. Intel would make the chip and Busicom would get the firm to programme its ROMs for the advanced calculators. The result of this design insight was that the Busicom calculators needed only four chips, a ROM, a RAM, an input/output IC and, centrally, a microprocessor. Hoff now needed to *invent* the only one of these proposed chips not to hand, the microprocessor.[83]

He then, with Frederico Fagin, did just that. With the blessings of Moore and Noyce, they and their design team produced a silicon chip one-sixth of an inch by one-eighth, containing 2250 microminiaturised transistors. The small size and high density are not the essence of the achievement – rather it is that the combination of elements including the in-built logic circuits created a programmable general purpose device, the minuscule CPU of a computer. Not only would it drive the Japanese calculator, but dozens, thousands, of other devices as well. Manufacturers would be able to use the CPU chip off the shelf, Intel simply executing their instructions in making specific ROMs.[84] Ted Hoff: 'The actual invention of the microprocessor wasn't as important as simply appreciating that there was a market for such a thing.'[85] This is something of an understatement, for Hoff's appreciation of a possible market masks the essential creative leap that he made. He was engaged in systems engineering – all the elements he used were already to hand – and there is nothing of the 'eureka' about his achievement. But in this he was like all the other *inventors* discussed here. They too were systems engineers; they too had all the elements to hand. They, like Hoff, were engaged in technological performances against a background of scientific competence, technological performances which were transformed by the operation of a supervening necessity into *inventions*.

The third transformation: the 'law' of the suppression of radical potential

Reiss of Northrup makes an IBM CPC programme itself, 1951

Our entire concern with semiconductors has been a necessary digression needed to illustrate the operation of the 'law' of the suppression of radical potential in computers. The development of the computer is the supervening necessity that creates the microprocessor. However, these two technologies are not, as is often claimed, in lock step. In fact, between the *invention* of the computer and the *invention* of the microprocessor is a 21-year gap. It is this gap which allows for the final argument in our contention that the computer's radical potential has been constrained. The slowness of the computer industry in adopting microelectronics joins the 'big-machine' design philosophy and the suppression of programme language developments (both discussed in the previous chapter) as the third major factor in the operation of the 'law' as it applies to computers.

The assumption here is that the radical potential of the computer is as a widely diffused device on the scale beginning to be achieved by the microcomputer of the 1980s. The very fact that computers were conceived of and marketed as a species of large inaccessible capital goods is, then, a most effective constraint. But the question is, were they necessarily so conceived and marketed? Or, could the micro-processor and the cheaper computer have been produced earlier?

Prima facie, the answer to these queries must be yes. Ted Hoff discovered nothing new – which probably contributes to his lack of widespread fame. Had the computer been seen as a machine with wide potential applications, then the active search for a microprocessor would have had a powerful supervening necessity and – since there was no advance in fundamental scientific knowledge between the discovery of the transistor effect and Hoff's CPU – there is every reason for hypothesising its entire developmental process could have been telescoped; but the reverse occurred.

During the 1950s, computers were enormously expensive valved machines in, largely, military service. For all that visionaries might dream of the new computer-based world, for all that some scientists and the entire media might talk of electronic brains, nobody was seriously interested in fully exploiting the device by making it smaller and cheaper. Ten years after its ideation and five after the first machine had run there were but 250 computers in the entire world.[86] 'For the first two decades of the existence of the high speed computer machines were so scarce and so expensive that man approached the computer the way an ancient Greek approached an oracle.'[87]

Atanasoff abandoned the ABC, a small-scale device, and Mauchly did not take it up. The Manchester Baby Mark 1, the first fully electronic machine to run a programme, is inscribed in the history as nothing more than a method for demonstrating the efficacy of the Williams storage tubes and was dismissed in favour of its bigger successor MADM. Pilot ACE, although bigger than these other two machines, was still small enough to be wheeled around and good enough to be the first commercially manufactured computer (the English Electric Deuce). Nevertheless, it too gave way to a monster successor, ACE, and was only built 'with the object of demonstrating the competence of the team as computer engineers.'[88] Whirlwind spun off a baby but that too was a neonatal fatality. Perhaps most significant of all, the IBM CPC was not developed, despite the evidence that it could have been fruitfully refined into a comparatively small-scale fully fledged substitute. Instead IBM opted for devices of much greater capacity and expense. There are, then, numerous moments in the incunabula period when the decision to make a greater number of smaller machines could have been taken but was not.

The same failure to think small can be seen with the next generation of the large machines which prevailed. Centrally by no stretch of the imagination was there a rush to replace valves with smaller transistors, for all that hindsight makes the valve a powerhungry, failure-prone heater.

It is now therefore necessary to mesh the history of solid-state devices to the history of computers since, as is demonstrated above, computers were not sufficiently important to semiconductors directly to figure in their development.

We must immediately discard the thought that these events were conditioned by computer scientists' ignorance of the advances in solid-state electronics.

On 3 October 1948, within months of Bell's announcement, Jack Good wrote in a letter to Turing:

Have you heard of the TRANSISTOR (or Transistor)? It is a small

crystal alleged to perform nearly all the functions of a vacuum tube.
It might easily be the biggest thing since the war. Is England going
to get a look-in?[89]

Nobody got a look-in, or rather looked in. Having established with
Colossus and ENIAC that valves worked, the small group of early
computer builders, their minds primarily focussed on storage systems,
did not explore the new technology. Transistors were clearly too
unreliable. It will be recalled that the builders of SEAC, which, with
its fellow SWAC, was the only one of the incunabula to use solid-state
diodes (the transistor functions as a triode), eschewed the most
modern form of those in favour of a more tested type. This was in
1953, by which time the Manchester group had designed a small
prototype computer using transistors.[90]
   During this period also, Bell Labs was exploring the transistor's
potential for computers. This work, to say the least, did not benefit
from Bell's well-developed publicity machine. A completely transistorised device was built which could multiply two 16-digit binary numbers
in 272 seconds. The Bell scientist, J.H. Felker, in an obscurely titled
1952 article 'Regenerative Amplifier for Digital Computer Applications', concluded that

> Transistor performance equal to that of vacuum tubes can be
> obtained in computer applications without sacrificing any of the
> obvious transistor advantages, such as low-power operation,
> ruggedness and extremely small size.[91]

The machine was real enough for its photograph to appear but the rest
is silence. A summary of early Bell Labs computers published in the
Bell System Technical Journal eleven years later traces the development of the Stibitz electromechanical machines up to the Mark VI
which was put into service in 1950 but makes no mention of a
transistorised device.[92] Perhaps the best clue as to the fate of this
machine is in Felker's acknowledgments – 'Some of the work described
was carried out under the sponsorship of the Navy Bureau of
Ordnance.' If the navy paid for and obtained the use of the Multiplier
and Control Unit, it is no wonder that the article concentrates only on
the transistorised memory system.
   The commonly made assertion that IBM put transistors into a
commercially available machine in 1955 comes from a business journal
report of that year and although a fine early example of 'revolutionary'
hype, is not true.[93] IBM's new machine for 1955 was the 705 featuring
'Magnetic core storage, increased speed, simultaneous reading and
writing of tape, direct memory transfer, a flexible accumulator, a new
flexible card reader, and a record storage unit' but no transistors.[94]

The supervening necessity that put transistors in computers was what had driven advances from the very beginning – nuclear research. It is true that by the mid-1950s one transistorised computer, the Bell Naval Ordnance machine, and a number of transistorised prototypes existed – at the IBM Military Products Division, MIT and Manchester University. And that there were a number of proposals on the boards, Philco's **Transac** (to be marketed as the 2000), Univac's **LARC** (for the *Livermore Atomic Radiation* Lab) and IBM's **Stretch** for Los Alamos. The military were not interested in reducing the computer size by using the smaller transistor. On the contrary, the effect of the transistor upon these machines was to make them even more enormous. Freed from the physical constraints imposed by valves, IBM's 7030, significantly nicknamed Stretch, for example, used 150,000 transistors. It was 75 times faster than the 704, twice as fast as LARC and capable of 400,000 instructions per second.

The American government funded the computer industry's investigation of the transistor, just as it had, in effect, funded the first commercial computers, but these military developments ignored the transistor's main advantage, its size. The results were often, in commercial terms, tentative. Stretch, for instance, was delivered in 1961 and worked at Los Alamos for a decade but did not yield the expected 7030 production model. Only a few were sold, even after the price was reduced from $13 to $8 million.[95]

IBM were in the business of selling machines at this point because of a consent decree in 1956 which held its leasing policies to be restrictive.[96] The leasing base restricted not only trade but also innovation and the huge success of its valve machines restrained IBM. 'New products destroy this leasing base, and thus, IBM is naturally hesitant to introduce new products until long after other firms have introduced similar products.'[97] And since the loyalty of IBM's lessors was already working to limit the ease with which even the largest firms, with even the most advanced products, could enter the market, the entire industry settled early into a certain technological conservatism. Philco, for example, did introduce the first transistorised computer, the 2000, in 1958 but it could not withstand IBM.

IBM, though, was never successfully to market a fully developed transistorised computer. The following, taken from a BBC documentary *The Chips are Down* transmitted in March of 1977 and one of the most influential 'information revolution' texts, is typical of the pastiche of error and distortion that has grown up around the history of the transistor/computer interface.

In the early 1950s, the switches in computers were valves. Each one was hand-made and expensive – around £5 each at today's prices – and the world market was dominated by huge American manufacturers. But in 1948 William Shockley invented the transistor, for which he was to get a Nobel prize. It murdered the valve industry with a rapidity that was brutal. (from script as published)[98]

One can at least agree that William Shockley did win the Nobel Prize.

The computer industry was no more eager to use the integrated circuit than it had been to use the transistor. Within IBM

> There were projections saying it would take all the sand in the world to supply enough semiconductor memory in order to satisfy IBM's needs so many years out. . . . It was just too great a risk to commit the total corporation to integrated circuits for computer logic.[99]

IBM, reflecting the prejudice against semiconductors noted above, were happier, however marginally, thinking about ICs for memory rather than for logic. The Atomic Energy Commission (AEC) again broke the bottleneck by funding Control Data Corporation (CDC) – a company formed when many of the original ERA people, including William Norris, broke away from what was now (Univac) Sperry Rand – to build a machine two and a half times as fast as Stretch, 1 million instructions per second (1 MPS). The result was the **CDC 6600** and IBM wrong footed.

IBM had been persevering with transistors, despite the unfortunate Stretch experience. In the early 1960s, a new series was developed and introduced in April 1964. The **System/360** used a hybrid integrated circuit/transistor design – rather as the SSEC was a compromise between electronic and electromechanical technologies in the late 1940s. But the logic of the separate components, present in the 701 to allow for a measure of customer configuration as well as increased rentals, was now carried further. The 360 series was the first not only to allow users to update all parts of the system, including the CPU, but to do so without needing to junk software and data. At the outset of its career, though, the 360 reflected, on the one hand, the CDC challenge and, on the other, company fears of the IC. These uncertainties were fuelled by the $5 billion investment the new line occasioned. The adaptability of the machine helped make it a major hit, some 30,000 being sold before 1970, at a rate initially of as many as 1000 a month. IBM built five new factories and increased its workforce by a third to cope.[100] However, although the 360 saw off both GE's and RCA's forays into computers, it was no match for the more up-to-date CDC 6600, the first of which was delivered in September 1964. To neutralise the threat, in the best tradition of Watson Sr (in his cut-throat cash register days at the beginning of the century), IBM announced the 360/91, a competitive I MIPS machine.

CDC was to claim successfully that this was a ploy and the subsequently publicised production difficulties of the 360/91 were actually its design period. CDC thereby lost sales while the computer market, hooked on the 360, waited for the comparable (and compatible) machine from IBM. The ploy cost IBM its software house and about $75 million in an out-of-court settlement with CDC; which result encouraged other firms – already given a foothold by the 360 peripherals market – to take the giant on and led to a restoration of a greater measure of competition in the mainframe industry in the 1970s.[101]

By 1965 there were still only 31,000 computers worldwide – and that could be a somewhat inflated figure – and the semiconductor industry had failed to take over the computer business with the transistor before the IC was available. The IC, despite the conservatism at work, was to be another matter. A host of firms – excluding IBM which was initially constrained from entering this market by another Justice Department attack – built minicomputers of increasing range and decreasing price essentially using the ICs to reduce size while maintaining the computing capacities of the previous generation of mainframes.

Two strands combine in the 'mini'. On the one hand mainframe manufacturers used the ICs with greater boldness than IBM to produce smaller machines. In 1966, for instance, Burroughs introduced two medium size machines, which, somewhat like the IBM 360s, were IC-hybrids. In 1968 CDC and NCR both marketed pure-bred, as it were, IC models.

More significantly some computer engineers began seriously to think about the possibilities of a smaller size machine, and did so without relying on the new IC technology. Ken Olsen, the member of Forrester's Whirlwind team at MIT who had built the baby to demonstrate the magnetic core memory, had then been deputed to control the SAGE production process at IBM's Poughkeepsie plant. Appalled by the inefficiencies of a large organisation, Olsen became convinced that he could do better by going smaller both in term of design and organisation. In 1957 he established the Digital Equipment Corporation and within three years he had produced his first machine, the *Programmed Data Processor*, **PDP 1**. At $120,000 it was not cheap, but it was a good deal cheaper than the competition. In 1963 Olsen produced the first minicomputer, the **PDP 8**, a transistorised machine, the size of a large filing cabinet, costing a mere $18,000. The PDP 8, with its 4K of memory, was the computer industry's first automobile (to use Noyce's analogy), as opposed to jet airplane. It was a commercial hit and between 1965 and 1970 Digital's sales went up 9-fold while its profits soared 20-fold.[102]

The PDP 8 vividly illuminated a real need for smaller computers and by 1969, IC minis were available for as little as $8000. Two years later one count revealed 75 different firms making minis, but the mainframe industry continued to ignore the logic at work. In 1969 IBM, for instance, having obviously decided that the world was not going to run out of sand, introduced a complete IC memory for the 360 system. This was the sort of computer industry decision that pushed Noyce and Intel into the memory business. Only in the early 1970s did IBM begin to replace the System/360 with the **System/370**, machines having both memory and logic on integrated circuits.[103]

All this activity meant that for the first time in the decade from 1965 a real relationship between semiconductors and computers emerged and the growth encouraged the founding of Intel and its rivals, specialised semiconductor firms which looked primarily to the computer industry for business. Computer production began to increase. As Year 25 of the computer age approached, there were 70,000 machines in the US, and, by 1976, the first successful year of the microcomputer, there were 200,000, 40 per cent classified as minis. (The inappropriateness of the term 'mini' for these devices is often remarked on as an example of how quickly the computer has developed but, on the contrary, 'mini' celebrates the deliberate design philosophy which made of the giant a standard and downgraded everything else.)

The IC was not produced for the computer. Like the transistor, no one quite knew what to do with it, although TI did build a very small computer in the early 1960s (rather as Bell built a radio or Fairchild a television) to demonstrate its potential. The eventual importance of ICs to computers did not lead, however, to a concentrated research effort on the computing applications of microelectronics. The normal military necessities were in abeyance, the post-Sputnik arms race having been deflected by the hot war in Vietnam. The computer industry – despite the growing market for minis – saw ICs primarily in terms of memory devices and, until the 1970s, failed to act as a driving force behind semiconductor R&D. Nobody envisaged the possibility of the microcomputer, although there can be no doubt that had the full potential of the IC as a logic circuit device as well as a data store been seized upon earlier, such a small, potentially more radical machine could have been built easily. It takes 12 years from the transistor effect (1947) to the planar process (1959), the essential technical break-through; and then 10 years from that to the *invention* of the microprocessor (1969); and a further five years (1974) before anybody thought to use a CPU to build a personal machine.

It is significant for this thesis that the *invention* of the microprocessor occurs in a firm dedicated to computer memory solid-state electronics, not a computer company. It is also significant that the entire process in both semiconductors and computing takes until the mid-1970s to mature, giving society decades to prepare for such impact as the micro-machines would make. If there was to be a 'revolution' here, this particular history would have to be very different. As TI's chief executive, J.F. Bucy, said in 1980, 'Looking back we probably should have started on microcomputers earlier', not least, perhaps, because his firm had actually built one.[104]

And so with the CPU, the pattern established with the transistor and the IC repeats itself.

> Robert Noyce recalls that when Intel introduced the microcomputer [actually the microprocessor] late in 1971, the industry's reaction was 'ho hum'. Semiconductor manufacturers had made so many extravagant promises in the past that the industry seemed to have become immune to claims of real advances. Besides, the big semiconductor companies – Texas Instruments, Motorola, and Fairchild – were preoccupied with their large current business, integrated circuits and calculator chips. [our brackets][105]

Among the first devices to contain a microprocessor was a digital clock which could be also arranged to sound like an electric piano. Ted Hoff:

> Some of the other stuff seemed really weird to us at that time. We were surprised when people came to us with ideas for slot machines, and some other people wanted to use our microcomputer [sic] to automate cows. They wanted to automatically record how many times the cows drank water, or something like that and correlate it with milk production.[106]

Meanwhile Intel had fulfilled its raison d'être by introducing, in 1970, the **1103 RAM** chip, which was to affect the size and expense of minicomputers by substantially reducing the cost of memory. It worked and worked reliably. By the time the original 4004 micro-processor gave way to the 8-bit **8080** in 1974, the reputation of the microprocessor had improved sufficiently over its predecessor's solid-state devices for it to be accepted. The 8080 did what it said it would do, reliably. The computer was finally ready to open its heart to the chip.

The entire development of the transistor was unrelated to computers; but now as we move from prototypes towards the *invention* in solid-state electronics, computers become significant and our digression ends.

Phase four: technological performance – production, spin-offs, redundancies

Jonathan Titus invents the PERSONAL COMPUTER, July 1974

A last question now suggests itself; even if it were true that the computer could have developed as a smaller and more accessible device earlier (as is argued above), is it now the case, when such small machines do exist and are comparatively widely diffused, that the opportunity for disruption is no more? Has the potential really been contained?

The development of the home, or better, desk-top microcomputer was not primarily determined by technology but rather by the range of factors which, we are suggesting, is usual in these circumstances – partially articulated needs, legal and economic constraints on unfettered innovation, research traditions and established industrial practices. Balanced against the supervening need for the micro-computer (in so far as any such need has been established) has been its potential for disruption with the result that the steady march of the years has eased the introduction and diffusion of the device.

The microprocessor was specifically designed for calculators and by the mid-1970s some three-quarters of a million had been sold for that purpose.[107] More widespread uses involved domestic oven timers, industrial control processes and, above all, video games.[108]

Space War, reputedly created by student computer enthusiasts at MIT in the early 1960s, was the Ur-video game, a widely shared cult among those with access to the mainframes at America's universities. Nolan Bushnell was one such Space War aficionado who combined his electrical engineering studies at the University of Utah with summer work at an amusement park.[109] These twin experiences sparked Bushnell to conceive of democratising Space War, making it as accessible as an electromechanical game in an amusement arcade. In 1972, he founded Atari and launched **Pong**, a microprocessor-driven toy which attached to a domestic television receiver. In 1974, Pong was featured in the catalogues and stores of Sears Roebuck, America's largest retailers. Video games were soon selling at a rate of 7 million units a year. Electronic game arcades flourished. Atari's revenues reached $40 million annually, and Bushnell sold to Warner Communications for $30 million. By 1980 Atari was retailing $100 million worth of games and low-level home computers and at the end of 1982 there were video games in 15 million American homes. It was a spin-off that happened in advance of the parent device – for, in that year, less than one in fifteen of those homes also possessed a personal computer.

Intel refined the 4004, producing in 1972 the first 8-Bit CPU, the

8008 and following that with the best-selling 8080 in 1974. These developments and the flood of Pongs created a general atmosphere in which the microcomputer, like a ripe fruit, was bound to fall off somebody's tree. The idea had become so obvious that even a few bold engineers within the computer industry were beginning to grasp it.

Digital and Hewlett-Packard dominated minicomputers, having a 55 per cent market share between them. They had developed a subsidiary educational market worth, by the early 1970s, about $40,000,000. Digital's educational division chief, David Ahl, became convinced that a machine even smaller than the PDP 8 mini would sell well in this area. He moved to the development group and produced plans for a $5000 computer for schools, but Olsen's firm, now the biggest private employer in Massachusetts, had become too big to continue to think small. The company felt that if one needed less computer than a PDP 8 one simply shared its time with other users, exactly the argument used by mainframe supporters in resisting the mini a few years earlier. Ahl left and eventually founded *Creative Computing*, one of the many magazines catering to the owners of the sort of machine Digital turned down.[110]

In 1975 Hewlett-Packard took much the same line as Digital choosing to ignore a crude single board microcomputer built by one of its more junior programmers, Steve Wozniak, in his spare time. It used a new $20 CPU, the MOS Technology 6502, and had 4K of RAM. Wozniak called his board the **Apple I** and he knew better than his employers there was a market for it.

Radio-Electronics had run a story in July 1974 outlining the design of a machine called the **Mark 8**, billed as 'Your Personal Minicomputer'. The Mark 8 was the brainchild of Jonathan Titus, a New York engineer then working for his PhD at the Virginia Polytechnic Institute. The article described a computer based on the Intel 8008 chip and invited readers to send away for a 48-page instruction manual, for $5.50, on how to build it. For a further $47.50 circuit boards were available, but the buyer had to obtain, from Intel and others, the chips required. All told the machine would cost the hobbyist about $250 to assemble. *Radio-Electronics* sold 10,000 copies of the blueprints and about 2500 orders were received for boards. The first home computer, with almost no memory and no means of storing its programmes after use, was probably assembled by between one and two thousand people. It spawned the earliest hobbyist clubs.

The next Intel chip, the 8080, led, within six months of the Mark 8, to a far more sophisticated machine – the **Altair 8800**. A group of three air force scientists had founded, with one civilian, a company in 1969 to produce, in their spare time, electronic gadgetry of various kinds, none of which did too well. Captain Edward Roberts bought out his faltering partners and began to sell a sophisticated electronic calculator kit. This remained a profitable business until 1974, when semiconductor firms, exploiting CPUs, marketed equally sophisticated, fully assembled but cheaper calculators. Roberts decided to move from calculator kits to computer kits. *Popular Electronics*, *Radio-Electronics*'s rival, having refused to publish Titus, was now looking for its own computer project. The editors had felt that the Mark 8 was more of a demonstration than a real machine and so laid down some requirements for their computer kit – it really had to be able to compute and it had to cost less than $400. Roberts's Altair 8800, with sixteen expansion slots, was heralded in the January 1975 edition of *Popular Electronics* as 'World's First Minicomputer Kit to Rival Commercial Models'. Like Titus's Mark 8, it was not a very accessible device, but it sold well to other enthusiasts, people who knew enough about electronics to make the 8800 work and it spawned a whole new world – more clubs sprang up, the first computer shop opened in Los Angeles in July 1975, the first magazine *Byte* appeared in August, the first world Altair Computer Conference was held in Albuquerque in March 1976.

Stephen Wozniak had failed to graduate college but he had built a transistor radio while in grade school and, at thirteen, a transistorised calculator which won a Bay Area science fair prize. In 1971 Wozniak made a small logic board out of ICs with one friend and with a new acquaintance, Steven Jobs, who was still at the high school in Cupertino Wozniak had himself attended, he began to manufacture a profitable line of illegal sound emitters used to make free calls on telephones. Wozniak and Jobs created a cash flow of several thousand dollars out of these **blue boxes**.

Five years later Wozniak was working for Hewlett-Packard and Jobs, who had also dropped out of college and had spent some time in India, was now working for Atari. The two natives of Silicon Valley met up again at 'Homebrew', the local Altair club. Wozniak built the Apple I with promptings from Jobs and, planning to sell it to their fellow enthusiasts at 'Homebrew', they unveiled the first Apple there in July 1976. The proprietor of the local Byte Shop was present and subsequently ordered 50 boards. At this point Hewlett-Packard passed up the Apple and Jobs sold his minibus and Wozniak his two Hewlett-Packard programmable calculators to pay for the manufacture of the orders. Some 175 were made and sold mainly through Bay Area shops or from Jobs's garage at $666 each.

Flushed with this success, Wozniak began to design and build the machine that was to become **Apple II** while Jobs approached his boss, Nolan Bushnell, for capital. Bushnell passed Jobs on and eventually he made contact with Mike Markkula who had been Intel's marketing manager and had just retired from the firm a millionaire. In October of 1976 Markkula became convinced that Wozniak's new machine was a potential winner and he added $91,000 to the $5000 Jobs and Wozniak put up to found the Apple Corporation. Markkula raised a further $660,000 from, *inter alia*, a Rockefeller company as well as obtaining a $250,000 line of credit from one of America's biggest banks. Wozniak created the first industry standard disk-drive in a burst of furious energy over the Christmas vacation of 1977 while riding high on Apple's first successful year.[111] (Ahl's plan for Digital's micro had not been helped by the failure of the proposed disk-drive on his prototype a few years before. By 1977 there was a proliferation of drives available using a new 5¼ inch standard floppy disk.)[112]

From the first 10,000 to respond to the original suggestion of Jonathan Titus, through the cult created by Altair, to the establishment of Apple, the computer industry took almost no notice of the burgeoning world of micros. The industry was used to machines costing a good deal more than a few hundred dollars and remained basically uninterested in the potential market these pioneers were unearthing. IBM, thinking as small as it could, showed in 1975 a 48k prototype which weighed only 50 lb, but the matter was not pursued. Commodore International, though, was an established electronics firm and it introduced the **Pet**, a true microcomputer, in 1977. A chain of electronic supply shops, Radio Shack, also produced a viable computer, the **TRS-80** for as little as $499.[113]

By the late 1970s the attractiveness of what was then still called the home computer was self-evident. Apple, on its launch as a public company in December 1980, was valued by the market at $1.2 billion and a number of firms, some from the mainstream of the industry, some – like Atari – from the games end, were becoming involved. Eventually IBM itself was to enter the fray with the **5100** which some considered a much more expensive alternative to the Altair and then, most seriously, in 1981 with the **PC**. It sold 35,000 of them that year.[114]

The number of computers in US homes had gone from a few hundred in 1975, the year of their introduction as a factory-assembled item, through to 20,000 a year later. In 1980 the industry passed the billion dollar sales mark and by 1982 there were just under a million machines. Current estimates of total numbers are curiously hard to come by, which, from the point of view of 'information revolution' hype is not without significance. Rather than computers, which are used to tally the numbers of everything else, the computer industry seems to prefer public relations experts to tot up its own successes. More sober estimates suggest that by 1984 there were 9,000,000 micros worldwide and a further 3/4 of a million mainframes and minis.[115]

The driving force, in this first decade of the home computer, still remained with those, like Roberts or Wozniak, who appeared to emerge periodically out of a garage (heavily backed by lines of credit) with a new wonder. Constant advances in microprocessor capacities rendered these machines quickly obsolete if not in terms of their actual use, certainly in the minds of potential buyers. Volatility is used by information revolutionaries to account for the bankruptcies and disasters the industry has suffered and there is some justification for this. In the bruising struggle to establish the standard machine mistakes are easily made, especially when one's opponent is IBM with its Svengali-grip on customers undiminished in this new area. Nevertheless, despite the plausibility of this explanation for occasional failures, within the boom there was a real problem. The home computer had little or no place in the home.

It rapidly became clear that the appetite to replace one's cookbooks, balance one's chequebook, keep track of one's stocks or spend happy evenings recalculating one's mortgage over different terms and interest rates was limited – and no amount of hype was going to change that. In the home, the computer's essential value – beyond games – was only as a computer; a device, rather like von Neumann's IAS machine, upon which to study – or better, play with – computing. A semantic shuffle was then accomplished. The home computer became the personal computer, which, while it could exist in the home, also had a function as a tool in the workplace.

We are still in the midst of the diffusion of this machine, the PC, but the limits of its potential can now perhaps be clearly seen and these lie beyond the issue of the failure of individual firms or even the operation of the depression of the early 1980s. Altair's Roberts sold out for $6.5 million worth of stock in a company which was itself bankrupt by 1979. Osborne, producer of the first portable machine, was also bankrupted. Texas Instruments pulled out. Atari failed to keep up with the games and could not balance this failure with success in the cheaper end of the computer market. The division might have been generating nine figure revenues in 1980 but within four years it was losing so much money it had to be sold. The experience of Coleco, a toy manufacturer, with its **Adam** is a further significant marker. Here was a machine which did everything that devices twice as expensive did and yet it performed so poorly that it had to be relaunched within months of hitting the American market. Coleco's profits currently depend on the Cabbage Patch line of dolls. By 1985, even Apple was laying off workers and the market had gone so soft that a general realisation of its limits was dawning.

Cost is no longer the major constraint, utility is. A study of twenty Greater New York computer-owning families, funded by an educational software house, discovered that the computer was underused and had no effect on the social organisation of the families. It was, for the children interested in it, a substitute for television, reducing but not replacing the amount watched – in short, not the beginnings of a brave new world but just another gadget.[116] Since the machines have no clear purpose, their design flounders. The Apple II survived for more than a decade because, like the IBM 360 before it, it was updatable. Apple seemed unable to profit from this and constantly sought another, incompatible and less adaptable, model to build on this initial success. The results were variable. Other firms were equally adept at failing to see what was really driving the market. In a tradition best exemplified by the automobile and film industries, the computer manufacturers convinced themselves, despite any evidence to the contrary, that what they had to sell was what the public really wanted to buy.

This picture contrasts most vividly with the hype of the information revolutionaries for whom the home computer was a crucial step in the creation of the wired city, the global village or what-have-you. It seems obvious that the personal computer will replace the electric typewriter, the calculator and spreadsheet and, in a more limited way, personal data banks and equipment for graphical work. This is to say, PCs are useful for people who use spreadsheets but will not create a need in those who do not work with figures. The same is true of certain types of graphical work and filing and all typing. For graphic artists, filers and typists the PC is a boon. For others, it is not. Since typing is the most widespread of these functions, and one which computing dramatically facilitates, word processing especially seems secure and therefore ensures the very wide diffusion of the PC. (It should be noted that the first IBM typewriter to have word processing capabilities, albeit of a limited type stored on magnetic tape, was introduced in 1964 but, being rather user-unfriendly, was not a success.)[117] It is fair to hypothesise that these basic alphanumeric and graphical functions will be an integral part of the standardised machine that is emerging and will be readily accessible in human terms.

The need for 'computer literacy' is a complete illusion, a skill only necessary for a transient moment in the development of the PC – in effect nothing more than the central plank of a campaign to sell the otherwise useless device to bemused parents as a necessary educational tool. There are, of course, computers in our children's futures, but no special language will be required. Without prejudice to the supposed intelligence of the fifth generation of machines, one thing must be obvious – they will be smart enough to talk a (nearly) natural human language – not **Basic**.

All this is not to say that networking, remote accessing of databanks, remote banking and other services, specialised programming of all sorts, amateur programming and (most of all) game playing will not have their place. It is to say that those places will be comparatively minor and will not amount to anything like a revolution. However, it is also to say, in essence, that the computer, in this personal setting, will help professional people perform professionally; but it will not make writers of non-writers, accountants of non-accountants and so on. The PC has been forty years coming and it is not about to change very much.

How then has this South Sea Bubble, this Railway Mania so come to dominate the discussion of these things? The debate is all but confounded by the rhodomontade. To read constantly that today's $300 microcomputer has more computing capacity than the ENIAC (which was not a computer, of course) or that if a Mercedes Benz was a chip it would now cost $300 (or, alternatively if a Rolls Royce were one it would only cost £70) does more than boggle the mind.[118] It creates a smokescreen of disinformation about our societies' major technological thrust.

One basic element in all this is the seductiveness of graphs. Nothing captures the essence of the hype surrounding computing so well as the dramatic growth curve graph which peppers the pages and video-screens of this literature. The graph's parameters are so chosen that the line of 'progress' in whatever form it might be, rises with rocket-like precipitousness towards the top of the page as the present or near future is approached. Most of these graphs are simply absurd – transistor sales from nowhere to millions, with no account taken of the place of such sales in the total electrical component market. This is, perhaps, the *locus classicus* but other diagrammatic 'truths' surely stand revealed as equally needing what the linguist Labov has called, in another connection, 'the withering interjection' viz: 'so what?'

So what that the dollars to compute millions of instructions per second have fallen between 1960 and 1980? So what that in 1950 1000 tubes took up a cubic foot whereas in 1956 10,000 transistors occupied the same space. So what that by 1958 1,000,000 could do the same thing? (Note that the date in each of these latter cases is typically more than a little in front of itself, but let that be.) The number of components per circuit has gone from 1 in 1959 (actually it should be 1957 or even 1953) to 262,144 in 1978.[119] So what, even, that the computer rental business grew by 78 per cent a year between 1954 and 1964.[120]

Another element of hype is the supposed youthfulness of the chief actor – Jobs was 22 when he sold the Apple concept to the 32-year-old Markkula. Noyce was 29 when he took over Fairchild and so on. But others, Watson Sr, Watson Jr, Shockley, Kilby, Mauchly, Olsen, Dummer (to take a few at random) were older. In fact, in the world of computers, as in other worlds, there is a mix. The ages of those writing the cheques and offering the lines of credit to enable exploitation of the whizz-kids' innovations is less than prominently inscribed in the records. And one can further question the supposed significance of the fortunes that have been made – so what that Bushnell earned $15 million from Atari or that Jobs was worth $162 million overnight when Apple went public. Has not such wealth always been the reward of entrepreneurship – 'this boundless drive for enrichment' – under capitalism?

The amazing youthfulness of certain players and the wonderful growth of ROMs, RAMs and dollars is not evidence that the world is in flux; rather revealed is just a bemused obsession with miniaturisation, youth and profits in the minds of the graphmakers and other commentators. The question as to what the actual societal impact of the machine has been is simply begged. The emblematic instance of this is that throughout the literature of the information revolutionists the suggested inevitable social upheaval seldom makes enough difference for the secretary in the new electronic office not to be a woman and her boss a man.

We have described what is claimed as the slow growth of computers from 250 worldwide in 1955, 31,000 in 1965, 70,000 in 1968 (US alone), to nearly 10 million by 1984, 90 per cent of them PCs. A generation has passed. The question can be asked, 'Are these numbers untoward?'

In 1899 in France, then the world leader in auto manufacture and not to be overtaken by the US until 1906, there were 1672 vehicles. Ten years later, in 1909, there were 46,000.[121] In the US, passenger vehicles alone went from 8000 to half a million in the first decade of the century and from half a million to 2 million in the second and from 2 million to 23 million in the third, 35 years after the *invention* of the automobile. 35 years after the *invention* of the computer there are not yet 10 million of them. So if Mercedes (and all other cars) were indeed chips, by 1930 American roads – with a third of the cars – would have been a great deal emptier than they were. Our purpose here is not to do a little mind boggling of our own, but rather to point out that the penetration of this device, as compared with automobiles (or televisions) is by no means fast. A basic fact about the microcomputer is that despite its enormous success there are far fewer of them than the industry expected, less than half as many as the 20,000,000 predicted a decade earlier for this date.[122] And with the motor car came a complete remaking of the entire environment, paved streets, gas stations, garages. Where is the equivalent impact with the computer?

> The computer in its modern form was born from the womb of the military. . . . It is probably a fair guess, although no one could possibly know, that a very considerable fraction of computers devoted to a single purpose today are still those dedicated to cheaper, more nearly certain ways to kill ever larger numbers of human beings.

What then can we *expect* from this strange fruit of human genius?

We can expect the kind of euphoric forecasting and assessment with which the popular and some of the scientific literature is filled. This has nothing to do with computers *per se*. . . .

We can also expect that the very intellectuals to whom we might reasonably look for lucid analysis and understanding of the impact of the computer on our world, the computer scientist and other scholars who claim to have made themselves authorities in this area, will, on the whole, see the emperor's new clothes more vividly than anybody else. . . .

It is not necessary to credit computers for accomplishments with which they have nothing to do. They can be realistically credited with having made possible some easing of the lives of some people. . . .[123]

Thus Joseph Weizenbaum, one of the computer scientists most clearly to see through the emperor's new clothes. His conclusion, it would appear, most happily accords with the historical record here presented. The development of the computer is of a piece with the development of television, subject to the same phases and exhibiting the same pattern. The schema of this work thus holds good across two of the four crucial technologies of the so-called 'information revolution' and reveals that there is no revolution at work. Now we will demonstrate that the pattern holds as true for a celestial technology as it does for these terrestrial examples.

LITTLE BIRD OF UNION AND UNDERSTANDING

Phase one: scientific competence

Sir Isaac Newton defines an ARTIFICIAL SATELLITE, 1687

(i) 'The birthplace of rocket navigation'
In the Fifth Definition of the *Principia* – 'a centripetal force is that by
which bodies are drawn or impelled, or in anyway tend, towards a
point as to a centre' – Newton described the launching of a projectile
from the top of a mountain so high that the air would no longer
impede its progress. Newton surmised, it would:

> never fall to the earth, but go forwards into the celestial spaces and
> proceed in its motion *in infinitum*. And after the same manner that a
> projectile, by force of gravity, may be made to revolve in an orbit,
> and go round the whole earth.[1]

Only two difficulties stand in the projectile's way – its speed would
have to be in excess of 5595 metres per second to achieve orbital
velocity and it would need to be made of heat-resistant materials to
avoid being burnt by the friction of the atmosphere; but, basically,
once these requirements are met, Newton's example would, and indeed
does, work (albeit by gravity alone).

The origin of the rocket, the device which was eventually to
demonstrate the truth of Newton's hypothesis, is lost. It emerges as an
instrument of war at the battle of K'ai-Feng-Foo in 1232 and within
three decades it is being used in Germany. By 1327, there were
military rockets in England but, over the following century, they were
slowly supplanted by the increasingly accurate cannon. It was the
shock of the rockets fired by the Sultan of Mysore's army at the battles
of Seringapatam in 1792 and 1799 that recalled British military
attention to the device. Sir William Congreve, working at the
Woolwich Arsenal, designed an iron cylinder 3½ feet long by 4 inches
in diameter which was mounted on a 15 foot pole. These were used

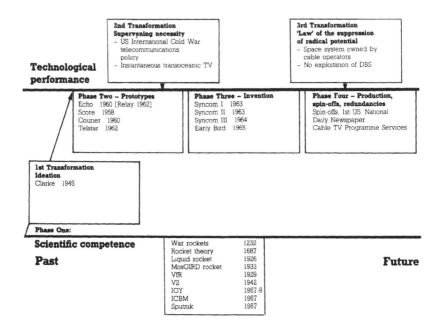

Figure 7 Geostationary communications satellites

against the Napoleonic invasion fleet at Boulogne. Their inaccuracy was such that the fleet was unharmed, although the town burned. Following subsequent actions against besieged cities – the only targets apparently big enough to ensure a hit – complaints were heard as to the brutality of the weapon.

The 'rockets' red glare . . . bombs bursting in air' seared itself on American consciousness at about this time, during the War of 1812. In opening the way to the rasing of Washington and in the attack on Fort McHenry, the British made effective use of their rockets. These successes, though, did not ensure the popularity of the technique on either side of the Atlantic, even after the American William Hale improved the Congreve model by introducing spin stabilisation instead of the long stick.

It is a very uncertain weapon. It may indeed spread havoc among the enemy, but it may also turn back upon the people who use it, causing, like the elephant of other days, the defeat of those whom it was designed to protect.[2]

The advantage that rockets had over cannons – that they were, comparatively, simpler to make and had greater range – was again wiped out in the nineteenth century as it had been in the fourteenth. Cannons caught up and military interest declined, except in situations where cannon could not be transported. Rockets were reserved 'for mountain and forest warfare, as in Abyssinia, in 1868, and in Ashantee, in 1874; and are very valuable for this purpose, from their extreme portability and moral effect.'[3]

Igniting a combustible substance inside a tube requires no great theoretical knowledge nor, indeed, a very high level of technological wherewithal, as civilian fireworks and the minor US craze for workable rocket models attest. The notion, put on paper by Jules Verne in 1865, that an extremely large cannon might fire a projectile which could escape from the earth's atmosphere was not too unlikely an extension of these realities. (That Verne should have determined Tampa, Florida, as the site for the start of his imaginary rocket journey *De la terre à la lune* is, on the contrary, uncanny.) The rediscovery and development of rocketry this century was thus, in the first instance, in the hands of amateurs – readers of Verne, enthusiastic for space travel and, paradoxically, uninterested in potential military applications.

> Russia is the birthplace of air navigation, as is manifested by documents on flights in an air balloon in 1731 by Ryazan official Kryakutnoi, who anticipated the French brothers Montgolfier by fifty-two years.
>
> Russia is the birthplace of the airplane, as is manifested by the patent issued in 1881 to A.F. Mozhaiskii, who anticipated by a quarter of a century the foreign inventors, the American Wright brothers.
>
> Russia is the birthplace of rocket navigation. The famous scientist K.E. Tsiolkovskii in 1903 published in the magazine Scientific Review the first part of his work 'Investigation of World Spaces by Reactive Instruments'* in which for the first time in the world he established and proved the feasibility of flight in cosmic space. Only nine years later in France and sixteen years later in the United States were works published on this subject. (* 'Reactive Instruments' = *reaktivbymi priborami*, i.e. rocket engines)[4]

Not all the translated Soviet rocket literature displays such chauvinistic 'firstism', although it must be said that the claims made for the rocket pioneer Tsiolkovskii, like those put forward for the television pioneer Rozing, are not without merit. Tsiolkovskii, a schoolmaster who died a state pensioner and a member of the Soviet Academy of Sciences in 1935 aged 78, was inspired by Jules Verne and outlined all the basic elements of the modern rocket, calculating the necessary mechanics of its flight and presciently suggesting liquid oxygen and hydrogen as its fuel. The 1903 article, not widely known until the eve of the First World War, marks the *ideation transformation* of the device. But it was left to another avid reader of Verne's, the American R.H. Goddard, to launch a liquid fuel projectile – to *invent* the contemporary rocket in 1926.

Goddard, whose first such rocket rose only 4 feet, was head of the physics department at Clark University. During the First World War, funded by the government through the agency of the Smithsonian, he was using powder fuels, although he had understood the possibility of a nuclear-powered engine by 1907. At war's end support ceased but Goddard continued his work publishing 'A Method for Reaching Extreme Altitudes' in 1919. By the mid-1920s he was in agreement with Tsiolkovskii about the need for a liquid propellant.

After the intervention of Lindbergh, money for a new series of experiments was raised from private sources – the Guggenheim Fund and the Carnegie Institution – and the work that would eventually produce the first liquid-powered rocket began. (Goddard received no further official backing until, in 1960, after his death and after the Russians had launched Sputnik, his widow and the Guggenheim Foundation were given $1,000,000 from the armed forces and NASA for the use of 214 of his patents. Subsequently NASA named an installation for him.)

The situation in Russia, where enthusiasm for space travel was encouraged by the revolution, was somewhat different. In 1924 a society for the study of interplanetary flight was founded out of a 25-strong informal group at the Air Force Academy. That year one of Tsiolkovskii's followers, F.A. Tsander, came to Lenin's attention at a meeting of the Moscow Province Committee of Inventors held in the city's university, where he read a paper proposing that a rocket could be designed to melt its own unwanted aerodynamic parts (wings, etc.) for fuel after it left the atmosphere. He explained that:

This creates a practical possibility of reducing the craft's full weight of 10,000 kg to 500 kg (the weight of small land-based airplanes) by the consumption of combustible material, and fully guarantees attainment of the enormous velocities required to overcome gravity.[5]

(Tsander also made notes on a winged space vehicle that would be powered by jets in the atmosphere and return to earth by glide-landings.)[6]

Another Tsiolkovskiian, Yu V. Kondratyuk, suggested an alternative scheme wherein the rocket would discard various stages as the fuel they contained was burned up, leaving only a capsule to make the space flight. A magazine devoted to these studies appeared during the 1920s and all this theorising resulted in the publication of N.A. Rynin's nine-volume *Space Travel* between 1928 and 1932.

On 17 August 1933, some five months after Tsander's death, a successful liquid-fuel launch was achieved by the MosGIRD (acronym for the Moscow Reactive Propulsion Study Group) which had been established as a main research centre and where courses on jet propulsion and rocketry were offered. The rocket was 2.4 metres long and weighed 19 kilograms. It rose 1500 metres.[7] But in the event neither the Russians, nor the Americans, were to dominate the field up to and during the Second World War.

Hermann Oberth, a Rumanian, was, like Goddard and Tsiolkovskii, inspired by Verne and wrote a theoretical work in German, *The Rocket Into Interplanetary Space*, which was published in 1923. (There were other texts. For instance the French scientist Esnault-Pelterie published on space exploration in 1913 and again in 1928 coining the term *'astronautique'*.)[8] Oberth became president of the unofficial German counterpart to GIRD, the *Verein für Raumschiffahrt* (The Society for Space Travel – VfR) in 1929 and served as technical consultant on Fritz Lang's film *Girl on the Moon*. By 1931, members of the *Verein*, who numbered nearly 900 including Esnault-Pelterie, were regularly launching rockets from a *Raketenflugplatz* on the outskirts of Berlin. One launch, for the purposes of a UFA newsreel, concluded with the rocket igniting the roof of a near-by police station and by 1934 the activity had ceased. Wernher von Braun, who joined the club in 1929, explained in a co-authored history of rocketry the circumstances of this decline and its sequel.

> The problems of the rocket pioneers were compounded by the increasing objections of the Berlin police to rocket flights within the city limits. In desperation the VfR group sought support from the German army. After a demonstration flight of a **Repulsor** (the liquid-fuelled rockets tested by the VfR in 1931) at the army proving grounds at nearby Kummersdorf in the summer of 1932, the army invited Wernher von Braun to do the experimental work for his doctor's thesis on rocket combustion phenomena at Kummersdorf.[9]

Rockets were again playing their historical role. As in India 150 years earlier, they were aiding the side with, if not the less developed, then certainly (because of Versailles) the more constrained, military technology. And, apart from avoiding the restrictions of the treaty, there was also the further push of a German military tradition which found expression in such giant artillery pieces as **Big Bertha**. The Berlin amateurs might have been desperate but the army needed little persuasion. It is possible that the Lang film had made them aware of rockets, really for the first time since Prussia had abandoned the weapon in 1864; but on the other hand, the Weimar staff officer, Colonel Karl Becker, responsible for ballistics research had included a section on rockets in his standard work on ordnance published in 1925.

Captain Walter Dornberger was given direct responsibility for the rocket programme and a static test stand was built on the army range at Kummersdorf about 25 km south of Berlin. Dornberger was explicit as to the practical thrust of the proposed work.

> The value of the sixth decimal place in the calculation of a trajectory to Venus interested us a little as the problem of heating and air regeneration in the pressurized cabin of a Mars ship. We wanted to advance the practice of rocket building with scientific thoroughness.[10]

Dornberger and his superior officers had visited the *Raketenflugplatz* and were impressed by the work, although appalled by the lack of documentation. Dornberger was particularly taken with a 19-year-old member of the club who had just graduated from the Berlin Technische Hochschule, Wernher von Braun. Von Braun topped the captain's list of possible technical assistants and when the army went ahead with its rocket programme in the autumn of 1932, von Braun was hired. While a civilian employee of the Weimarian army, he began work on his PhD at the Friedrich-Wilhelm University.

German rocket research from the outset envisaged the large projectile and the rocket-powered aircraft. The basic specification was to update Big Bertha by doubling the range from the gun's 78 miles to the proposed rocket's 156 miles and increasing the explosive payload 100-fold from 22 lb to 2200 lb, one metric ton. However, in order to facilitate rail transport, it was to be no bigger than 9 feet in diameter and 46 feet long. A series of devices, designated *Aggregat* (Assembly) 1, 2 etc., were designed and tested. The **A-1** weighed 330 lb and blew up on its stand. The **A-2** rose 6500 feet.

In 1937, von Braun was installed, as technical director over a staff of 80, at Peenemunde, a desolate Baltic spot at which a large research centre was built. The **A-3** achieved an altitude of 7½ miles and an 11-mile range. The *Fernraket* (Longrange Rocket) **A4**, after a series of initial disasters in 1942, was reliable enough to be put into production. It became the '*Vergeltungswaffen Zwei*' – **V-2** (Vengeance Weapon #2; the Luftwaffe's pilotless buzzbomb, the Fi-103, was designated **V-1**).

The V-2, as per the original specification, was 46 feet long and 5½ feet in diameter, had a range from 180 to 210 miles and travelled at 3500 miles per hour. The first of them, built largely by enslaved victims of the Nazis as were all German arms, fell on Chiswick and Epping, 9 September 1944. Against England, 1115 were launched, killing 2754 people and injuring 6523 more. On 8 March 1945 a single V-2 killed 110 and injured 123. 2000 more were used in Europe; but the vengeance weapon, which was overall less destructive than conventional bombing, did not turn the tide of war and the assault ended in the spring of 1945. By that time von Braun and his team were designing an **A-10** multistage intercontinental missile which would carry a V-2 warhead across the Atlantic.

Von Braun's work was unique in that it concentrated on large projectiles. Other secret military developments in the 1930s, such as those in Britain, eschewed this scale in favour of smaller tactical devices. These led to a variety of rocket weapons, anti-aircraft, air-to-air and personal. The Russians, also concentrating on tactical devices, produced the powder-fuelled **Katyusha** rockets which were, for example, an important contributory element to the Soviet victory at Stalingrad. Some experimental rocket-powered aeroplanes were built but never made operational. After the Hot War, the Russians found themselves in a markedly inferior technological position vis-à-vis the West. Even without their pioneering interest in space travel, it was to be expected that they, technologically disadvantaged as the Germans had been two decades earlier, would turn to rocketry as an area where comparatively primitive devices could be made to yield major results.

Peenemunde is today within the borders of the German Democratic Republic and it is often suggested that the Russians (who captured it in 1945) simply walked off with all the V-2 know-how, but this is not the case. In February of 1945 von Braun, with scores of his key people and their families, left the area and headed south to find the Americans – which they did by April. The Russians secured the services of only one key V-2 scientist, Helmut Grottrup. (Grottrup had been briefly arrested by the Gestapo, as had von Braun and Magnus von Braun his younger brother, in 1943 because their continual talk of space-travel was interpreted as treachery. It is likely that the episode was a ploy in Himmler's attempt to obtain control of the rocket programme but, much as Himmler himself had resisted Goebbels' efforts to take over Jewish policy in the late 1930s, the army had the rocket scientists released and the programme as a whole escaped Himmler's embrace.)[11]

For reasons of personal ambition rather than political convictions, Grottrup refused the American contract which von Braun and the mass of the Peenemunde technicians accepted and crossed over to the Red Army. Grottrup's first task in the Soviet Zone was to reconstruct the V-2 assembly line which he did by September 1946. This period terminated abruptly the following month when he and about 200 lesser technicians were transported to the USSR. There he supervised the launching of a series of V-2s and designed a variant which, although much the same size as the German original, was twice as heavy and had a detachable warhead, 27 per cent more thrust and totally different internal arrangements.

The Germans in the USSR spent the late 1940s creating ever more powerful rockets, each of which was further away from the V-2, on paper: the last so planned was to have a warhead of 6600 lb and a range of 1800 miles. These rockets were never built and slowly the Russians let the Germans go. Grottrup was returned to the DDR, among the last to leave, late in 1953. There is some evidence that Grottrup's plans were more sophisticated than those drawn up during the same period for the Americans by the von Braun group, but the Russians exploited 'their' Germans far less effectively. They had neither the wherewithal nor the inclination to pamper and, more than that, there seemed to be a genuine feeling among certain of their experts that the Soviets, as the pioneers of rocketry, knew enough to go it alone.

Prominent among those holding this view was Sergev P. Korolyov, eventually the chief designer of the Russian space programme. Korolyov, a Ukrainian, had visited the elderly Tsiolkovskii at his home in Kaluga and had succeeded Tsander as head of MosGIRD in 1932. An aircraft designer under Tupolyov, he was responsible for the first rocket-powered plane prototype during the war. After the war Korolyov and his fellows felt they were abreast of the Germans in theory and basically took from the Grottrup group, on a piecemeal basis, only production know-how rather than design ideas. A secret Commission for Long-Range Rockets was established in 1947. Although the primary thrust was towards the intercontinental ballistic missile (ICBM), which the Kremlin saw as vital to its interests (for a revamped V-2 could only at best devastate Finland and West Germany), there was also an interest in space science unmatched in the West. For instance, a series of V-2s carrying 14 dogs was launched between 1949 and 1952 to explore life support systems. It is against this background that the return of Grotrup and his men must be seen.

Building on these foundations, the Russians made good progress. They had an extensive background, a measure of German production skills and the technology, telemetry apart, remained rather basic. The drive to an ICBM, although obviously not publicly acknowledged, was behind the intensive development programme instigated in 1947. The need to carry a heavy A-Bomb conditioned the design philosophy; a big missile was needed. By the mid-1950s the prototypes of the **RD-107** and **RD-108**, which were to serve as first stage boosters until 1971, were being developed in the Gas Dynamics Lab in Leningrad. There was public talk of rocket engines with 1.1 million pounds of thrust and the Russians exploited the fundamental ambiguities which allowed a warhead-carrying device also to be used for more peaceful purposes. The ICBM came automatically disguised, as it were, as a space rocket. A 1951 report in the Soviet military newspaper *Red Fleet* picked up by a German press agency stated:

> that a moon rocket has already been worked out in the Soviet Union. It is said to be 60 meters long, and to have a maximum diameter of 15 meters, a weight of 1000 tons, and 20 motors with a total power of 350 million horsepower.[12]

This ambiguity between missiles and space probes is the mark of twentieth-century rocketry. It was what, ostensibly, got von Braun into trouble with the Gestapo and it was stridently exploited by the Soviets throughout the Cold War. For instance, in the English language *Soviet Review of World Events*, a professor wrote in 1954:

> During World War II, we all saw the frightful consequences of rocket weapons used against the civilian population of Britain. It is the duty of every scientist working in the field of rocket techniques to prevent the achievements of human genius from ever being used for such purposes again.[13]

Be that as it may, a device with 350 million horsepower nevertheless had a number of potential uses apart from going to the moon. It could deliver a warhead to any terrestrial target; or it could fulfil the terms of Newton's scheme by lifting a projectile – a satellite – into orbit.

> There is little difference in design and performance between an intercontinental rocket missile and a satellite. Thus a rocket missile with a free space trajectory of 6,000 miles requires a minimum energy of launching which corresponds to an initial velocity of 4.4 miles per second, while a satellite requires 5.4.[14]

The above 1951 Soviet report listed among the plans being investigated for the moon journey, take-off and landing by the rocket on an 'outer-station', a satellite free of the earth's gravitational brake. This was another vision of Tsiolkovskii's to be found in his 1895 short story, 'Dreams of Heaven and Earth':

> an imaginary earth satellite, like the moon, but brought arbitrarily closer to our planet, to a point barely outside the limits of its atmosphere, that is, about 300 versts from its surface, constitutes if its mass is very small, an example of a gravity-free environment.[15]

In November 1953, the president of the Soviet Academy, A.N. Nesmeyanov, announced at a meeting of the World Peace Council in Vienna that a satellite was possible. Six months later, Nesmeyanov's organisation established a special permanent commission to study 'interplanetary communications', as space travel is called in the USSR.

The Russians had just exploded an H-Bomb and there was a general, and increasing, supervening necessity of Cold War military competition. A small group of geophysicists meeting at the home of Professor van Allen in the US in 1950, seeking in part to combat these developments, agreed that it was now appropriate to instigate a third worldwide observational and experimental programme of international scientific cooperation. The previous two, in 1882-3 and 1932-3 had achieved considerable results and so a committee was set up, in 1952, to invite participation and plan the International Geophysical Year (IGY). The Russians did not respond. The IGY was to run from July 1957 until December 1958, the next period of maximal sunspot activity. The coordinating committee, at a meeting in Rome, October 1954, recommended 'that thought be given to the launching of small satellite vehicles'.[16] It was already six months past the deadline for acceptances into the IGY programme, so that same meeting was stunned to receive, via the Russian Embassy in Rome, word that the USSR would after all participate.

Eight months later, on 29 July 1955, President Eisenhower announced that the US would launch a satellite during the IGY. The following day, armed with Korolyov's assurances, the Soviet Union made a similar announcement. Leonid Sedov, the chair of the Academy's 'interplanetary communications' commission, stated a few days later, in Copenhagen, that the Russian satellite would be bigger than the one proposed by the Americans and would certainly be orbited in time for the IGY. Korolyov and his group, who kept a close watch on their American opposite numbers, calculated how much payload the best-known potential American launcher, **Viking**, could put into orbit and they knew they could beat it. In June 1957, Academician Nesmeyanov stated publicly that the device and its rocket were ready and would be launched within a few months. That same month, the Russians released film of three dogs in sub-orbital flight. The Soviet IGY official presenting the footage said, 'I would like the British correspondents to inform the British Society of Happy Dogs about this because the Society has protested to the Soviet Union against such experiments.'[17]

The Russian intention to launch an orbital device was communicated to the Chair of the American IGY Committee, as per the international agreements made for the Year. In July, the Soviet magazine *Radio* contained the following announcement:

The Institute of Radio Engineering and Electronics of the USSR
Academy of Sciences asks radio amateurs to report on the
preparations for the reception of signals from satellites launched in
the USSR . . . to report the radio signal data and to forward the
magnetic tapes with the recorded signals to the following address:
Moscow, K9, Makhovaya ul., 11, IRE AN SSSR.[18]

In August, the Soviets claimed that they had successfully tested an
ICBM and Krushchev said it could be targeted anywhere in the world.
In September, Radio Moscow broadcast that the launch of the
satellite, designated **Sputnik** – travelling companion – was imminent.
On October 1st, Sputnik's transmission frequencies were published.
On October 4th, Sputnik was successfully launched.
In the USA 11½ years earlier, in its first major investigation, what
was then called the Rand Project had predicted the effect of such a
feat:

Since mastery of the elements is a reliable index of material
progress, the nation which first makes significant achievements in
space travel will be acknowledged as the world leader in both
military and scientific techniques. To visualize the impact on the
world, one can imagine the consternation and admiration that would
be felt here if the US were to discover suddenly, that some other
nation had already put up a successful satellite.[19]

(ii)   An experimental world-circling spaceship
The Americans were as wrong-footed by the Russian satellite as they
had been by the British rockets 143 years earlier. In general, the
United States treated all Soviet announcements of technological plans
and intentions with a grain of salt. As we have seen, the Soviets
claimed primacy in *inventing* radio and television as well as aeronautics
and space travel, and the rhetoric of the times rendered these claims,
although not in fact without a measure of justice in some instances,
intemperate. Historically, in long-range rocketry, Russian theorising
had not produced devices in the metal and, after the war, the
Americans knew that they, and not the Russians, had secured the
supposedly crucial German expertise.
There was, in addition, considerable doubt about the feasibility of
the ICBM, the necessary parent of any satellite. For instance, no less
an authority than Vannevar Bush told the Special Senate Committee
on Atomic Energy in 1945:

The people who have been writing these things that annoy me have
been talking about a 3,000-mile high-angle rocket shot from one
continent to another carrying an atomic bomb, and so directed as to

be a precise weapon which would land on a certain target such as this city. I say technically I don't think anyone in the world knows how to do such a thing and I feel confident it will not be done for a very long time to come. I think we can leave that out of our thinking. I wish the American public would leave that out of their thinking.[20]

Such influential opinion, which reflected a certain war-weariness and a confidence that the only potential enemy, the Soviets, was *hors de combat*, led to a period of restrained experimentation. The Americans, like the Russians with the Katyusha, had a range of effective battle-proved rockets, such as the **Bazooka**. In 1945 the latest of these was still in the testing stage at the Jet Propulsion Lab of the California Institute of Technology. The **WAC Corporal** was fired by the army at the White Sands Proving Ground, New Mexico (near the site of the Trinity atomic bomb test), in October 1945 to an altitude of 23,500 feet.[21]

This was the state of American rocketry found by the 118 German V-2 scientists who, following von Braun, had contracted themselves to the United States army. Their odyssey had begun in the spring of 1945 with the decision to flee Peenemunde. Like the computer scientist Konrad Zuse, the group wandered south into the Harz mountains, to Garmisch-Partenkirchen and Oberammergau. There a series of confusing encounters with Russian, British and American military personnel took place. The Peenemunde technicians were clear that their objective should be – since they 'despised' the French, feared the Russians and fretted that the British could not afford them – a form of agreement with the Americans. Some were required to launch, under British auspices with Russian and American observers, the last V-2s to be fired from German soil. The American army, for its part, with a minimal amount of bureaucratic ineptness, accepted von Braun's offer of service, crated up a vast haul of enough V-2 parts to build about 100 rockets and shipped the group, via France, Boston and the Aberdeen range, to rendezvous with the parts at the White Sands site. The war with Germany had been over 18 weeks when von Braun and the first six scientists reached Boston. Reuniting the families took rather longer, but within two years of the war's end Peenemunde had been reassembled, men, families and machines, in New Mexico.

At first, the Americans had much the same plans for 'their' Germans as the Russians did and at the test site between April 1946 and September 1952 64 V-2s were launched, none with dogs. Given the attenuated interest in rocketry, ordnance had determined that instant multistage rockets, for which the V-2 launchings provided the know-how, were a cheap way of continuing development. The first **Bumper-WAC**, created by placing a WAC Corporal on top of a V-2, rose

244 miles over White Sands in February of 1949. Such experiments created missiles whose range was too great for the proving ground, so a new site was established on a swampy lagoon-fringed Florida beach, Cape Canaveral, about 100 miles across the peninsular from Jules Verne's Tampa launch pad.

As in Russia, there was a certain difficulty about using the Germans, but the American army kept the team together and well provided for, moving it in 1950 to a chemical warfare plant, the Redstone Arsenal, in Huntsville, Alabama. The arsenal had been engaged in testing liquid propellants in 1945 but had since closed. Von Braun's first task was to build a surface-to-surface missile with a 500 mile range – the **Redstone** – which was being tested by the summer of 1953. Thus just as Grottrup was being returned to Germany by Korolyov, having not built any devices, von Braun launched his first American rocket, one larger and more powerful than the V-2.

The Germans, fundamentally happy with their American lot, were becoming naturalised; and the Americans, less inhibited by the immediate past than the Russians, were placing them more and more in the forefront of US rocket technology. So it was that the von Braun team stayed on the job. Of the original 118, 8 being dead, 64 were still at Huntsville when it became a NASA installation, the Marshall Space Flight Center. However the V-2 cast a long enough shadow over von Braun's career to deny him leadership of the space programme to which his scientific and managerial abilities had most significantly contributed. He was never to head any American agency.

The Russians, despite their H-Bomb on the one hand and the animadversions of Senator McCarthy on the other, were still not being taken seriously enough to constitute a supervening necessity which required that the Americans put their ICBM house in order. Of course, the Cold War necessitated ICBM feasibility programmes but initially, given the German advantage (as it might be termed) and a general sense of superiority, the Americans were prepared, unlike the Russians, to wait for the lighter H-Bomb and a concomitantly smaller rocket to carry it. By the mid-1950s rival long-range rockets proliferated among the American army, navy and the newly independent (as of the summer 1947) United States air force. Von Braun's contribution, for the army, was to stretch Redstone into the **Jupiter**, while the UASF was developing an almost identical **Thor** as well as **Titan** and **Atlas** and the navy had **Viking**. Each of these rockets could lift a satellite as well as warhead, as had been theoretically understood in the US since at least 1946.

The first US satellite proposal, which called for a single-stage rocket, had been made by a naval study group in October of 1945 and a figure of between $5 and $8 million had been determined as the probable development cost of the test vehicle alone. The navy approached the

Army Air Force, but was rebuffed. Instead the airmen asked Project Rand to do a parallel study. (Rand – from Research *and* Development – was initially a subsidiary of Douglas Aircraft. Its secret mission was 'to wage intercontinental warfare by any and all means' and it became an autonomous non-profit research institution in 1948.[22])

The Army Air Force's request led to Rand's very first report, 'Preliminary Design of an Experimental World-Circling Spaceship', produced, because of the existence of the navy's rival suggestion, in 12 weeks (and quoted above, page 235). The writers began by rejecting the navy's single-stage proposal in favour of a multistage solution which would be able to orbit a 600 lb payload. They estimated it would take five years (i.e. 1951) to achieve. However, the report acknowledged the ambiguity of rocketry by indicating that such a satellite would be too small to carry an A-Bomb and therefore the usefulness of the exercise in terms of strategic weapon development was limited. In fact the satellite was not a weapon at all and that, given military policy, rendered it unfundable. Rand, trapped by this, sought to escape by appeals to history:

> In making the decision as to whether or not to undertake construction of such a craft now, it is not inappropriate to view our present situation as similar to that in airplanes prior to the flight of the Wright brothers. We can see no more clearly now all of the utility and implications of spaceships than the Wright brothers could see flights of B-29s bombing Japan and air transports circling the globe.[23]

This was followed up by a supplementary report 'The Time Factor In The Satellite Program', which took such arguments further by calling attention to the usefulness of rocket technology to a more backward potential enemy.

> The possibility of constructing a satellite has been well publicized both here and in Germany and the data of the Germans are available to various possible enemies of the United States. Thus, from a competitive point of view, the decision to carry through a satellite development is a matter of timing dependent upon whether this country can afford to wait an appreciable length of time before launching definite activity. . . . Since the United States is far ahead of any country in both airplanes and sea power, and since others are abreast of the United States in rocket applications, we can expect strong competition in the latter field as being the quickest shortcut for challenging this country's position.[24]

The country, confident and prosperous under Eisenhower, was

nevertheless prepared to wait. The interservice committee on guided missiles said in 1948 that 'neither the navy nor the USAF has yet established either a military or scientific utility commensurate with the presently expected cost of a satellite vehicle.'[25] By 1949 Rand was reduced to pushing for a satellite as 'an instrument of political strategy' whereby it was hypothesised that information obtained by satellite observation would, when conveyed to the 'ruling hierarchy' in the Kremlin, cause 'suspicion, disruption and possible purges'.[26] (Lest this seem too far-fetched it has been well pointed out that not dissimilar results occurred with *the impact that Sputniks I and II had upon the "ruling hierarchy" in the United States in 1957*. (italics in original)[27] But, of course, in the latter case, the ruling hierarchy was confirmed in its ideology. The possibility of similar confirmations occurring among communists has always proved difficult for American opinion to grasp.)

The satellite programme was moribund until the geophysicists began calling for such devices in 1954. However, the international attention this interest involved did nothing to assuage the rivalry of the services. Now, in addition to the naval and USAF schemes, the army also had a contender in the form of an adapted Jupiter C (for Composite) vehicle. Von Braun prepared a report on 'Project Orbiter', 'A Minimum Satellite Vehicle Based on Components Available from Missile Developments of the Army Ordnance Corp', i.e. a ready-made system. In January of 1955 the Department of Defence set up an eight-man civilian committee (two chosen by each of the three services and two by the DoD itself) and instructed it to decide among these proposals by August of 1955 – which it did, four days after Eisenhower publicly announced that the US would orbit a satellite for the IGY.

The committee told the air force to concentrate on Atlas as an ICBM and not get diverted by satellites, and, bowing to naval scorn which described von Braun's Jupiter C as 'cluster's last stand', opted for the navy's own **Vanguard** proposal. Vanguard consisted of a stretched Viking, an **Aerobee** second stage and a solid-fuel device with a 40 lb satellite on top. Von Braun was promising only a 15 lb payload and his design was certainly lashed up and less elegant than Vanguard. Further, the Jupiter was a more closely guarded military secret than the Viking, less suitable for a 'peaceful' application. Yet it is likely that the general hostility of the aeronautics industry to the army's tradition of in-house weapon development was among the most significant factors underlying the DoD decision. The naval scheme emerged out of the heart of the military-industrial complex, with all the usual procurement paraphernalia, whereas the army's came from its émigré German labour force sequestered in a wholly owned and operated factory in Alabama.[28]

The sequel is as dramatic an illustration of the limitations of technological determinism, even under conditions of extreme international ideological competition, as these pages afford. Vanguard was badly designed and the corrections took until the spring of 1956 to effect. Nevertheless the secretary of defence, Charles Wilson, ordered that 'the Redstone and Jupiter missiles will *not* be used to launch a satellite'. (italics in original)[29] Moreover, the secretary said, the army group was to abandon ICBM research altogether and concentrate on devices with no more than a 200-mile range. Thus there would be no wasteful duplication of effort. If no rockets had existed, this would have been a rational proceeding; as it was, the order ignored the fact that the army, whatever the plan called for, was closer to an ICBM than either of its rivals, and was therefore closer also to a satellite launcher.

In response, when it became apparent that the navy had nowhere to test Vanguard, the army refused to interrupt the enforced conclusion of its Jupiter programme on the Cape. A whole new navy complex had to be built at Canaveral and radio tracking stations constructed across the Caribbean and down to Santiago, Chile. At the time Sputnik orbited, Vanguard had yet to be tested. On 6 December 1957 Vanguard failed to leave its pad. 'Oh, what a FLOPNIK', headlined the London *Daily Herald*.[30]

The irony is that von Braun's rocket had been successfully tested 15 months earlier, travelling a distance of 3335 miles and reaching an altitude of 682 miles. The suppression of the Jupiter C's radical potential was carried to lengths that bordered on the bizarre. Prior to the first launch in September 1956 the Pentagon became so concerned that von Braun might simply attach a satellite without authorisation that it ordered General Medaris, the ranking officer on site at Cape Canaveral, personally to make sure the nose cone was loaded with sand, satellite-less. Subsequently Pentagon officials inspected Huntsville, checking there were no live (i.e satellite-ful) rocket stages stored.[31] It was apparently the very success of the German outsiders which prompted Wilson's order forbidding further work, thereby preserving the integrity of the decision privileging the navy. (One cannot but suppose that Senator McCarthy, who died in May of 1957, would have had a ready explanation for these events – they must have been the work of communists in the Department of Defence.)

The government attempted to shrug off Sputnik I – 'a neat scientific trick' said Wilson, by now the ex-secretary; 'a silly bauble in the sky', announced one of Eisenhower's aides and the president himself professed that Sputnik 'does not raise my apprehension one iota'.[32] The administration had not, after all, reacted publicly to the announcement of a Soviet ICBM by Krushchev in August, 1957 and Sputnik I, a lot smaller than an H-Bomb, was only a radio transmitter which did nothing more than announce its own presence.

Sputnik II, launched on 3 November 1957 was, militarily, quite another matter. It weighed 1120 lb and was complex enough to carry a life-support system for the dog Laika. Clearly it could have been hoisted aloft only by a very large rocket indeed. It is possible that Korolyov altered the order of the Sputniks, moving the Laika flight up from third to second, skipping the launch of a smaller, more scientific package, specifically to make this understandable to the Americans.[33] Ex-Wehrmacht General Walter Dornberger, von Braun's old V-2 boss now at the Bell Aircraft Company in Buffalo, certainly took the point. 'There can be absolutely no doubt now that the USSR has the means for sending an atomic or hydrogen warhead anywhere in the world.'[34]

On the day Sputnik I was orbited, the new secretary of defence, Neil McElroy, happened to be visiting Huntsville. Von Braun told him that he had two Jupiter Cs stashed away and could orbit a payload within 60 days. His commanding officer, Medaris, suggested 90 days might be more feasible. The go-ahead came shortly after Sputnik II when the Eisenhower administration finally comprehended both the scope of the propaganda defeat it had suffered and the reality of the Soviet ICBM threat. In the event it took von Braun 85 days to transform a Jupiter C into the **Juno One Space Carrier** by the single addition of the solid fuel rocket to which the payload, **Explorer I**, was attached. Explorer I was launched 31 January 1958, wiping out the ignominy of the Vanguard fiasco. With Juno, von Braun achieved his lifelong ambition to build a space craft; and with Explorer, Professor van Allen's hopes for the IGY were also dramatically fulfilled because the satellite revealed the belts of trapped radiation surrounding the earth which now bear his name.

Beyond the short-term response of unleashing von Braun and the Juno One, the more sustained American reaction to Sputnik II was a 10-fold increase in expenditure on space activities in five years. Five years after that, by 1966, the peak year, NASA – which had been established in 1958 to break through military rivalries – was costing nearly as much as the whole Department of Defence. The UK was losing 21.3 per cent of its science postgraduates to the US in 'the brain drain', and the heavens had gained 522 man-made objects – among them a number of devices dedicated to telecommunications.[35]

The development of the launching rocket constitutes the crucial element in the ground of scientific competence for the communications satellite for, without it, obviously, there could be no technological performance in this area. But we have seen how the technology of the satellite, a spin-off of missile research without a clearly defined military purpose of its own, was delayed (as we would expect) because of the operation of the 'law' of the suppression of radical potential. The ambiguities of the long-range rocket as both satellite launcher and warhead delivery system contributed to this suppression as did the particular configurations of weapon production modes and interservice rivalries in the United States, not to mention Russian strategic concerns and national pride in the 1950s. Nevertheless and for whatever reason, as with television and computers, satellites were possible a full decade or more before they were *invented*:

> [B]ecause of the German V-2, the implications of the development of control theory and practice and the development of powerful radiotelemetry – it became clear after the war (in 1946) that, with the same *Aggregat* and a perfectly plausible second stage, Peenemunde could have orbited a payload as early as 1942.[36]

Clearly then, von Braun and Grottrup could have done the same thing for the Americans and the Russians in 1946 had there been a perceived supervening necessity for a communications or other satellite; and by 1945 ideas for a whole range of artificial space bodies including a geostationary communications device were, like the rockets to launch them, already at hand.

The first transformation: ideation

Herman Oberth invents a COMMUNICATIONS SATELLITE, 1923

In Oberth's *The Rocket into Planetary Space*, Arthur C. Clarke acknowledges there to be the first reference to a communications satellite, one relying on mirrors reflecting sunlight. Clarke suggests, though, that his outline of a chain of three manned geostationary radio satellites, published in the October 1945 edition of *Wireless World*, is the first proper articulation of the idea.

> In any event [he wrote in 1962] (until *Pravda* trumps me) I think I can claim priority for the first detailed, specific technical exposition of the global comsat (sic) system, with particular reference to synchronous orbits.[37]

Pravda would almost certainly let the matter stand with perhaps a *sotto voce* plea for Tsiolkovskii's proposal for an orbiting launch pad since, in 1956, the Russian journal *Radio* duplicated Clarke's proposal without acknowledgment or, significantly, claim of priority.[38] Before Oberth, Tsiolkovskii and Clarke, in an *Atlantic Monthly* short story, in 1869, Everett Hale dreamed up a brick-built navigation satellite for determining longitude.[39]

Clarke had been the treasurer of the British equivalent of GERD and VfR, the British Interplanetary Society. Before the war, the BIS mustered about 100 members of whom the truly committed numbered about 10. They met weekly in pubs, cafes and each other's flats and by 1938 had produced and published plans for a three-man lunar rocket. Work in the metal was less grandiose, not least because the 1875 Explosives Act rendered the establishment of even a preliminary *Rakatenflugplatz* in London totally illegal. The group continued to meet during the war, and on one occasion had just heard a speaker (blinded by his own side's smokescreen) dismiss tales of large German rockets as pure Nazi propaganda when the building they were in was shaken by a V-2 explosion. After the war, the society, with Clarke in the chair, no doubt suppressing only with difficulty its memory of the 9277 British V-2 dead and wounded, moved to confer honorary membership on Wernher von Braun as soon as his address was discovered. It also captured G.B. Shaw for its rolls and engaged in public battles with Professor Tolkien and Dr C.S. Lewis. But by this time it had made its great contribution to the conquest of space – in the form of Clarke's ideation of the communications satellite.[40]

Although it is possible, by suitable choice of frequencies, and routes, to provide telephony circuits between any two points or regions of the Earth for a large part of the time, long-distance communication is greatly hampered by the peculiarities of the ionosphere, and there are even occasions when it may be impossible. A true broadcast service, giving constant field strength at all times over the whole globe would be invaluable, not to say indispensable, in a world society.

Clarke, in the *Wireless World* version of his idea, outlined in general the possibilities of rocket developments and the various periods produced by different orbits. He went on:

It will be observed that one orbit, with a radius of 42,000 km, has a period of exactly twenty-four hours. A body in such an orbit, if its place coincided with that of the Earth's equator, would revolve with the Earth and would thus be stationary above the same spot on the planet. It would remain fixed in the sky of a whole hemisphere and unlike all other heavenly bodies would neither rise nor set. . . .

Let us now suppose that . . . a station were built in this orbit. It could be provided with receiving and transmitting equipment . . . and could act as a repeater to relay transmissions between any two points on the hemisphere beneath, using any frequency that will penetrate the ionosphere. . . .

A single station could only provide coverage to half the globe, and for a world service three would be required, though more could be readily utilized.[41]

Clarke thought his space stations would have to be manned and serviced by shuttle but, in the event, developments in electronics allowed for smaller unmanned devices to perform exactly the function he outlined. However, the elegance of his case was not overwhelming. He had proposed what was to be known as a high-orbit system, the one presently in use in the West for the commercial telecommunications network. There are other possibilities involving lower non-synchronous orbits and in the years after Sputnik the advantages of Clarke's proposal, simpler ground stations and fewer satellites, were not conclusively established.

The satellites for the high-orbit solution would have to be lifted 22,375 miles and they would therefore be smaller than low-orbiting types, given the payload capacities of the first generation of space rockets. It was possible that reaching the synchronous orbit would reduce the weight of the satellite by 90 per cent and if they were smaller, they would be able to do less, carry fewer signals. They would need complex engine and gyro systems to get them to and keep them on their stations in the sky. Such machinery would shorten their working life and increase the possibility of failure, so the advantage of only three satellites could become a liability since a malfunction would remove a third of the net. Finally, bouncing two-way voice messages through 44,750 miles of space and back again would involve 540 milliseconds' delay which might prove unacceptable to users.

In 1954, no less a scientist than Bell Labs' director of Research Communications, J.R. Pierce, made an alternative and seemingly more viable proposal. Pierce, who had coined the name transistor and wrote science fiction under a pseudonym, discussed the technicalities involved in a communications satellite network and suggested the low-orbit option was the most practical.[42] The first details of the scheme were published in 1955 and in the years that followed Pierce's group at Bell Labs developed the suggestion until it became AT&T's favoured space segment plan. Broadband capacity for television as well as telephony, facsimile and data could be handled by a 180 lb satellite which in turn could be put into an orbit with apogee of 3450 miles and a perigee of 575 by the Delta configuration of air force's equivalent to Jupiter, the Thor missile. Such an orbit could give good visibility between Western Europe and the North-Eastern United States if it were inclined at 45° to the Equator.[43]

Against these ever-hardening realities and the never-diminishing power of the phone company, Clarke could, in the late 1950s, offer little; although Hughes, an aeronautics firm, were becoming interested in the practicalities of his scheme perhaps as a way of keeping the telephone company out of space. But the rockets to lift a payload comparable to the one proposed by AT&T to synchronous orbit did not apparently exist and nor did the satellite itself. The high-orbit idea, the low-orbit alternative, and a number of split-the-difference variations were all canvassed in the US at this time. The Russians too were obviously concerned with satellite possibilities for communication and observation from 1954 at the latest. They had been made aware of Hermann Oberth's world-circling mirror idea in 1926 as a device with which one can 'observe and photograph inaccessible countries'.[44] Oberth himself, now 66 years old, was brought from Germany to Huntsville by von Braun in 1955 and still considered the mirror scheme reasonable enough to spend some of his time in the Research Projects Office refining the calculations for its orbit.[45] The possible uses of space for a whole range of tasks (including destruction) were well established before the first intercontinental rockets flew, the idea of the geostationary communications satellite among them.

Phase two: technological performance – prototypes

The US navy commandeers the moon as a COMMUNICATIONS SATELLITE, 1954

For communication purposes a satellite is any heavenly object off which radio signals can be bounced. Established long-range electronic signalling had used the ionosphere, which is impermeable by radio waves, as a sounding board since 1901. Indeed, the discovery of the ionosphere itself was a result of successful wireless telegraphy experiments beginning with Marconi's transatlantic transmission of that year.

Marconi had articulated a law for distance in radio: 'I find that, with parity of other conditions, vertical wires 20 feet long are sufficient for communicating one mile, 40 feet four miles, 80 feet sixteen miles, and so on.'[46] This insight was the heart of Marconi's *invention*, for radio waves had been demonstrated by Hertz and Lodge to prove the correctness of Maxwell's electromagnetic theory years before. In 1879 D.E. Hughes, of the microphone, had shown a radio, a **coherer** or primitive receiver, to Stokes, the president of the Royal Society, who 'rather pooh-poohed all results' and told him that his experiments revealed induction, which was well understood, rather than radiation. (Nevertheless Hughes was made an FRS that same year.)[47] Marconi used a coherer to detect his radio signals which was a development of one designed by his professor at Bologna, Rhighi, and took up a suggestion made by Loomis in 1872 that aerials attached to kites would probably allow for transoceanic signalling.[48] Marconi's main contribution in 1895 was, then, not the *invention* of radio but rather the discovery that Loomis's hunch was right: the taller the transmitting mast, the further the signal radiated. His Russian competitor Popov independently produced a coherer of identical design later that year

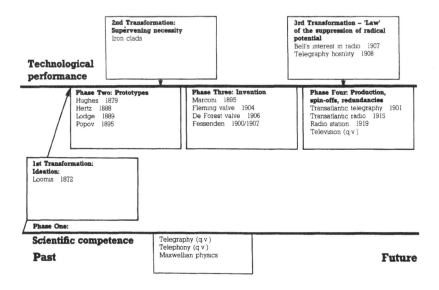

Figure 8 Radio

but used it for detecting the natural radio waves produced by distant thunder storms. (Soviet claims for Popov are more tendentious than those made on behalf of, say, Rozing.)[49] The signal sent to Marconi in Poldhu, Cornwall, from St Johns, Newfoundland, on 12 December 1901 obeyed his law, since a distance of 1800 miles was achieved with kite-born aerials at around 1000 feet; but, until he actually did it, there was some considerable doubt as to the possibility. In the late nineteenth century, faced with the knowledge that radio waves travelled in straight lines like light, scientific opinion was uncertain as to their efficacy for signalling along the curve of the earth.[50]

(The 'law' of the suppression of radical potential can be noted in the sequel to Marconi's transmission. The Anglo-American Telegraph Company held the telegraphic monopoly from the government of Newfoundland and exercised its rights in having the Marconi people thrown off the island. The government of Canada found another haven for the experimenters.)

To save the Maxwellian paradigm, Heaviside and, independently, A.E. Kennelly revived a suggestion made in 1882 to explain changes in the earth's magnetic field and hypothesised the existence of a spherical shell of ionised air surrounding the earth which was impermeable by radio waves. The ionosphere would allow both Marconi's distance law and the straight line hypothesis to coexist. The ionosphere was not experimentally verified until 1925, but it was utilised from 1901. In 1918 the US army transmitted a signal to France and the Marconi Company bounced one around the world from its station in Caernarvon to Sydney. Commercial transatlantic wireless telegraphy was introduced in 1920. Radio telephony developed in parallel, Bell making experimental transmissions to Honolulu and Paris in 1915 and a public service being established between AT&T and the GPO in 1927. By the early 1930s the world radio-telephone network was in place. Short-wave bands were also specifically allocated for broadcasting overseas and most 'developed' nations established such services during this decade.

The extensive use of ionospheric waves was both boon and bane to Clarke's communication satellites. The boon was that the ionosphere is an extremely volatile sounding board, subject to atmospheric and celestial interference, varying with day, night and the seasons, and virtually useless during periods of sun-spot activity. There was a need for something better. The bane was that the ionosphere presumably invalidated the scheme because communication between earth and satellite would be impossible; the waves could not get through it. Clarke dismissed this, stating 'with certainty' that 'shorter wave lengths are not reflected back to the Earth' and he was right.[51] There is a radio window for waves between 8mm and 20m which was imperfectly understood in 1945 but was subsequently demonstrated and used.

Project Diana, 1946, offered concrete evidence of the viability of Clarke's concept. In this experiment the US Army Signal Corps bounced a radar signal (i.e. very short-wave) off the moon. By 1954 the navy had developed a Communication by Moon Relay (CMR) system which it used to link Washington to Honolulu and which was in operation between 1959 and 1964. In 1959 Bell Labs used the moon as a sounding board to create a voice circuit between Holmdel, New Jersey and Pasadena.[52] In 1961 Jodrell Bank used it to talk to Australia. But all celestial bodies, even the moon, are too far away for viable voice links because of the time lapse involved in sending and returning the signal across the vastness of near space. A closer object would avoid this problem.

The simplest solution was Oberth's mirror. NASA launched a 100 foot metallised balloon, designated **Echo I**, on 12 August 1960, using a Delta rocket. This object is the first man-made prototype in satellite technology. At an altitude of about 1000 miles it was automatically inflated and was used, on its first pass, to transmit a recording of President Eisenhower between Holmdel and an experimental site in the Mojave Desert. In 1962 Echo I reflected a poor-quality television image between California and Massachusetts.[53] Although simple in principle Echo required extremely complex amplifying and detection equipment on the ground. MASERS (microwave amplifiers, see chapter 2 above) were used and an enormous hearing tube 20 feet across was built by Bell to spot the faint signals being bounced back.[54] Echo II, a more rigid version, was orbited at 920 miles in July 1962. When this series of mirrored balloon experiments was concluded in 1963, the Bell Lab horn, in effect the prototype ground station, was assigned to radio astronomy work. The most valuable balloon so far has been **Pageos** (*Pa*ssive *Geo*detic *S*atellite). It was orbited at 2600 miles in 1966 and was the basis, four years later, of an extremely complex survey to correct the world's map.[55] Echo I re-entered the atmosphere and burned up in 1968.

Over Christmas 1958, previous to the Echo series, the DoD had orbited atop an Atlas rocket a smaller metal device, **Score** (*S*ignal *Co*mmunications by *O*rbiting *R*elay *E*quipment) – built by RCA. Rather like Sputnik, it could only transmit, in its case another recorded Eisenhower message which played on command from the ground. Score worked for 12 days.

It was obvious that active repeater satellites would reduce the need for enormous and elaborate receiving stations because the device would do some of the amplification itself. However the emerging understanding of the complex radiation fields of the earth raised some questions as to the viability of electronic equipment exposed for long periods in such an atmosphere. In consequence, the passive experiments continued beyond the launchings of the first repeater satellites.

Courier I-B was an Army Signal Corps' satellite which could cope with twenty dash-dot teleprinter signals at once, but not in real time. It stored the signals it received and played them back on subsequent passes. Orbited on 4 October 1960, it weighed 500 lb and was effectively powered by its 20,000 solar cells for 18 days.

All these experiments revealed the viability of space-side communications and the United States began actively to evolve policy for institutionalising the new system. Previous experience, as ever, was invoked. As far as telecommunications policies were concerned, the relevant particulars were encapsulated in a statement made just after the war by Truman's Communications Policy Board:

> The U.S. almost alone among the nations of the world relies on privately owned telecommunications companies and it should be the policy of the U.S. Government to encourage the health of these common carriers as a vital national asset.[56]

In the spring of 1959, the American tradition of state-aided entrepreneurship in telecommunications was confirmed for the new technologies of space. A House Committee felt that enough was known to mandate the immediate creation of a 'useful worldwide communication system based on the use of satellites' and that although nobody 'was prepared to envisage a point in time when Government assistance in the form of providing launching vehicles . . . and actual launching operations would not be required', nevertheless the aim should be the 'complete commercial operation of the system'.[57]

It was on this basis that AT&T took the proposal Pierce had been working on for six years to the FCC, formally requesting permission for a system with 50 low-altitude satellites in polar orbit and 26 ground stations to be co-owned between itself and the foreign telecommunications organisations with which it currently shared transoceanic cables and radio links. The petition caused an immediate bureaucratic problem since, if the FCC agreed to the plan, it would in effect be also suggesting to NASA that it launch the Bell devices. A compromise was formally arrived at whereby NASA continued with its passive experiments but agreed, if the FCC (among others) had given permission, to investigate low-level systems as long as no military or industrial effort was duplicated and, in the words of NASA's first administrator, Keith Glennan, 'only so long as is necessary to assure that timely development of a commercially feasible communications system will be completed by private industry.'[58]

Thus, as a NASA memorandum put it, 'the traditional policy of conducting international communications services through private enterprise subject to government regulation' was facilitated.[59] Yet, beyond the self-evident contradiction of NASA's central role, the whole issue was also part and parcel of an extremely intense international competition in space.

In 1961, Kennedy had asked Johnson, following the first Russian manned space flight (and the Bay of Pigs incident the same month) whether there was 'any other space program which promises dramatic results in (sic) which we could win'; and Johnson, in preparing the response which became the message 'On Urgent National Needs' (the 'man to the moon' announcement) sent to the Congress in May, included the development of a space communications system. The unstated problem was that such a system like the telephone, was seen as a species of 'natural monopoly', and, if private, a 'natural' sphere of expansion for the telephone company and the other American common carriers.

The Kennedy administration moved immediately to prevent this 'natural' development. It cancelled the proposal that NASA seek $10 million of private R&D for communications research, a Republican suggestion which in effect was nothing but an order to

NASA to have AT&T do the work. Instead, despite affirmations of the role of private enterprise, the research effort was to be funded by government. There was to be no repetition of the design stage competition that had allowed the Soviets to steal a march with Sputnik, and even AT&T's labours were not to go to waste.[60] On 27 July 1961 NASA agreed to launch the Bell-financed satellite designated **Telstar** which had been in planning since the mid-1950s, and a day later it contracted RCA, the maker of Score, to build it another almost identical device but with a deeper projected orbit; and four weeks after that it signed an agreement with Hughes Aircraft to provide an experimental synchronous device. Given that NASA itself was in the middle of an extended experiment with passive devices, this meant that all the possibilities were covered. Clarke's idea was to have its day in court. AT&T, perhaps because of the sudden welter of competition, paid NASA's entire launch bill five months before the event. It was on notice that it was not automatically to extend its 'monopoly' into space.

AT&T was first in the metal. Telstar, containing 2528 transistors (and not one IC), was nearly 3 feet in diameter, weighed 170 lb and was powered by 3600 solar cells. It was launched into elliptical path, its perigee 593 miles and its apogee 3503 miles, on 10 July 1962. It took less than 2¾ hours to circle the globe. Beside demonstrating the essential superiority of the Bell proposal, the satellite was designed to monitor Van Allen radiation. This last was increased that same month 100-fold by another in a series of detonations in the outer atmosphere, by the USA, of nuclear devices. Three had been burst in 1958 and now a 1.5 megaton bomb, poetically designated **Starfish**, was exploded 250 miles up. (Irresponsible space experiments were in vogue. In one, a cheap alternative to the entire telecommunications satellite programme was sought. Operation West Ford was designed to create a sort of Saturnian ring with 350 million tiny copper needles off which radio signals could be bounced without fear of celestial interference or terrestrial jamming. On reaching orbit the canister with the needles failed to open and the experiment was abandoned, to the profound relief of the world's astronomers, among others.)

Telstar ceased to operate on its 1242nd orbit in November. It was restored, through clever manipulation of its radio command signals, briefly in February 1963, but then died. It had worked and in the words of the Bell report written at the time:

> Design of a second generation Telstar satellite could be approached by Bell Telephone Laboratories with confidence as an engineering project. Where uncertainties exist, they have to do with the conditions existing in space.[61]

With Bell, their traditional partner, involved, the British and French Post Offices had constructed ground stations as their part of the Telstar experiment. The British opted for a parabolic dish which was erected in Cornwall not far from Poldhu on Goonhilly Down. The French had built a copy of the Bell design at Pleumeur-Bodou in Brittany. The American ground station was erected by AT&T within a natural bowl at Andover, Maine. The design was similar to that used for the Echo receiver but the listening aperture was increased from 400 square feet to 3600 square feet. The horn, which was covered with an air-inflated radome to protect it from Maine's annual 90 inch snowfall, was 90 feet long and 68 feet in diameter at the mouth.

Henry Thoreau, in 1854, wrote: 'We are in great haste to construct a magnetic telegraph from Maine to Texas; but Maine and Texas, it may be, have nothing to communicate.'[62] Echoing the thought, Ed Murrow, in contemplating the communications satellite, said 'The issue, gentlemen, is not how we deliver it but what our delivery has to say.'[63] Neither Thoreau nor Morrow would have been much comforted by Telstar. The Vatican, with more hope than truth, called the machine 'a little star that harms nobody and has the virtues of union and understanding'; but neither virtue was much in evidence.

Telstar was a shot fired in a species of technological war in which Russia was, though the most obvious official enemy, the least important practically. The real function of the bird was to allow AT&T to defeat its American rivals and bring the free world's cable system to its knees. The communication satellite was the only available means of transoceanic live television, and much was made of this in publicising the experiment, but TV was as tangential to the basic issue as Soviet competition.

Satellites were one among four systems available for general intercontinental telecommunication. Radio telephony had been, as described above, in use since the late 1920s but was of poor quality and extremely unreliable. The other two methods were not so disadvantaged.

In 1849, Dr O'Shaughnessy (after, Sir William O'Shaughnessy Brooke) laid a bare iron rod beneath the waters of the Huldee in India and sent a telegraphic signal from bank to bank, a distance of 4200 feet.[64] To get from the Huldee to the 1600 miles that separate Newfoundland from Ireland was a task which took sixteen years. In 1854 a New York, Newfoundland and London Telegraph (subsequently Atlantic Telegraph) Company was formed by a retired American businessperson, Cyrus Field. By 1856 the company's line had crossed the Gulf of St Lawrence and reached land's end in Newfoundland, 1000 miles at the cost of about $1000 a mile.[65] The main task of crossing the ocean was put in hand and in August of 1858 the two continents were joined. Queen and president exchanged telegrams – 'It is a triumph more glorious, because far more useful to mankind, than was ever won by conqueror on the field of battle', said Buchanan.[66] Then within the month the cable was dead, leaking.

Slowly capital was reassembled. It was now that Willoughby Smith, who was to establish the importance of selenium to television while laying this very cable, became involved. He was the chief electrician of the Gutta-Percha Company, makers of the best available natural latex insulating material. With an improved cable, Brunel's enormous and previously ill-fated *Great Eastern*, the world's biggest ship, was commandeered. Finally in July 1866 a permanent link was made between Heart's Content on the western shore and Valentia in Ireland.

> Gooch, Heart's Content to Glass, Valentia, 27 July, 6 p.m.: Our shore end has just been laid and a most perfect cable, under God's blessing, has completed telegraphic communication between England and the Continent of America.[67]

In the words of Cyrus Field's son, 'The heart of the world beats under the sea'.[68]

By the end of the 1920s there were 21 transatlantic cables and 3500 under other waters. From 1879 to 1970 the cost dropped from $100 per message between Great Britain and the USA to a maximum of 25 cents a word. In 1957, the year of Sputnik, the British Cable and Wireless Company carried 491 million words on its worldwide system. But despite the size of this traffic, neither the telegraph cables nor the ionospheric radio telephony net duplicated the ease and efficiencies of a national telephone system. The third alternative to the satellites did.

The super-powers might have been in the heat of a space race, but the world's telephone engineers were using post-war electronic developments to solve the problem of ionospheric communications in a more prosaic way. With MASERS it was now possible to build voice-carrying cables using repeaters, amplifiers, of such durability and sensitivity that they could be placed under the oceans. AT&T had established itself as the leading American entity in international voice communications by its 1927 radio telephony agreement with Britain. It joined the GPO and the Canadian government in 1952 to plan a telephone link across the Atlantic. Ninety years after the first Atlantic telegraph cable, 27 years after the inauguration of the Atlantic radio telephone, at a cost of £16 million, with 51 repeaters (built in the US and the UK) and a capacity of 36 channels, the wire was opened for business in September 1956. Its success fuelled other plans – a second cable from the US to France, a third from the UK to Canada, the first leg in the upgrading to voice of a worldwide Commonwealth system first established for the Empire in the nineteenth century. Clearly, had the space race not occurred, the late 1950s and early 1960s would have seen a cat's cradle of transoceanic voice cables being laid. AT&T was hedging its bets and was big enough both to press for its own satellite system (at a development cost of some $50 million) and also to commit $178 million to international telephone cable development in the same year.[69]

As for the space side of the bet, even the greater expense of the low-orbit system with its proliferation of satellites and complex tracking ground stations was to AT&T's advantage. American telephone company profits are pegged to investments and therefore the more expensive Telstar turned out to be, the better.

In the West, the satellites were, and would remain for the foreseeable future, an American technological preserve; but also for the Europeans, although cables were the more accessible technology, the potential ground station business, especially the complex one created by the low-orbit satellites, was attractive. Hence the fuss when the GPO ground station at Goonhilly picked up only a poor indecipherable image on the first Telstar television transmission, while the French at Pleumeur-Bodeau received it perfectly. A triumph was announced for French technology (for all that it was in this instance identical to American). Amid much publicity, a great leap forward was claimed by French salesmen pushing their version of the Bell ground re-ceiving system against the British variant. The business of ground stations was estimated to be worth $200 million over a couple of decades.

In the meanwhile, the Americans put on a show. The first Telstar image, which Goonhilly missed, was of the Andover randome, a US flag in the foreground with the 'Star-Spangled Banner' over. (It should not be forgotten that AT&T quit show business in 1926.) The British replied in kind with pictures of the Goonhilly control room, without flags and anthems, and the French transmitted eight minutes of variety on videotape. Telstar, with its rapid orbit period, was visible to both Atlantic shores for less than half-an-hour at a time so the era of specially created live international television began with brief shows. The programme transmitted on 23 July 1962, the first of what was to be a shortlived genre, was like nothing so much as that screened by the brothers Lumière in the Grand Café on the Boulevard des Capucines, 28 September 1895. 200 million people watched a series of animated postcards. The delivery system might have been a technological wonder, but the live image of a native American turned out to look just like his filmed image and the audience promptly returned to their national television fare. Thoreau, contemplating the first international communications link 108 years before, had been prescient:

We are eager to tunnel under the Atlantic and bring the old world some weeks nearer to the new; but perchance the first news that will leak through into the broad flapping American ear will be that Princess Adelaide has the whooping cough.[70]

However limited the significance of these cultural events, Bell, in cahoots with the television industries of two continents, had made Telstar a household word.

Relay I, the satellite built for NASA by RCA, was never to achieve such fame but, unlike Telstar, it outlived its planned life and, because of its higher orbit and longer passes, was responsible for the more serious exploitation of the new technology. International television saw the first live transmissions of actual events, i.e. occasions not specially and solely produced for this purpose, e.g. Sir Winston Churchill's acceptance of honorary US citizenship, the funeral of Pope John XXIII and the coronation of his successor, the launch of the **Mercury** spacecraft designated **Faith 7**.

Relay had been placed in a deeper – more 'medium' – orbit, its perigee 227 miles and its apogee 1109 miles more than Telstar's. It was just under 3 feet long, weighed 172 lb, with a twelve telephone/one TV channel capacity. Relay's effectiveness, and the rapidly increasing power of the booster rockets available, forced Pierce and Bell Labs away from the low-orbit random solution. (In the event, because it was thought to offer greater security, only the US military were to build a low-orbit system. Launched in batches of six to eight satellites at a time, **IDSCP** – *I*nitial *D*efence *S*atellite *C*ommunication *P*rogramme – was completed by the summer of 1967 and was used during the Vietnam war.) AT&T's $50 million investment included another Telstar which NASA launched on 7 May 1963 into an elliptical path with an apogee of 6717 miles.[71] Relay II was orbited on 21 January 1964 and was visible over the Atlantic for as long as 70 minutes. It was the first communications satellite experiment to be received outside of North America and Europe, in Brazil on a smaller 30 foot dish built by ITT.[72]

The most extensive exploitation of the medium-orbit system has been by the Russians. Given their geographical situation and the contiguity of their primary political sphere of influence, they did not need the worldwide capabilities of either low or high systems. In April 1965 **Molniya (lighting) IA** satellite was launched into an orbit which came as close as 375 miles to the earth but, over the USSR, was at 24,780 miles for eight of every twelve hours, thereby reducing the number of satellites required to establish a constant capability. In May television transmission was demonstrated with a film from Vladivostok about fishermen in the Sea of Japan. A second Molniya was launched in October. Over the next two years some 25 ground stations were built and on 3 October 1967 the launch of Molniya IF, the sixth of the series, inaugurated the entire system designated _ **Orbita**.[73] The Russians were to claim that Orbita had greater capacity than the geostationary system eventually *invented* for the West and there is little reason, given the fundamental point that lower orbits permit greater payloads, to question this assertion.

Russia had been behind Europe and America in building long-distance links for telephone and television use. With Orbita she skipped this stage of development altogether and there is perhaps no better example of the satellite's centralising tendency than the fact that only two of the first 25 stations could transmit. Orbita, without microwave or cable competition, is the mainstay of Soviet media distribution, allowing not only for the creation of a more homogeneous televisual culture across the internal empire but also enabling the simultaneous editions of *Pravda* and other publications to be produced. In addition, the system allows for broadcasting to the rest of Eastern Europe, a necessary bolstering of the electronic curtain. This alternative to the West was required since, for instance, Poland took part in the Telstar experiments.[74]

These then are the direct prototypes of intercontinental geo-stationary satellite communications systems, but in addition there are a variety of other uses to which the technology that created space communications can be put and these too should be considered among the prototypes leading to the technological performance of Clarke's idea.

Abandoning plans for a 2000 mph spyplane, the USA instead launched **Tiros I**, *T*elevision *I*nfra *R*ed *O*bservation *S*atellite and **Midas I**, *Mi*ssile *D*efence *A*larm *S*ystem in 1960. (**Midas 4** orbited the canister with the 350 million copper needles.) Techniques were developed to recover packets dropped by satellites, the first being from **Discovery 13** and **14** and almost certainly including photographic film. A year later **Samos**, *S*atellite *a*nd *M*issile *O*bservation *S*ystem, was orbited and a year after that, in **Kosmos 4** at 7000 lb the Soviet response was on station.[75] (Rockets have been associated with surveillance techniques since 1904 when Alfred Maul attached a camera to one. By 1912, Maul's camera rose to 2600 feet, was gyroscopically stabilised and took a full (8 × 10 inch) plate, but it was also redundant. Kodak's first camera specifically designed for use from an aeroplane was marketed in 1915.[76]) In January 1972 the US orbited **Big Bird**, an 11-ton satellite over 50 feet long. Such a device could contain a camera with a 5 foot aperture and a focal length of 20 feet which was theoretically capable, if the weather stayed out of its way, of a 330 lines/mm resolution (cf the cinema's 35mm with its 30 lines/mm), rendering visible any object bigger than 6 inches, from 100 miles up.

Surveillance techniques have been applied to geological surveys of ever-increasing complexity and observations. The **ERTS** (*E*arth *R*esources *T*echnology *S*atellite) series begun in 1972 and, renamed **Landsat** by 1975, produces a picture of the entire globe every 18 days – images showing water, minerals, vegetation and the works of man which, with the 1000 shots taken by the **Gemini** and **Apollo** missions, are available commercially from the United States government.

Weather observation was part of Vanguard II's mission in 1959, the raison d'être of **Tiros I** and, from 1964, NASA's **Nimbus** series was also designed for this task. The navigation aid proposed by Hale in 1869 became a reality with **Transit IB** in 1960. The Russians have matched all of these functions, for instance launching their own meteorology series, the **Metoras**. By 1972 the generic Kosmos series anounted to 542 objects, including, perhaps, some carrying bombs and others capable of destroying fellow satellites in orbit.

The crucial determinant of all these developments was the rocket. Once that was to hand all the rest, including communications satellites, followed in short order. NASA had explored the low option and a variant. Now, while Congress took up the matter of who was finally to own the system, NASA awaited the *invention*, the most radical of the communications options, of a synchronous satellite that would perform the task Clarke had assigned it seventeen years earlier.

The second transformation: supervening necessity

BBC TV, by facsimile transmission, obtains film of Her Majesty's arrival in America, October 1957

Satellites are the least radical of the four central technologies here discussed. They are enormously capital intensive and difficult of access, two strong prophylactics against disruptive potential. But they are also the least radical in another sense. They do nothing that other technologies do not do; as delivery systems designed to accommodate pre-existent transmission modes, they are as much redundancies as they are spin-offs. They can, therefore, be classed as duplicates of transoceanic telephone wires or as spin-offs from Cold War necessities, built not out of need but simply because military equivalents were in place and there was no reason not to have civilian versions as well. Such a view could point, as is done below, to the chronic underutilisation of the international commercial satellite net as clear evidence that there was no meaningful need at all.

But it would not explain at least two factors which constitute valid supervening necessities. Although not sufficient of themselves to have produced the R&D the system required, these forces were certainly strong enough to cause international civilian communications satellites to be derived from their military parents. The two comparatively minor factors highlight the satellite's only non-redundant aspects – the need for the USA to match its post-war 'imperial', as it were, role with an 'imperial' communications system which it owned or dominated and the advantage accruing to the television industry in having instant pictures available transoceanically.

On 21 October 1957, seventeen days after Sputnik, the BBC screened film of the Queen's arrival in the USA at Staten Island. The 15-second shot had been sent across the Atlantic frame by frame, using existing press wire picture facilities. It had taken nine hours.

The elaborate proceeeding which gave the BBC its royal film is symptomatic of a form of disease among television journalists from which their print colleagues have increasingly suffered ever since the telegraph – the idea that news is somehow a perishable commodity, like tomatoes. It cannot and will not wait. This was not always the case. Before the telegraph, it was a valid journalistic scoop to correct ones' rivals with 'Late News Of The Peace of Westphalia' and the like. After the telegraph, late news became, for no very logical reason, no news at all. The rush to be first has infected print reporters, illustrators, photographers, radio broadcasters, cinema newsreel producers and now television personnel.

The need is, in fact, internal to the media, of a piece with the desire noted above on the part of sports producers to develop an instant replay device (see page 96). The logic of professionalism, bred over the century and a half since the telegraph, meant that news personnel, rather than their audience, were the ones addicted to getting the pictures ever faster. It would be difficult to quantify the advantages to national life of the 15 royal television seconds on the day of the arrival as opposed to the many minutes available the following day: but, as had been said before, a supervening necessity need not be widely grounded nor utilitarianly sound. The perception, the perception of an influential few, is enough.

The actual availability of instantaneous channels of communication has done very little for international understanding, except in times of war. The fact that the radio was an open-eared device did not create, unless in circumstances that were exceptional, a transnational audience. (Reith's objection to dance music on Sunday forcing the British public to tune into foreign stations a mite less dour would be an example of such exceptional circumstances.) National broadcasting services exercised, in almost all situations, a mesmerising effect on their national audiences, even where language was not a barrier. Despite the valiant efforts of the shortwave services, the BBC's motto, 'Nation Shall Speak Peace Unto Nation' is therefore (unless construed as a specific instruction to the English *in rem* their Celtic neighbours) almost devoid of meaning. The need for instantaneous television could not have been seriously considered as a general cultural one, a demand from the mass audience, at least not if the experience with radio was a guide.

The world television audience demanded American television programming, but that demand did not require instantaneous transmission – liners or stratocruisers were quite fast enough to bring *I Love*

Lucy half-hours to European and Latin American shores. Vice-President Johnson may have greeted Telstar as 'a dramatic opening of a new frontier for international telephone, international television, and most of all international understanding', but the shows it facilitated were a bust. The existence of the capability created no new programming and worked to speed up the transmission only of material – the Tokyo Olympics, President Kennedy's funeral – which would have been internationally distributed anyway.

Of course, no band of broadcasters, however addicted to instantaneity, could have secured the billions of dollars required by the space programme. But in the context of the technological phase of the Cold War that developed in the late 1950s and early 1960s, broadcasters played their own small part in seizing the moment. The moment was there to be seized not just because of the Space Race. In America it was realised that the traditional hegemony of Britain and, to a lesser extent, France over the international cable system, a hegemony the emerging transoceanic telephone wires did not threaten at all, could be overthrown with a satellite system dominated by the USA. The economic significance of British domination has been highlighted by Schiller.

> The decisive role played by the British worldwide communications network – both its control of the physical hardware of oceanic cables and its administration and business organisation of news and information – which held the colonial system together, promoted its advantages, and insulated it from external assault, had not escaped attention in the United States. It was against these finely spun, structural ties that an American offensive was mounted.
> Conveniently, the attack could avail itself of the virtuous language and praiseworthy objectives of 'free flow of information' and 'worldwide access to news'.[77]

America had been chafing at this particular bit for over a century.

From the dawn of the age of electronic communications, governments have believed that vital national interests are involved in the development and ownership of telecommunications systems. This belief exerts a tremendous influence on all aspects of telecommunications. It justifies limitations on programme importation, such as those behind which British television has flourished for nearly 40 years. It allows the American broadcasting and telephone industries to operate internally without any international competition. It governs the radio jamming policies of the Eastern Bloc. It accounts for the desire for autonomous systems exhibited by all new nations. The incorporeal nature of these ephemeral messages has not prevented them from attracting the most protectionist of policies and, in the case in hand, the American national telecommunications interest accounts for the existence of essentially parallel international satellite and transoceanic cable telephones.

One of the earliest debates on these issues was occasioned by the presentation to Congress, in January 1856, of a bill to charter, in concert with the British government, Cyrus Field's transatlantic telegraph. There were objections to the scheme.

> The real animus of the opposition was a fear of giving some advantage to Great Britain . . . there were those who felt that in this submarine cable England was literally crawling under the sea to get some advantage of the United States.[78]

In reply to Senator Hunter of Virginia who asked 'What security are we to have that in time of war we shall have the use of the telegraph?', William Seward, whose bill it was, could only pray that 'after the telegraphic wire is once laid, there will be no more war between the United States and Great Britain.' In Britain, the parallel bill caused Lord Redesdale to ask why, since the cable began in Ireland and terminated in Newfoundland, both colonies, the US had to be involved at all! Both bills, however, passed.[79]

Out of these arguments, the concept of vital national interest, by which states mould telecommunications, emerged. In 1875, while explaining to Congress why a French company had been denied permission to land a cable six years previously, President Grant enunciated what had rapidly become the established position: 'I was unwilling to yield to a foreign state the right to say that its grantees might land on our shores while it denied a similar right to our people to land on its shores.'[80]

This principle of co-ownership and reciprocity effectively updated the notion of inviolate borders for the age of international electronic communications. It has stood at the foundation of the entire edifice of national and international policies. Grant's price of admission to the French was that American telegraph companies be allowed to share in the French cable. With that genius for blabbergab that has never quit American life, this essentially restrictive practice became known as the 'open shores' policy. It means the exact opposite.

Seward's transatlantic cable bill not only provoked the beginnings of American policy towards international telecommunications, it also confirmed the 'tradition' of private ownership for such purposes. Senator Bayard of Delaware queried the appropriateness of a private company being involved:

> It is a mail operation. It is a Post-Office arrangement. It is for the transmission of intelligence, and that is what I understood to be the

function of the Post-Office Department. I hold it, therefore, to be as legitimately within the proper powers of Government, as the employing of a stage-coach, or a steam-car, or a ship, to transport the mails, either to foreign countries, or to different portions of our own country.[81]

Congress had already refused a decade before to acquire Morse's telegraphic patents because its domestic profitability was in doubt and, since there was no reason to believe the transatlantic wire would be any more of a money-maker, Bayard's was a view out of step with this crucial precedent.

In 1866, the Congress had passed the Post Roads Act which limited government involvement in the internal telecommunications system to 'aid in the construction of telegraph lines and to secure the government the use of same for postal, military and other purposes.'[82] Government aid to enable private profit, the 'tradition' informing Eisenhower and Kennedy communications satellite policy, was in place; but it had been questioned, especially in the years around the First World War. Serious populist political moves were made to nationalise telephony, including AT&T, and for a time during the war the company was in effect reduced to being an agency of the Post Office.[83] In 1919 the navy nearly established its own monopoly over radio because this 'effective national instrument' depended on British owned patents.[84] Yet in each of these areas, despite the fact that the newer technologies were profitable, the model of non-involvement with the unprofitable telegraph was eventually reaffirmed and private entities either restored (AT&T) or established (RCA) (see below pp. 302, 351).

In the 1856 debate, Seward allowed that the United States had 'no dominions on the other side of the Atlantic Ocean'. Clearly a worldwide cable system of which the transatlantic wire was the first link would be of greater benefit to Britain and her empire. British imperial interests, which included a monopoly on the gutta percha produced by trees in British Malaya, ensured her dominance of the system. So strong was the hold that even the transformation to Commonwealth in the 1930s and the loss of empire in the aftermath of the Second World War did not seem to affect it. Despite this continued British dominance, the crucial supervening necessity, Sputnik, did not immediately push the United States to a civilian satellite communications system which would dislodge the British – rather it led to a man being placed on the moon and a whole slough of military applications wherewith to confront the Russians. Nevertheless the advantages of satellite communications to United States interests outside of this central confrontation were not ignored. Johnson, translated from the chair of the Senate Space Committee to the Vice Presidency-elect, strongly endorsed this option. A staff report to his committee said: 'To construe space telecommunications too narrowly would overlook its immediate and rich potential as an instrument of U.S. foreign policy.'[85]

The civilian machines were, then, spin-offs which would not have been realised in the metal had not they afforded the US an opportunity, denied for 100 years and now required by her changed international position as leader of the West, to gain the ascendancy of a worldwide telecommunications network through superior technology. Also satisfied was a further and comparatively insignificant supervening necessity for instantaneous transoceanic television pictures.

Phase three: technological performance – invention

Williams, Rosen and Hudspeth invent the GEOSTATIONARY SATELLITE, 1962

The technological performance NASA had contracted Hughes Aircraft to perform in the agreement of August 1961 was simply to *invent* a geostationary satellite, **Syncom**. Since all supervening necessities, both major and minor, had already occurred, the *invention* took place contemporaneously with the development of the most immediate prototypes, low-orbit repeater satellites such as Telstar.

Syncom had to be small enough to fit upon a known rocket and light enough to allow that rocket to lob it 22,375 miles up. Thereafter the satellite had to find and keep an exact station. It had to be capable of repeating at least one telephone conversation. Hughes had been chosen for this task because it had been investigating the possibility of Clarke's idea even as AT&T was campaigning for the other solution. It had told a congressional committee in 1961 that it would have a 300-voice channel device ready within months, but NASA's Goddard Space Center had cautiously stipulated a less ambitious scheme to be completed in a year and a half.[86] Thus commissioned, Hughes' initial plan was to orbit a comparatively simple machine that would weigh no more than 33 lb and would be unable to carry television. It assigned three of its designers, Williams, Hudspeth and Rosen to the jigsaw puzzle and they made good progress. They decided to spin the device gyroscopically for stabilisation. They also decided on elegant light-weight hydrogen peroxide and nitrogen-powered manoeuvring gas-jet engines for fine adjustments when the orbit was achieved. But the capacity to carry television became vital, because of the publicity generated by Telstar's transmissions, and Syncom had to increase its weight so that it could do likewise. To achieve high orbit now required that the satellite be attached to an extra solid-fuel rocket. These designs should, theoretically, get the satellite to where it had to be and keep it there.

But, unlike their competitors who had military models to go by, Williams and company were pioneers. **Advent**, the army's first geostationary device contracted to GTE, had only been put on the drawing board in 1960.[87] In 1961, James Fiske, the Bell Labs' president, had laid out for a professional audience some of Hughes' and GTE's problems. To get the rocket from Cape Canaveral at 28½° of latitude north to the equator required that it make a left-turn. No guidance system then in existence could accomplish such a feat. Further, the high-orbit solution would delay speech by nearly a third of a second while the wave travelled up and down from the satellite, a distance of 44,750 miles, and Bell research on very long terrestrial telephone lines showed that such a time lapse was just twice as long as telephone users found bearable.[88] In addition, no contemporary rocket could lift the 180 lb of a Telstar or Relay to such an altitude, and all the extra engines required so to do exponentially increased the chances of failure.

This attack was but a strand of the general strategy being used by the international common carriers in the emerging debate on who was to own the new system. Led by AT&T, and with the support of the FCC, these carriers took the 'traditional' view, the one that would most effectively suppress the technology's radical potential. They held that, as the most experienced transoceanic telecommunications entities, they should own this new potential competitor and profit fully from that ownership. That enormous sums of public money were necessary to develop the system for them was, according to this 'traditional' view, of no account.

The domestic telephone and telegraph companies (AT&T apart) took a slightly different position, suggesting the system be owned by all authorised common carriers as the 1934 Communications Act defined them, which of course included themselves. Nevertheless many common carriers, those who also had manufacturing interests, did not want the principle of general common carrier ownership overly to extend AT&T's power. NASA had chosen a common carrier, RCA, whose interest in space antedated AT&T's and who had given Score its voice, to duplicate Telstar with Relay. For these companies, constraining AT&T became a question of denying the superiority of the low-orbit system. RCA, despite low/medium-orbit Relay, argued that even if the high-orbit method took longer to develop it would be economically more viable. GTE went so far as to suggest that: 'A random system (i.e. AT&T's) could discredit us before the world as a leader in space communications if Russia establishes a stationary satellite system.' (our brackets)[89]

The aeronautics companies, who regarded space as their bailiwick, had a different view again, agreeing with the principle of private ownership but demanding that they be included. Hughes, an aircraft manufacturer, had communications satellite plans sufficiently advanced to secure the NASA contract for the high-orbit experiment and Lockheed had been formulating its designs as long as had AT&T. Lockheed worried that Bell would get into the missile business.

The Justice Department examined all the ownership proposals for traces of 'monopoly':

> The Justice Department firmly believes that a project so important to the national interest should not be owned or controlled by a single private organisation irrespective of the extent to which such a system will be subject to government regulation.[90]

Not for the first time around a new technology, indeed as with the telephone itself, AT&T's aggressive position needed to be balanced by more defensive public relations pronouncements (see below page 350 for further examples of this tendency). Its chief engineer James Dingman said:

> Hard as it may be for some to understand, our sole interest is in the earliest practicable establishment of a worldwide commercial satellite system useful to all international communications carriers and agencies here and abroad.[91]

Of all the players, only Philco broached a radical solution, proposing that the US build a system which then be handed over for the UN to operate. (Philco, it will be remembered, wanted an 800-line television standard in 1939.) All these insurances against disruption were occasioned by the most conservative potential application of the technology, since nobody in government nor industry conceived of a satellite that would be directly receivable in the home.

Despite AT&T's somewhat beleagured position in the debate, its criticism of the Hughes' scheme highlighted the fact that the synchronous satellite was a communications device and Bell Labs, the acknowledged authority on telecommunications, was against it. In advance of Advent, Hughes had no guarantees that even Syncom's basic concept was going to work. In May of 1962, as Congress prepared to act on satellite communications, the DoD cancelled the Advent project. Faced with these uncertainties, both political and technical, Hughes began to cool off to the point of withdrawing.

In telecommunications, geostationery satellite is the systems engineering *invention par excellence* with nothing to 'discover' and seemingly little or no place for intuition or emotion even of the 'will to think' kind Shockley celebrates. It is significant to note, therefore, that Hughes's design team was deeply involved in its work. Williams was actually prepared to back his technological hunches by offering his own savings to the management so that the project might continue and, to that end, he burst into a vice president's office and slapped a cheque for $10,000 on the desk. Hughes stuck with Syncom, as the row about institutional frameworks intensified on the Hill.

Against the somewhat diverse views of industry, those positions first deployed in connection with the telegraph (and first defeated in 1857) were once more being put forward. Liberal democrats, lead by Estes Kefauver, bolstered by the public money poured into the space programme, put yet again what might be called the 'Post Office' argument, and claimed the system for the public. Not only was the technological debate about orbits overshadowed by the Soviet presence, the political discourse was also affected by the possibility that the Russians might at any minute inaugurate either a high- or low-orbit system and that it might prove as attractive to the world as one, in the words of Senator Russell Long, 'owned by the American monopolies'.[92] The radical potential of communication satellites as disrupters of the established international telecommunications system had already led to the duplication of low-orbit machines. It was now about to be institutionally constrained even before the synchronous model was *invented*.

Unlike the similar rows that had taken place about the telegraph, the telephone and radio, the argument about space communications perforce involved other nations, nations which had, whatever their ideologies, tended historically to accept the 'Post Office' argument. International communications involved governments. Even the US government, despite its 'tradition' of leaving communications to private enterprise, had attended the first international radio frequency allocation conference in Berlin in 1906 where an International Radiotelegraph Union had also been established. (The United States had instigated the model of such organisations when in 1863, at its suggestion, the Postal Union was formed.) The Radiotelegraph Union merged, in 1932, with its telegraphic opposite number to create an International Telecommunications Union (ITU). Frequency allocation agreements, dating in their latest form from a 1947 Atlantic City meeting, made no provision for space communications. As early as 1955, the US Department of State began to prepare for the revision of these regulations at an ITU Radio Administrative Conference to be held in Geneva in the winter of 1959. In the interim all space telecommunications, civilian and military, had to be conducted so as not to offend against the established users of the spectrum. The ITU in 1959 agreed to a 13-band allocation for three years and determined to reconvene at an Extraordinary Administrative Radio Conference in 1963 to make more permanent arrangements.

Aside from the ITU, the United Nations was another brake on unfettered national activity in space. On 20 December 1961 the

General Assembly accepted the report of its Committee on the Peaceful Uses of Outer Space which sought the early establishment of a satellite communications system to be 'available to the nations of the world as soon as practicable on a global and non-discriminatory basis'.[93]

Kennedy was as aware as anybody of the dangers of 'monopoly' (i.e. the phone company) both internally and from the foreign policy perspective. He had acted to prevent AT&T buying the system when he cancelled the Eisenhower proposal for NASA to seek $10 million for R&D from private resources; but nevertheless he agreed with the 'tradition' of keeping the government out of enterprises which could be profitable to the citizenry. The executive branch therefore developed a compromise proposal in which a new entity would be created to run the space system. It was not, as Kefauver and his faction wanted, to be organised along the lines of the TVA, i.e. a governmental entity; rather it was to be a private company somewhat like the original RCA wherein radio interests were given control. In the proposal for the exploitation of the satellites all the common carriers were to have control but were to share the new company with the public – which obviously could include electronic manufacturers and the aerospace industry. Thus the public's tax dollars were to be returned in the form of an investment opportunity and a new government sponsored company, in some sense legally representative of the entire tele-communications industry, could present itself internationally for business untainted by a history of involvement in American capitalism.

The neopopulist proponents of the 'Post Office' position were sufficiently opposed, despite the fact that the company would have to obtain State Department approval for any foreign contractual agree-ments, to stage a filibuster against the bill. Public ownership became associated with high-orbit schemes simply because private ownership was connected to AT&T and the low-orbit solution. In the midst of the debate, Telstar flew, but it made no difference. The 'law' of the suppression of radical potential had already worked to such good effect that even AT&T itself had given up its hopes of domination, such was the hostility to it from all quarters (except the FCC which, as is not uncommon, behaved more like industry advocate than regulator). AT&T supported the president, already claiming the space-side channels were only to supplement existing means of communication, and it continued actively to pursue its transoceanic cable plans. On 10 August 1962, faced with the need to have America's house in order for the upcoming Administrative Radio Conference, the Senate began to consider the bill. After four days of filibuster, it voted, for the first time in 35 years, cloture. The bill then passed on August 17th and in the House on August 27th. It was signed into law as 'The Communica-tions Satellite Act' on the 31st. The preamble states:

> In order to facilitate this development and to provide for the widest possible participation by private enterprises, United States participation in the global system shall be in the form of a private corporation, subject to appropriate governmental regulation.[94]

Title III establishes a Communications Satellite Corporation (COMSAT) in which, 'Fifty per centum of the shares of stock . . . shall be reserved for purchase by authorized carriers' and at 'no time shall any stockholder who is not an authorized carrier . . . own more than 10 per centum of the shares'.[95] It immediately became a blue-chip stock and the second tooth of the new technology had been drawn.

In December 1962 Relay I gave a further (redundant but) successful low-orbit demonstration. Fourteen weeks later Syncom I, the *invention*, was ready to go.

At less than half Relay's weight, 79 lb, Syncom was nevertheless capable of television. The Delta booster pushed it into the sort of orbit the Russians would choose for the Molniyas some years later, lifting it 150 miles. From that point it travelled out to its apogee of 22,375 miles where the extra solid-stage rocket was designed to fire – to kick the orbit out, as it were, into a circle. The engine ignited on a command from earth but immediately thereafter Syncom went dead. It was lost for a week. When rediscovered by astronomers it was found to have achieved a perigee of 21,3754 miles, an apogee of 2,823 miles and a period of 23 hours 45 minutes. The basic manoeuvre had worked and it was probably an exploding gas-jet engine that had killed the bird. It was not an auspicious demonstration of the superiority of the high-orbit option.

Syncom II, with specially designed peroxide engines, as opposed to the shelf-items with which its predecessor had been equipped, was launched on 23 July 1963, nine weeks after Telstar II. Syncom II made it into a circular orbit but not on the equatorial plane. Instead it described a 'figure-of-8' movement across the globe. Nevertheless it was visible for long periods of time and it drifted until both California and Nigeria could see it. An American warship in Lagos harbour signalled Paso Robles, 7700 miles away, via Syncom, a satellite distance record. The orbit, though inclined, could be adjusted to an exact 24-hour period, checking out the gas-jet engines, and by August 15 daily seconds-long pulsings of gas had placed it on station.

That fall, under the Atlantic, AT&T inaugurated the third telephone cable in which it had an interest. In January 1964 RCA's Relay II was launched for NASA.

On 19 August 1964 Syncom III was placed on exactly the right equatorial orbit by a Delta with three solid-fuel rockets strapped to its sides, its normal thrust of 170,000 lb thereby increased to 330,500 lb. The extra power gave it the wherewithal to make the left-turn south of the Cape.

A day later, in Washington, recalling the desire for a worldwide network set forth in the UN resolution three years previously, an 'Agreement Establishing Interim Arrangements For A Global Commercial Communications Satellite System' was opened for signature. The word 'commercial' was the only significant addition. Ten nations, as well as the Vatican City, signed.[96]

Syncom III reached its station over the International Date Line on September 11th just in time to transmit the Tokyo Summer Olympics to America and dramatically illustrate the advantages of a stationary device. The images were carried to Europe from the USA in hour-long bits on Relay II. European television companies, in the spirit of the BBC's experiment with the facsimile transmission system, were seeking ways and means to be first with the Olympic pictures, including the chartering of jets equipped with film-developing labs and the like. They gave up such plans for ever.[97] The geostationary satellite was *invented* and was obviously superior. Three more nations signed the interim agreement and the battle to contain the threat of the new technology – a threat seriously increased by Syncom III's effective existence – moved to the international arena.

The third transformation: the 'law' of the suppression of radical potential

AT&T, PTT and Deutsche Bundespost open a TRANSATLANTIC TELEPHONE CABLE, 1965

Satellites are the most centralised technology here discussed, the most difficult of access, yet nevertheless their history reveals in a very clear-cut way the operation of the 'law' of suppression. Above has been detailed the developments which prevented AT&T from dominating the American interest in satellite communications either technically or institutionally. The disruptive potential to the delicately balanced internal market was thus contained. Now we must examine the attempts made by the world to prevent that American interest itself from dominating so that a similar containment of potential disruption could occur internationally. The subsequent suppression of domestic satellites within the US and the general failure, into the mid-1980s, to exploit direct satellite-to-home possibilities, with which we will also deal, illustrates not so much the institutional structures of containment but more the suppression of the technology's radical heart. We begin internationally.

(i) The communications satellite corporation shall act as the manager
In 1963, the Extraordinary Administrative Radio Conference met in Geneva and determined wavebands for space communications. The

American delegation was 43-strong but the support of the Soviet Union was needed to beat off a French proposal that broadcasting transmissions, radio or television, from any space object be banned. Enough spectrum for 9000 voice channels and 4 TV circuits was agreed. The permanency of the arrangement was questioned by Israel for fear that this would enshrine the USA's and USSR's 'first come, first served' dominance. Despite this, the allocations were adopted as permanent. Comsat, which had been incorporated in March 1963, had something concrete to sell.[98]

Comsat's directors, led by the ex-chair of Esso and including three appointed by AT&T, were at odds with the US government's vision of a utility company and thought the RCA model of autonomous entrepreneur more appropriate. They were now running a free enterprise which would simply establish and operate a system more extensive and cheaper than any other. The rest of the world would therefore negotiate with Comsat to use it. This was the plan they presented at meetings with the Europeans, the Canadians and the Japanese in the summer. They were disabused of this vision of their own business when the Europeans, who had established CETS (Conference Européenne des Telecommunications par Satellites) as a forum for achieving a joint position, told an American governmental group after the Geneva radio frequency meeting that they simply would not deal with Comsat.

In the European response in the early 1960s one major theme was the protection of the generation of voice-quality cables then being laid, whose technology was not a complete American monopoly. TAT I, the first Anglo-American voice cable, had been joined by TAT II in 1959 and TAT III in 1963. The British and their Commonwealth partners were working on **Compac**, the first transpacific cable, which would run from Vancouver to Auckland and Sydney via Fiji, while the French and German Post Offices, with AT&T, planned TAT IV for 1965.

The chairman of the British space development organisation said, 'It by no means follows that the US communication system will necessarily be the best solution for others'; and in the British parliament it was stated that:

> there is now a growing feeling that . . . we shall finally end up by starving the transatlantic cable of telegraphic communications from America and assisting Comsat to get off the ground, and that Britain will merely end up renting a line from the Americans.[99]

The British Postmaster General replied: 'The only way of preventing an American monopoly in this sphere . . . is to join in partnership with the United States and so secure the right to influence the course of events.'[100]

Yet he spoke in the heat of the technological moment, for there need have been no course of events to influence – no viable Comsat at all – as a careful examination of the background to the Satellite Communications Act and the continued interest of the American international common carriers in cable revealed. The Europeans could have just maintained their refusal to deal with Comsat and to plan cables with AT&T. By denying AT&T the system, and in creating an alternative entity which could not, because of AT&T, have monopoly powers, Congress had let the entire world off the hook. It had also negated the United States's best shot at achieving a dominated global telecommunications system. It was the seductiveness of the technology, rather than the competitive reality, that fed the feeling of entrapment in Europe and elsewhere. Most believed that by the 1970s cables would be obsolete and that belief conditioned two needs – first, to constrain the development of the satellites until the expected obsolescence occurred, cables having a life of 23 or more years; and second, to make sure the world system would be in a substantive way international. Behind these needs were the sort of nationalistic worries that had animated certain members of Congress in 1854, but now the technological boot was on the other foot. So however mesmerised they might be by space, for the Europeans the principle of national autonomy over telecommunications was too important to be swept aside, especially by a pushy American corporation.

There was, at least among the majority of the nations concerned, no overt hostility to space research *per se*. Telstar had allowed the British and Americans to synchronise their watches to within 10 milliseconds and NASA had launched a British scientific satellite **Ariel** (or UK I) and a Canadian one, **Alouette**, in 1962. The British were particularly pathetic in the face of this dependency, the parliamentary secretary to the Ministry of Science claiming: 'We believe our programme is bringing in far more in the way of scientific knowledge than by sending men or even dogs up into space.'[101]

In February 1964, when next Comsat approached CETS, in Rome, it was in the company of officials from State and the FCC and now the skeleton of a plan evolved. Obviously Comsat would have to manage the system, as the only organisation with access to the requisite technological wherewithal held by NASA, but it would be subject to an international committee of control. CETS price for this was a promised transfer of technology among the member nations of the committee so that research, development and manufacture in the long term would be more dispersed. A British Foreign Office official said, 'We have to make sure that no doors are slammed in the face of European industry.' Further, provisions for new members would have to be made involving a changing structure of ownership; and finally the whole deal would have to be an interim one pending the maturation of the system. These points, which were largely acceptable to the US government, were hammered out over the next several months, resulting in the Interim Agreement, which would operate through 1969, being open for signatures in August 1964.

Comsat handed over control of its projected system to an International Telecommunication Satellite Consortium, Intelsat as it came to be called. Intelsat initially allowed for at least 5 more European states to join and so set its investment requirements and concomitant ownership shares on a quota basis, assuming the participation of 19 industrialised nations (or rather 18 nations and the Vatican City). Determined by telephone usage, the United States began with 61 per cent; Britain came next with an 8.4 per cent interest, but all were to surrender pro rated portions of these quotas to other nations as they joined. The United States was not to fall below 50.6 per cent, however. By 1969, when these arrangements ran out, 70 nations had initialled the Interim Agreement.

Intelsat was governed by an executive and representative body called the Interim Communications Satellite Committee – ICSC. The ICSC, which met once every two months, comprised a growing number of nations or representatives of groups of nations determined by quota. In 1967, for instance, there was one representative for the Philippines, Thailand, Singapore, India, Indonesia, Malaya, and New Zealand (some 595 million people at this point) and one for the original signatory, the Vatican City (population 855).

The arrangement did work well in a crucial particular – the ICSC determined the rates and therefore prevented any undercutting of the cables. Even the dominant Comsat voice was influenced in this regard by the fact that one-third of its board were American carriers, who had large interests in cable. Intelsat functioned internationally as Comsat functioned domestically, as the agent containing the disruptive potential of the new technology.

Beyond this central achievement Intelsat was also a success in that a majority of nations, including therefore many from the non-aligned group, joined. There can be no doubt that it met the basic requirements of the UN resolution and went to considerable lengths in its pricing policies to favour the disadvantaged – although this fact must be tempered by the observation that it persisted with the technology with which it began, comparatively small satellites requiring huge and expensive ground stations. Intelsat was prepared to treat circuits on its satellite transponders somewhat as a supermarket treats loss leaders, the real business being in ground stations. Few later signatories were on the emerging voice-cable net and selling the satellite alternative (and building the attendant ground stations) dovetailed with these (mainly) new nation's desires for the trappings of statehood in so neat a package that some might call it a neocolonialist model. In the last year of the agreement's life, Cameroon, Guatemala, Côte d'Ivoire, Jamaica, South Vietnam and Luxembourg joined.

But the Soviet Union remained aloof, taking as dim a view of Comsat as it did of the project to release millions of copper needles into the ionosphere. The hostility was transferred to Intelsat, which was seen as simply a Comsat cover. Using arguments that had been made against the corporation in the congressional debate, a Soviet spokesman asked:

It is well known that private enterprise in the U.S.A. has a long experience of discriminating against Americans themselves, so how can we expect this enterprise to be unbiased and just on the international scene?[102]

They further objected to the weighted voting system in the ICSC and the fact that Intelsat was independent of UNO. In 1968 this animosity was translated into a rival scheme, Intersputnik, which was passed to the UN (as Intelsat had in fact been) by the USSR, Bulgaria, Cuba, Czechoslovakia, Hungary, Mongolia, Poland and Rumania. The attractiveness of Intersputnik's 'one nation, one vote' formula was somewhat diluted by the Soviet invasion of Czechoslovakia and the inherent technical drawbacks of Orbita as a global system. Intersputnik, with a second generation of Molniya IIs joining twenty Molniya Is, was in existence by the early 1970s, the same time as Intelsat emerged in a permanent form. (The Soviets were to orbit geostationary satellites in the late 1970s.)

Apart from non-participation in Intelsat (which was limited to the communist bloc), other troubles emerged during the period of the Interim Agreement. Governments were allowed to nominate bodies to represent them on Intelsat and thereby at the ICSC as well. Most chose their telecommunications ministries or post offices; some, the USA, Japan and Italy for instance, sent other organisations. The result was that internationalising Comsat's system required that it be handed over to Intelsat which was 61 per cent American and which 61 per cent was represented by Comsat. Comsat, as its earliest negotiating position indicated, was singularly ill-equipped to cope with the diplomacy this entailed and proved to be incapable of behaving with restraint. For instance, the agreement specified that Comsat was to have control of the day-to-day management of the system, including research and procurement, ensuring that: 'contracts are so distributed that equipment is designed, developed and produced in the States whose Governments are Parties to this Agreement in approximate proportion to the respective quotas. . . .'[103]

Eleven European states out of Comsat's 18 partners had just established a multinational space research organisation and a rocket development team which were designed to create a space programme and which were ignored by Comsat when it let its first Intelsat contracts. Comsat exhibited more concern for its domestic constituency. It requested Hughes build a commercial version of Syncom, a Type 303. At the same time, it further contracted Hughes as well as AT&T, RCA, ITT and Space Technology Laboratories to submit design proposals for a complete system, either high or low orbit. It was not that the Europeans had illusions about their endeavours in space, rather that the nature of Intelsat required national technological inferiority to be massaged with due process. Comsat handouts spoke of Intelsat as a 'unique partnership for progress' but the reality was rather different. The American corporation did not even bother to insert in the District of Columbia telephone directory a telephone number for Intelsat which was secreted on half of the third floor of its Washington headquarters.[104]

Hughes claimed that high orbits could save some $200 million over the first few years of operation despite the shorter life predicted for the geostationary satellites. Syncom III was exhibiting all of the advantages and none of the problems suggested for such machines. Even after its peroxide engines ran out it remained on station and was still capable of working efficiently. If the Syncoms were replaced every three years, as against the projected span of 10 years for the Telstar-type, still massive savings could be effected in reduced launchings and much simplified ground stations. On 6 April 1965 NASA orbited the 303, to be known as **Early Bird** or **Intelsat I**. It weighed 85 lb and took 23 hours 57 minutes, give or take 57 seconds, to circle the earth. It was equipped with 240 voice circuits or, if these were cleared down, 1 television channel. Vice-President Humphrey exclaimed as Early Bird was demonstrated 'My goodness, now we'll be able to call everybody!' But he could have done that, more or less, already – there were 374 telephone lines available under the Atlantic. There was no rush to rent transponders on the satellite and after the free honeymoon was over the television networks decided they could live without the facility on a day-to-day basis since it was so expensive. It was to remain Intelsat's sole space-side asset until October 1966.

The decision to stay with Hughes had been made the previous year and the Intelsat II series had a similar specification to that of Early Bird, 240 channels, but a larger 'footprint' to cover more of the globe. One was partially lost when its motors failed to fire for 10 seconds, rendering it operable for only nine hours a day, but by 1967 three working machines were in place and Clarke's dream was a reality. The world was girdled with Intelsats. The Australians and Japanese, having fewer cable alternatives, made comparatively greater use of the Pacific bird than was made of the one over the Atlantic.

Comsat's partners had viewed these early developments with a degree of equanimity since obviously Hughes had a head start on everybody. However, with the Intelsat III series, patience was at an end. Comsat still failed to discuss a possible launcher contract with the Europeans and decided again for a Hughes satellite. The third generation was to have 6000 channels and be ready in 1970 and the FCC recommended that Comsat wait for it. The State Department, at the behest of Intelsat's other high tech members, had to persuade Comsat that it would be politically sounder to introduce an interim satellite with a greater percentage of non-US components; hence the procurement of a different series of Intelsat IIIs from another manufacturer (TRW), with only 1200 circuits, but twice as much foreign subcontracting (4.6 per cent) than had Intelsat II (2.3 per cent). And when the first four Hughes Intelsat IVs were built, with 9000 rather than 6000 circuits, they too contained non-American components, 26 per cent. To achieve this, Hughes were forced to train the foreign technologists in the secrets of the hardware so that they could return home, duplicate the work and send it back to Hughes. Throughout these crucial years when, to preserve its position, Comsat should have been eager to waste money in this way, instead it 'insisted that Intelsat purchased only the best equipment available at the least expense.' Comsat was unrepentant and, as early as 1967, was having to defend this logic. James McCormack, the company's chief executive, stated:

> Improvements can be made, but a fundamental restructuring will hardly commend itself to those who are interested in maintaining the pace of progress which has been enjoyed during these initial years.[105]

The Europeans contributed to Comsat's gaucherie by spending together less than 10 per cent of what the USA or the USSR were spending separately on space and by producing poor results. But nevertheless, there was activity; French, British (and Japanese) alternatives to Cape Canaveral had been established and rockets and satellites launched, albeit of a size and scope well behind those of the super-powers. And this was the rub; Intelsat was supposed to be an agency which positively discriminated in favour of the technologically disadvantaged and clearly, under Comsat's management, it was failing so to do. A high-level committee of Eurospace experts put it like this in 1967:

> The fundamental question is above all more political than economic in nature: do the European governments consider it necessary and desirable to devote the necessary resources to programmes in order to secure an independent long-term capability and not to remain

tributaries of American facilities or dependent on global systems in which the United States has a preponderant influence?[106]

The French and the Germans, acting on this question, inaugurated the first direct internal challenge to Intelsat by planning a communications satellite of their own, **Symphonie**. (All other non-super-power devices had been scientific, meteorological or navigational.)

Matching these moves during the period of Intelsat's Interim Agreement, there was considerable dissatisfaction among the rank and file, nations which had no technological pretensions but wished nevertheless to have some more meaningful say over the system. For many of these states satellite links were the only available means of international communications and they felt at the mercy of the Americans. In the ground station, they had acquired a symbol, but the US acquired 'modern and direct communication to many areas of the world which previously had none, thus enabling American business to better utilize its operations in many countries'.[107] As the end of the decade approached, and with it the termination of the Interim Agreement, the stage was set for a negotiation which would yield nothing in complexity and intensity to the other battles herein recounted.

Comsat had made some moves to protect itself. By 1969 nearly 10 per cent of its employees were foreign nationals, although only 2.5 per cent of its 641 managers were. The US share of Intelsat had dropped to 52.61 per cent. Thirteen per cent of research contracts were being awarded abroad, and this was increasing. The US Government insisted on more – that Comsat institutionalise the inherently schizophrenic quality of its position by dividing administrative from technical management within the organisation. But the basic facts remained – of the $323 million the system had cost so far, no less than 92 per cent had been spent in the US.

The Europeans came to the table in greater disarray than in 1964. For the states among their number with lesser high-tech pretensions, Spain for instance, the failure to effect a greater measure of technological transfer meant little. Protection of the cable system was more important to the UK than to some of its partners. Exacerbating these differences of position was the threatened collapse of the joint Eurospace programe. Europa rockets, with stages made in different countries, had achieved 11 straight launch-pad failures, causing Britain to announce its withdrawal to go it alone, as the French had already successfully done.

Outside of Europe, many of the developing countries with attenuated telephone systems (and therefore less power in the ICSC) wanted a 'one seat, one vote' system and some even suggested a truly representative assembly to manage the operation on UN lines.

It was expected to take only weeks or at the most months when Washington's largest international conference ever convened in February 1969 but, in the event, two and a quarter years were to pass before the 'Agreement and Operating Agreement Relating to the International Telecommunications Satellite Organisation', the Definitive Arrangements, were opened for signature. The time was spent in what has been described as 'tedious technical negotiations', but, although they may well have been tedious, the arguments were about power, not machinery. Needless to say Intelsat emerged with a very complex governance structure but also one in which Comsat's, and therefore the US's, role was much reduced.

Interest was still determined by telephone usage based quotas but the ceiling was placed at 40 per cent. Voting within the Board of Governors, which replaced the ICSC, was still weighted. Two new representative bodies, an Assembly of Parties (i.e. governments) and a Meeting of Signatories (i.e. agencies) were established on a 'one person, one vote' basis. The parties were to meet biennially, the signatories annually; but the former could only recommend to the Board of Governors and the latter can only act upon recommendations of the Board of Governors, wherein the power therefore still resides. The US fought for Comsat to retain a veto (via a 66 per cent vote of the Board which it could always prevent by denying the 40 per cent share it was expected to hold) but were forced to abandon the demand. Similarly Comsat's role as operator of the system was reduced from 'manager' to 'management services contractor' working to a small permanent Intelsat staff led by a director general. But even after 27 months of debate, this last was not final being subject to review in 1979 with the possibility that Comsat be replaced entirely. (This did not happen and Comsat still manages the system.)

It was not that the world was able, the Soviet-dominated bloc aside, to escape from the realities of US power; rather the Americans themselves had emasculated their own control of the system by allowing competing international carriers to continue to exist. The world needed to talk to America but did not need Comsat to do this. It could be argued that if Comsat had achieved a diplomatic triumph in its day-to-day operations, the result would have been much the same in the long run. The Definitive Arrangements reflected a reality in which Comsat and the satellites could be ignored, even by the most poverty-stricken state, and other American companies and cables engaged. The agreement therefore represented a considerable setback for US policy and those ambitions, expressed in the early 1960s, that substituting space for terrestrial technology might give America control over the world's communication system. In the debate, the US tried to move the agenda away from management issues altogether, arguing that Comsat had done, basically, a fine job and that recent reforms would solve outstanding difficulties. But efficiency was not the point; national interest was. Comsat was not only forced to divorce administrative from technical management, it was relieved of the former function altogether.

Other battles were lost. Comsat came to the table determined to kill the Franco-German bird, Symphonie, and to obtain the right for Intelsat alone to provide navigational and other specialised services. The US Government felt that Comsat's stand against so-called regional satellite systems could not be sustained and, instead, the best that could be achieved was to insist on compatibility, i.e. no Orbitas. The British resisted the navigational request since it would have affected traditional British dominance of such systems. The arrangements were open for initialling on 20 August 1971. Arthur C. Clarke, as the system's father, addressed the delegates of the 54 nations gathered to sign the agreement:

'Today, my friends, whether you intend to or not, whether you wish to or not, you have just signed far more than yet another intergovernmental agreement. You have just signed the first draft of the articles of Federation of the United States of Earth.'[108]

The kindest explanation is that, having missed the events of the conference, he cannot have read the documents.

In 1971, the Intelsat IV system was introduced using the Hughes third-generation satellites – 3750 voice circuits plus 2 television channels. But rivals were at hand. In 1973, a domestic satellite was launched for the Canadians by NASA and in 1975 Symphonie A was orbited for the French and German post offices (also by NASA) while the cables continued, an 845-voice-circuit TAT V having been laid in 1970 from the US to Spain. In 1976, the first Intelsat IV-A (6250 voice and 2 TV) was orbited, another compromise while waiting for the more advanced Intelsat Vs. The cables refused to fulfil the prophecy of obsolescence proclaimed so stridently a few years earlier. Not only were more cables being laid, but their capacity was increasing. TAT VI with 4000 circuits was operational the same year as Intelsat IV. Transpac (a second transpacific wire) with 845 circuits was also operational in 1976; and, following the failures of previously dependable American launchers, there was even talk of using a French booster for the Intelsat V series. Comsat had not supplanted the cables and had not even been able to maintain a space-side international monopoly. The 'law' of the suppression of radical potential had worked.

(ii)   The dictates of their business judgement

Comsat proved as attractive a stock as Congress had hoped, the 50 per cent available to the public being snapped up by 150,000 buyers. The value of their holdings, an average of 20 shares per person, doubled to $100 million. Yet Comsat was as unsuccessful in carving out for itself a pre-eminent position domestically as it was internationally.

It suffered a string of defeats, beginning with the denial of its claim to be sole proprietor of the American ground stations and concluding with the determination by President Reagan in 1984 that America's best interests would not after all be served by an international monopoly such as Intelsat upon which it served. Along the way, it acquired a variety of domestic competitors even in space, never achieved a technological edge over NASA, failed to obtain its own military business (such as the terrestrial carriers traditionally had) and saw its plans for a direct broadcast satellite (DBS) collapse. In just two decades Comsat went from licensed potential giant, the agent of America's 'imperial' communications ambitions, to being just another rich telecommunications player. In this decline is also described the containment of the technology Comsat sought to exploit.

The first site of defeat was the ground stations, then costing around $10 million each. Comsat argued that its mission required sole ownership of the terrestrial facilities associated with the system, but the Satellite Communications Act specifically allowed the FFC to license the carriers as well as Comsat to operate such stations. Comsat claimed that if it were denied ownership, it would be the only Intelsat agency without such control and this would reflect badly on its leadership position. The international ground station market had developed very much as expected. Technically, despite the initial problem, the British open dish worked efficiently and did not develop troubles as the radomes did, water collecting in their wrinkles with resultant reception problems. The British firms did well. Marconi built stations in Hong Kong and on Ascension Island; Plessey, GEC and AEI built the second dish at Goonhilly. The Japanese tied up with the Americans, Nippon with Hughes, Mitsubishi with TRW and Bell. Siemens built the German station, ITT the Spanish, RCA the first Canadian, Philco the Italian and Bell the first French. The station in Bahrein had been built by Marconi and was not owned by the local government but by Cable and Wireless. Comsat was not involved in this market and had not built most of the stations it was running. Nevertheless, as with many of the arguments it deployed in the 1960s, the supposed threat to its status that this entailed is somewhat hard to understand – for was it not the only Intelsat agency with a space-side segment to control, and did not this affect its leadership position in the organisation?

In 1966, after a fruitless negotiation between Comsat and the carriers (complicated by the carriers' ownership of Comsat), the FCC issued the Solomonic judgment that 7 of the 8 earth stations in question should be controlled on a 50-50 basis for three years. The Alaskan station was to be wholly owned by Comsat. (The issue was a bitter one not because of the operational profits accruing from ownership but rather the fact that profits were pegged to investments and that therefore the ground stations represented an opportunity to increase return overall.)

The FFC found against Comsat on another important issue. The military, who permanently used some 17 per cent of the carriers' terrestrial system, let bids for 10 satellite circuits to Hawaii and points west. (At the same time the Pentagon was pressing AT&T to undertake a second transpacific cable.) The FCC had decided that Comsat should be a carriers' carrier, which is to say an organisation having nothing to do with users, only carriers. Comsat, however, did bid on the Hawaiian circuits and undercut the carriers; but the FCC was adamant that Comsat not engage in competitive practices of this kind whatever the cost to the American taxpayer.

The military seemed to be committed to helping Comsat get off the ground, even perhaps allowing the impending passage of the 1962 Act to influence its cancellation of the geostationary Advent project. At one point, immediately prior to the cancellation, DoD was arguing on the Hill that Advent had commercial possibilities.[109] After passage of the Act, it began negotiating with Comsat for a complete geostationary system, although Philco had already been given the task of developing a low-orbit one upon the cancellation of Advent. There is a possibility that the entire DoD/Comsat negotiation was an exercise in covert government support for the fledgling company, since Comsat could offer nothing that Advent had not previously offered. The system could not even reach all points on the globe, since a mixture of equatorial and polar orbits are needed for total coverage. Had an arrangement been made, the international implications would have been profound. The European post offices would have had to deal with a private corporation, already a difficulty, engaged on presumably secret American defence work, probably an impossibility. The whole business delayed the go-ahead to Philco for the system that eventually became IDCSP (see page 283 below).[110] Comsat developed no direct military connection. Instead the military were soon spending over $20 million a year on leasing Intelsat satellite circuits via the carriers. Their appetite for overseas communication links was considerable, since these commercial arrangements were in addition to the Syncoms which NASA had donated to them.

Comsat's next defeat was over the issue of domestic satellites. In 1965 ABC proposed to the FCC a scheme for an equatorial geostationary device which would allow it to network its own programmes coast-to-coast. 1965 was the last year of black-and-white prime time and it was estimated that the process of colourisation then under way was going to cost between $250 and $275 million. Prime-time colour alone required $100 million. It is obvious that ABC, as the trailing third network and the last one to go to colour, would be most interested in obtaining the relief from AT&T's long line charges that the satellite promised.

All the usual paraphernalia which allows the FCC to be such an exemplary agent of the 'law' of the suppression of radical potential went into play. In March of the following year, the commission undertook to study the issue, which it did – for the next 7 years. It studied to such good purpose that, with the help of the Nixon White House, it in effect froze domestic satellite television distribution as effectively as television itself had been frozen during the period 1948-52 and, as in the earlier period, inaction allowed the players to regroup around the new technology.

Clearly, since international versions of the device were in existence, the delay from first request in 1965 to the launch of a domestic satellite in 1974 could not have been technological. Nor could it have been financial, since the American networks in 1965 spent about $55 million with AT&T on the long lines ABC was proposing to by-pass. It was estimated that a satellite could have done the job for $25 million.

The ABC request highlighted Comsat's impossible situation domestically. At this point Comsat had about $200 million of surplus capital and clearly from its public shareholders' point of view the application of these monies to meeting domestic needs was not only appropriate but necessary.[111] No other area of expansion was so obvious and over no other area could Comsat lay such good claim; but Comsat was constrained in its response. It was also about a quarter-owned by AT&T and therefore any proposals for domestic satellites, voice or video, threatened a substantial element of its biggest shareholder's annual revenues. Of course, if it did destroy AT&T's network television business, the compensation to the telephone company was built in through its ownership of its nemesis; but one quarter of $25 million is not as attractive as all of $55 million, especially as the terrestrial hardware, expensive microwave links and coaxial cables, would also be rendered obsolete. Comsat therefore petitioned the FCC for permission to establish a demonstration rather than a working domestic satellite system. There was much at stake for Comsat. Should it fail to extend its international space-side monopoly into the domestic area, then without doubt its 'leadership position' would be adversely affected.

AT&T petitioned directly for its own domestic satellite. It denied ABC's claims as to the potential savings involved and as only it really knew the actual costs of the terrestrial system, this was an authoritative counterclaim. Nevertheless, even AT&T allowed that a 35 per cent saving could be effected were it permitted to launch the system. By

1968, AT&T had revised its opinion, claiming that satellites would be just as expensive as terrestrial links, a position it held and which, of course, proved to be in the nature of a self-fulfilling prophecy.

The Philco Ford role, that of radical visionary, was this time played by the Ford Foundation under the guidance of ex-commercial broadcaster Fred Friendly. Already commercial television of the 1950s was beginning to be regarded as a Golden Age, while the current output, characterised by an FFC chair as a 'vast wasteland', was causing increasing concern. Friendly's own experience, when important live news coverage he was producing was replaced by reruns of *I Love Lucy*, was one of the more highly publicised occasions for this concern. The struggling educational (i.e. non-commercial) television stations had never been able to afford AT&T's longline prices and therefore had been unable to constitute themselves into a network. (The FFC mandated pricing policy of the telephone company, it will be remembered, was exactly designed, during the television freeze period 1948-52, to inhibit the growth of fourth networks (see above, page 80).) Without being a network, public programming, watched piece-meal across the nation, never achieved a sufficiently coherent audience to impinge upon the commercial rating system, which in turn meant that, since it could not demonstrate an audience, its political clout was reduced. It was estimated that NET, a main provider of non-commercial programmes, would have to spend $8 million a year with AT&T to create a network, just $2 million more than was budgeted for production. Friendly saw satellites as a way of both giving public television its network and substantially increased revenues:

> I mentioned to Bundy the idea of a constellation of satellites serving all broadcasters and operated by a non-profit organisation, the profits themselves being used in helping found an educational network.[112]

The scheme was sponsored by the Ford Foundation with technical advice from Hughes and presented to the FCC in the summer of 1966. It attracted considerable attention, being subjected to a hearing by a Senate subcommittee. Faced with a popular proposal that would in effect have taxed a most powerful interest group, the networks, and removed significant revenues from another, AT&T, the FCC remained mute.

By the end of 1969, the Republicans now being in the White House, the FCC was girding its loins to grant Comsat a temporary licence for a domestic satellite system. The Executive Branch requested an opportunity to study the question and quickly affirmed its hostility to a monopoly supplier. It suggested that the FCC license all-comers and so the commission, in its First Report and Order on this matter, asked:

What persons with what plans are presently willing to come forward to pioneer the development of domestic communications satellite services according to the dictates of their business judgment, technical ingenuity, and any pertinent public interest requirements laid down by the Commission?[113]

Thus the FCC announced what was to become known as the domestic 'open skies' policy. The networks were now paying AT&T some $75 million for longlines. Eighteen applications were received from 15 organisations. The Ford Foundation plan was dead.

In addition to reaffirming their separate bids, AT&T and Comsat (which it still owned in part) put in a joint scheme whereby AT&T would own and operate five ground stations, while Comsat held the space segment. This same year, 1971, the Canadian government let a contract for the world's first domestic satellite. Despite its Inuit name **Anik** (Brother), it was American built and American launched. By January 1973 it was operatonal, the first communications satellite to be owned by an Intelsat signatory outside of Intelsat control. (Symphonie was launched two years later.) Ironically, since Anik was orbited for social rather than market reasons, it was also destined to be the first US commercial domestic satellite. RCA was so impatient of FCC prevarications that it requested permission from it and the Canadian regulatory agency to use an Anik transponder. A further irony is that at least some of Anik's intended beneficiaries were less than pleased with it. An Inuit delegation to the CBC complained:

Hasn't the white man caused enough trouble already? Does he now have to send us a 'brother' who does not know our problems; a brother whom we do not trust, because we believe he is only going to hasten our downfall? We only ask that the white man leave us in peace, so that we can follow our ancestors. We do not need him.[114]

It has been suggested that Anik (obviously because of the RCA move) 'earned a return of capital investment that was virtually unprecedented in the telecommunications industry'.[115]

Within a decade, the Comsat Act was rewritten, removing the privileged ownership position of the carriers at their own request. They sold off their shares and relinquished their seats on the Comsat board. Satellite communications had not turned out to be such a threat that these protections were needed against it, and the carriers had determined that they were better off being free to compete openly. Thus, just as AT&T had quit another government creation, RCA, in the 1920s, so it now, and for the same sort of business advantage, abandoned its interest in Comsat.

Westar, the first US domestic satellite, was orbited in the summer of

1974. It had been 4 years since the FCC's 'First Report and Order', 11 years since Syncom 1. Moreover from the point of the processes of suppression, Westar was owned by that first beneficiary of America's laissez faire communications policies, Western Union, a telegraph company – although to survive until this great age it had needed considerable government help (see below, chapter 6 passim). It was to be a further decade before any of the networks moved entirely to satellite distribution.[116]

(iii) From ground station to home dish

Comsat was now no longer in direct control of Intelsat and a number of signatories had broken its monopoly of space. It had lost the battle to access the telecommunications market place directly and was denied the domestic satellite system. It had not even managed to contain NASA's continuing experiments with communication devices. NASA had specialised needs generated by the **Apollo** programme which meant that it did not abandon the tradition begun with the Echo balloons. The result was that, leaving aside secret military developments, NASA publicly unveiled a continuing series of devices which were in design and capacity always one step in front of the contemporary Intelsat machines Comsat was producing. The *A*pplications *T*echnology *S*atellite series, begun by the agency with the **ATS 1** in 1966, involved multiple function devices with communications, meteorological and navigation capabilities. In 1968 NASA launched the first despun antenna satellite, ATS 3 whereon the antenna spun round at the same speed as the main body of the satellite but in the opposite direction, maintaining a position constant to the earth – a necessary advance if massive satellites with huge antennae were to be built. ATS 3 could carry television and was available as a back-up to the Intelsat system. Prince Charles's investiture was seen across in America and Europe saw the Mexico City Olympics via ATS 3.

The ATS 6 launched in 1974 directly challenged Intelsat's plans for telecommunication aid to the Third World, which plans were at odds with the organisation's continued reliance on large ground stations. The NASA bird, with its 10-metre antenna, did not require multimillion dollar receiving dishes. It was placed over India and used by the Indian space agency as a development tool. The educational potential of satellite television turned out to be much like the educational potential of terrestrial television. The fad for such experimentation has passed, although Intelsat still has its dreams. In the 1980s, it announced **Share** (*S*atellites for *H*ealth *a*nd *R*ural *E*ducation), a 16-month free test and demonstration offered to nations in celebration of the organisation's 20th birthday. (Such schemes have a venerable history since the socially ameliorating effects of media hardware have long been puffed. The magic lantern was recommended 'for all educational purposes' in 1705.[117] Every new medium since then has been touted as a crucial educational tool, but the results have never matched expectations.)

ATS 6 made small ground stations possible, which had not been of much interest to Intelsat; and it showed the way to even smaller receivers, home dishes. It is in the failure to develop these last and their attendant DBSs that the final expression of the 'law' of the suppression of radical potential in connection with space communications can be found. And, as ever, the question of technological availability can be readily answered.

Britain's Defence Minister Healey and Secretary McNamara signed an agreement in Washington in November 1965 which involved Britain in the IDSCP (*Initial Defence Satellite Communication Program*) being developed by Philco. The British defence establishment was to get a complete satellite-based system to replace all British military communications links worldwide. The Johnson administration, behind the Tonkin Resolution of 1964, was escalating its commitment in South East Asia and was pleased to encourage a faltering Britain in maintaining its military presence east of Suez. Also the British, with their advanced MASER technology, had something technically to contribute to the IDSCP.

In 1965 the Signals and Research Establishment was set up at Christchurch, Hants. By 1966 it was working on small portable ground stations and **Idex** and **Scot** emerged. Both were a fraction of the size of contemporary commercial stations. Idex, which cost £35,000, weighed 1.3 tons but could be pulled by a Land Rover. Scot, an even smaller device 6 feet in diameter, was installed on *HMS Wakeful* and *HMS Intrepid*. The ground station had become a dish.

The permanent part of the 1965 deal was implemented in the **DSCS-I** (*Defence Satellite Communication System*-I) in the US and **Skynet** in the UK. Skynet had its own ground stations in Cyprus, Singapore and Australia, like so many electronic equivalents of the old coaling stations; but it did not prevent the withdrawal from east of Suez. The system used a synchronous satellite over the Indian ocean 30 feet high with 10,000-channel capacity. The DSCS-I/Skynet system was a perfect expression of American dominance of the sort that Comsat failed to attain on the civilian side. The *Sunday Telegraph* remarked, 'How long can we expect an independent foreign policy if we must rely on America to maintain our world wide defence communications?'

Clearly blocked as Comsat, the company spun-off a wholly owned subsidiary, Comsat General, to be freer than its parent in establishing a foothold in specialised satellite applications. This ploy, coupled with the 'open skies' atmosphere at the FCC, worked. Comsat General built a navigation system, **Marisat**, initially mainly for use by the US navy. As with the IDSCP system, small dishes were developed, 4 feet in diameter, for shipboard installation. By the early 1970s, at the latest, experimental satellites could be received on dishes under 3 feet across and with the military **Tascat** communications system 1-foot dishes were used. To receive the commercial domestic satellites then coming on stream required 30 foot (9 m) dishes, costing $80,000.

From the beginning home dishes have been promised, giving access to all television programes bouncing off any satellite within range. Prices for such a home receiving device have been quoted from £14 for the dish to £35 to convert the receiver. In advance of such technology, over a million Americans, as of the summer of 1985, had taken the matter into their own hands and erected dishes, at between $2000 and $5000, to obtain domestic satellite television signals not designed for home reception at all. It was estimated that as many as a quarter of rural homes had either no reception or reception so limited in quality and choice as to make this sort of expenditure worthwhile. About 35,000 dishes a month were being installed.[118] It was clear that many dishes were being sold within cable franchise areas, thus creating a real market in TV signals with actual alternatives.

Beyond some local ordinances forbidding the erection of home satellite receivers, the field was almost completely unregulated and the dish-owners were tapping into signals for which others, via cable, paid. The cable industry responded to the threat by planning to scramble its signals. HBO, the leading US cable pay television service, was the first to announce it would encrypt, and various schemes to collect from the dish owners were canvassed, although none at first sight seemed viable. In the 1984 Cable Telecommunication Act, the right to unscrambled reception via dish was confirmed, but unauthorised reception of a scrambled signal became a federal offence.[119] The battle between cable and dish in the United States took place in a technological arena where it was possible to build a dish with 'materials commonly found in the local hardwarestore; plywood, standard dimension lumber, threaded rods, nuts, bolts, washers, screen, paint and so on.'[120]

The fact is that the various delays in putting DBS into place have effectively destroyed it, as far as can be seen, for the near future. The National Association of Broadcasters attempted to smother the technology at birth at the FCC and in the courts and those delays alone might well have done the trick. In 1984, a 5-channel DBS satellite, operated by USCI Inc to serve the North-East and Midwest of the US, was marketed. It charged $40 a month for the service and $400 for the dish, but it managed to sign up only 10,000 subscribers and was bankrupted. Comsat, which was the first to announce active exploration of DBS in 1979, spent some $139 million before dropping out of a partnership intended to save USCI.[121] CBS and Paramount had previously decided there was no future in a DBS partnership with

Comsat itself. None of the other four proposed American services is in place and three potential operators, RCA, Western Union and Rupert Murdoch have withdrawn altogether. With more than a million dishes already scattered across the countryside receiving existing satellite-borne broadcasting and cable services, with the country 42.9 per cent cabled and with 15 million VCRs, the space for DBS was shrinking fast.

In Europe the British plan for a DBS to be run exclusively by the BBC became a risk-spreading consortium scheme to supply programming (via Unisat) to the British cable system, should that ever emerge. The consortium was disbanded in the summer of 1985. Of the 19 European satellites launched or planned, Unisat is one of 8 intended for direct broadcasting. It is hard to believe many, if any, of these will operate in this mode. For Europe, the other distribution means that have been adopted, notably again VCR and in certain countries cable, are less threatening to national autonomy than is DBS. One of the earliest demonstrations of international satellite television beamed President de Gaulle into an American schoolroom. President Johnson took this seriously enough to demand a right of reply. American commercial interests, using the state of Luxembourg as an umbrella, have had DBS plans promptly rebuffed by the rest of Europe.

The failure of DBS, its virtual redundancy, reflects one side of the coin of constraint. The persistence of transoceanic cables is the obverse. As significant as its failure to enter the domestic satellite market as a privileged player was Comsat's inability to halt the growth of voice quality cables after the Intelsat net was in place. The cables themselves, as we have pointed out, increased in capacity to match the satellites. **Cantat 2**, UK/Canada, had 1840 circuits. TAT VI had 4000. Further, the self-evident cost effectiveness of the satellites, anyway because of national interest not as significant a factor as it might have been, became less clear as cable costs fell from the $305 per circuit mile of TAT I to the $8 per mile of TAT VI and the real cost of satellites, launches, lost birds, shorter lives and terrestrial investments became better understood. Although economists believe that above 600 miles satellites must be cheaper, it turns out to be by not so great a factor as to overwhelm nationalistic considerations.

Comsat seriously resisted TAT VI, the 845-circuit wire to the Iberian peninsula, arguing that it was unnecessary and would divert traffic from Intelsat, but lost. The planning for TAT VI involved the European postal authorities (CEPT) and American carriers and government agencies but significantly excluded not only Comsat but the other governmentally appointed commercial space organisation, the Italian firm Societa Telespazio. It was clear by the early 1970s that the American carriers as well as their European opposite numbers were no longer under the spell of space. When the FCC suggested that

Intelsat IV must be preferred to TAT VI, the carriers, with the help of their correspondents on the eastern shore, declined to comply. The crisis occasioned by TAT VI was actually an Intelsat IV crisis, for it had been (and continues to be) the case that the satellite system had been consistently underused. During the first decade it had 'chronic excess capacity'.[122] So it was the viability of the next generation of birds, not the phasing out of cables that dominated thinking. The balance between the two technologies was maintained. Intelsat gave a more than adequate return to its investor nations (especially those few who had interests in the ground station market), but no cable interest was superseded.

In 1927 there were 10,000 telephone calls at $75 each between the US and the UK. By 1957 there were 250,000 and by 1961 4.3 million. There were 115 million message units leaving America in 1982, about 30 per cent originating in the New York area and 37 per cent of them destined for Great Britain.[123] By 1985 the cheapest cost for an initial transatlantic minute was $1.17 plus tax.

The projected growth of such business was described by proponents of the satellites as a 'nightmare' which only space technology could solve. It turned out to be no such thing, although different institutional arrangements, notably the appointment of AT&T as American agent rather than the creation of Comsat, might have yielded a profoundly changed result. In either case, the technology would have been subservient to the politics of international communication.

Intelsat enters its third decade with regional and national competitors in space, competitors under the waters and, by a decision of President Reagan late in 1984, potential competitors as providers of certain sorts of specialised circuits (business data transmissions) even over the Atlantic. (The FCC has five such applications for 'cream-skimming' the profitable international data business pending.) Its enviable 20-year record of political peace is threatened by an Israeli plan to launch a compatible national satellite against Arab objections. Intelsat has played its part in the reduction of international telephone costs and has raised expectations that any event considered news-worthy in any part of the world will be instantly available to television viewers everywhere. As a vehicle of American imperial communications ambitions it has been less successful. Britain still shares about half the world's cables with the US and has an interest in Intelsat a third as large as the US's 30 per cent.

The technological determinists are now no longer as enamoured of space as they were. Although the recovery of satellites became a civilian reality in the year 1984, the business remains somewhat chancy and the numbers of birds that can be stationed over the equator limited. The equatorial band, some 170,000 miles long, can accommodate only about 1800 satellites since if they are placed any closer together their signals will overlap. Finding the techniques to avoid the exhaustion of the band, which looked like being the fastest-ever depleted natural resource, is likely not to be necessary, since waveguides and fibre optics promise to take cable technology a quantum leap forward. An American entity, Tel-Optic, has received FCC authorisation to proceed with its $600 million scheme to lay two fibre optic cables, in partnership with Britain's Cable and Wireless, across the Atlantic. The wires, with a capacity of 13,000 voice channels or 18 television channels (280 MIPS), would be operational by 1989 and 1992 respectively.[124] Far from destroying the transoceanic cable, communications satellites themselves could be nothing but space litter by the end of the century.

Phase four: technological performance – production, spin-offs, redundancies

Time Inc rents a SATELLITE TV CIRCUIT, 1975

Understandable as Clarke's enthusiasm for them is, satellites have been the least significant of the four central technologies. Beyond their part in the falling international communications tariff and the commonplace of commonplace news pictures and special events on television, there are few tangible results of communications satellites impinging on most people's lives. Instant international facsimile service is useful but only for a few people. Business satellites of the sort pioneered, with not that much success, by IBM, Comsat General and Aetna Insurance in the late 1970s have a greater effect but again not on that many lives. Yet there are other satellite communication technology spin-offs which, although just as unrevolutionary, are of more general impact.

The appearance in America in 1982 of a new daily newspaper, *USA Today*, must count among the latter. By 1983 it was being printed simultaneously via satellite in 17 cities and was read by over 1 million people. It is in fact a national newspaper, America's first to be specially designed for that function. Other established papers are using space technology for multiple printings but are still maintaining their local or regional qualities; *USA Today* is the first true child of the marriage of press and bird. Just as the Soviet Molniyas enabled simultaneous editions of *Pravda*, so Gannet Inc used the same technology to produce *USA Today*.

In 1964, Clarke predicted that one result of satellite communication systems would be just such a paper. Like all technological determinists, he let his enthusiasm for the hardware blind his understanding of present realities and their impact on probable developments; 'Influential newspapers such as the London and New York *Times* will', he claimed, 'experience a great increase in distribution and immediacy' because of satellite remote-control of multiple dispersed presses.[125] Since the function of a newspaper is to reinforce every nuance and particular of its readership's image of the world, only when the view from New York totally matches the view from London in the minds of substantial numbers of people will Clarke's prophecy have the remotest chance of being realised. But he then went on: 'One of the first countries to benefit from this will be, rather ironically, the United States, which has never possessed a really national newspaper.' In this he was, nearly two decades before the event, correct. Needless to say America's first national daily, a miracle of technology, is, garish colours aside, essentially a traditional popular paper with a heavy emphasis on human interest materials. Its look and some of its format is heavily derived from television but, since television's news style is itself acquired from *USA Today*'s predecessors, the result is, culturally speaking, somewhat circular.

More significant is the part domestic satellites have played in the development of the American cable industry.

Wilkes-Barre, in the centre of Pennsylvania's depressed anthracite coal industry, was, in 1972, to add to its other troubles, a hurricane disaster area. On November 4th that year, in the midst of another storm, 365 cable subscribers saw the very first transmission of a new cable service, **Home Box Office** – HBO. For an extra fee above the one they normally paid to the cable operator, they received uninterrupted by commercials a hockey game and the feature film, *Sometimes a Great Nation*. The great notion of pay-television was not born that stormy night in North Eastern Pennsylvania. As we have seen, it is as old as television itself, since the earliest pioneers worked on TV specifically as a means of delivering movies directly into the home. Paying for transmission, whether over the air or through a wire, simply required scrambling the picture at one end and unscrambling it at the other. The technology for so doing was first demonstrated in 1931. More extensive field trials of pay-TV form part of Hollywood's sustained attempt to cope with the rise of New York television in the late 1940s (see above, pp. 76, 80 et seq).

Pay-television is not immediately helped by cable, except that Pennsylvanians, whose mountains had given birth to cable TV in America, had become used to the idea of paying for their television (see below, chapter 6). But balancing this advantage was cable's fragmentisation of the audience which meant pay-TV's chances of success were, if anything, reduced. At least a scrambled broadcast signal reaches out to every television set in the community, whereas with cable only a reduced pool of subscribers is accessed. Yet, despite these difficulties, the country's first effective pay-TV service was made possible by reconstituting the scattered pockets of cable-viewers into a national audience via satellite. Thus pay-TV in America is a spin-off from satellite technology.

HBO was the brain-child of a cable visionary, the man who had wired lower Manhattan, Charles Dolan. Dolan had virtually sold Sterling Manhattan Cable to Time Inc, and among his first moves in his new berth at the big organisation was a proposed pay-TV service for cable. HBO's official history quotes Dolan as having conceived the idea in the summer of 1971. 'My family and I sailed to France on the QE II for a vacation. On the way over, I wrote the first of several memos proposing a pay TV channel.'[126] Time invested $150,000 which would grow to some $30 million before the service broke even, 6 years later.

Time Inc's involvement with audiovisual media stretches back to the *March of Time* newsreels which began in 1931. It acquired its complement of television stations (now all sold to avoid 'conflict of interest' charges) and was responsible for the effective marketing of BBC product in the public television system during the 1970s. Its involvement with cable has been its most successful video enterprise, by far. It is now one of the largest owners of cable systems in the country with rising 2½ million subscribers in over 400 locations.

After two years of effort HBO had been sold to a mere 57,000 subscribers on 42 cable systems. This was a quarter of the way to financial break-even point and, although a solid enough achievement in the sorry history of American pay-TV, it was not sufficient to guarantee the business. Even if the 200,000 mark had been passed by the end of 1975, HBO would still have been a marginal enterprise.

The idea that, by using a satellite to reach these discrete (and by broadcasting standards minute) cable systems, an economically viable audience could be reconstituted occurred shortly after the ABC petition to the FCC. A number of cable industry entities were therefore involved in the 'open skies' bids of 1970 and 1971 and Dolan had mentioned the satellite option in his initial proposal to Time. By 1974, Westar was in business and the existence of a commercial domestic system affected FCC attitudes to the cable industry. Up to that point it had been, at the behest of the broadcasters, denying pay-cable permission for sports events and restricting the use of features, but it was now eager for the domestic satellite system to fly and therefore pleased to encourage pay-cable uses of it. Time spent $6.5 million dollars on leasing two satellite transponders from RCA for five years. The satellite pay-cable rights to the Ali-Frazier heavyweight fight in the Philippines were acquired and HBO went live on the bird on 30 September, 1975 with 'The Thrilla from Manilla', as the bout was billed.

The channel had been gaining subscribers at the rate of 15,000 a month. After the fight that figure doubled and by the year's end it had nearly 300,000 homes spread out over many small cable systems. Eight years later it has over 12 million and, on occasion, it achieves a bigger audience than any of the three broadcast networks. By the early 1980s each subscriber was paying an average of $10 a month for the service, about half of which went to Time; $60 million a month, nearly a quarter of a billion a year. Repetition (for 'convenience' most programmes are repeated more than six times a month) reduced the service's need for material and costs drastically. A 24-hour programme was sustained with about twenty new titles each month, with much of the material 'presold' as hit movies (and much more which failed to achieve theatrical release piggybackng, as it were, on the hits). Operating profits for video division (which include cable systems as well as these services) went from just over $20 million in 1978 to $160 million in 1982; but, by 1984, with its rate of subscriber acquisition finally levelling off and with expensive Hollywood contracts, production obligations based on the expectation of an ever-growing subscriber base, HBO was suddenly in difficulties. Some 8 per cent of its workforce, including the executives responsible for the exclusive movie deals were fired.

HBO must nevertheless be considered a hit. It acquired direct competitors such as Viacom's **Showtime** in 1978 and Warner's **The Movie Channel** in 1979; but by 1983, when these two joined forces, it was fairly clear that the market could only stand for two major film-based services.

Other satellite-distributed services were offered on a non-subscription basis – so-called 'basic' services which came into the home together with the retransmission of broadcasting signals for the monthly cable fee. A yacht-racing local television station owner in Atlanta, Ted Turner, leased satellite time in 1977 to offer cable services his undistinguished (that is to say normal smaller market independent station) programming of old TV shows and even older movies. He could, however, now claim a national audience, for within five years the station, WTBS, was reaching 26 million cabled homes. Armed with this audience, he still bought his programming as if for a local market but he sold advertising time on a national ratecard. The result was millions of dollars of increased profits. The 'superstation' was born. (It should not be forgotten that WTBS without the satellite is a highly profitable enterprise. With the satellite, Turner has used enhanced revenues to spin-off a flurry of loss-making ventures including a 24-hour news service and a music channel – a rival of Warner's successful Music TV (**MTV**) – which was off the air within weeks of its launch.)

Turner's emulators included services which were not broadcast anywhere, i.e. specially created for cable. **ESPN** (The *E*ducation and *Sp*orts *N*etwork, now entirely sports) reached even more homes than

Turner's superstation and was owned, eventually, by ABC. Most carried commercials, some did not; but the advertising dollars were hard to come by – $400 million in 1984 as opposed to $8.4 billion billed by the broadcasting networks. Some charged, others (at least at the outset) paid cable operators per subscriber. By 1984 there were 4 domestic satellites carrying at least 40 television services, many of which were only part-time and highly specialised and the vast majority of which were not profitable.[127]

Without satellites, cable television, split into thousands of small discrete systems, could never have established a national audience in the US, and without such an audience its ability to create any measure of original programming would have been reduced if not removed altogether. The degree to which cable's growth has depended upon these extra services is, despite the vigorous assertions of the cable industry, debatable. There are, however, some ramifications of the marriage of cable and satellite which are beyond dispute and cannot be ignored.

Dishes, when the FCC in its continued support of the domestic satellite system approved 4.5 m *television receive only* ones in 1976 (TVROs), sprouted at the headends (central distribution points) of cable systems. Like the Orbita stations in Russia, they could not transmit. There were 500 of them in 1977 and about 5400 by 1984. Their price dropped from $80,000 (for the mandated pre-1976 9 m version) to under $5000. It was this business within the cable industry itself which created dish production lines, making the devices subsequently also marketed to home owners – thereby contributing to the failures (thus far) of DBS and a major threat to ex-urban cable systems.

It has been our contention that, despite the efforts at every stage constraining the development of satellite communications, from the failure to exploit the V-2 to the institutional arrangements governing Intelsat, less disruption was threatened by this technology than by the others here considered. Satellites, because high-profile national interests as well as economic factors are involved, offer a very clear example of how the 'law' of the suppression of radical potential works, however inherently limited the power of disruption may be. As the systems of ownership and control slowly emerged, no national or international player was disturbed and few new players were discovered. The cost of this calm is simply that the full potential of the system and its spin-offs is denied. There are no worldwide tele-communications as cheap or cheaper than local ones; and direct broadcasting which would make the television receiver open-eyed as the short-wave radio is open-eared is unlikely ever to prosper. The Federation of the United States of the Earth will need more, much more, than balls of metal in the sky to facilitate its birth.

There now remains one element of the 'global village' to examine. A history of the television screen, in both its guises – receiver and video-display unit – has been offered. The above explains the history of space-side communications. Now all these elements must be linked – by the telephone wire. It is to the history of wired communication, specifically the telephone, that we now turn, both to complete the account of the central technologies of the supposed information revolution and to show that the model can be applied to a nineteenth-century technology.

COMMUNICATE BY WORD OF MOUTH

Phase one: scientific competence

Dr Hooke invents the TELEPHONE, 1665

The earliest practical telephone transmitter consisted of a diaphragm attached to a wire. The end of the wire dipped into a bowl containing an acid solution. An electrical contact was fixed to the bowl. As the voice vibrated the diaphragm so the wire moved. This created a variable resistance in the solution which was registered through the contact. The device was used by Alexander Graham Bell to utter the immortal words 'Mr Watson, come here I want you' on 10 March, 1876. Bell did not design this contrivance.

Its specification had been deposited in a caveat – 'a description of an invention not yet perfected' – in the Washington Patent Office nearly a month earlier on 14 February, 1876 by Elisha Gray, the co-owner and chief scientist of a Chicago telegraphic equipment manufacturing company.[1] That same day, some two hours earlier it would seem, although no record was kept, Bell patented an 'Improvement in Telegraphy' using electromagnets and a vibrating diaphragm of a kind he had been experimenting with for many months. For the past couple of years he had been in competition with Gray, both of them in the footsteps of many others, to produce a device which could increase the capacity of telegraph wires by allowing a multiplicity of signals to be carried simultaneously. It is perhaps no wonder then that evening at 5 Exeter Place, Boston, when it must have dawned on Bell that Gray's design might well transmit sound better than his own, that he spilt the acid on his clothing. Bell's patent – US#174465 – had been allowed but a week. It had been issued a mere three days before. Yet Gray's machine was clearly superior (and more fully described) than the one Bell had sketched in his deposition. Watson's was not the only help Bell would need.

More than two centuries of scientific inquiry and technological development preceded the extraordinary coincidence of that day in Washington.

In considering the ground of scientific competence for telephony, an examination of all that Bell and Gray knew or could have known of the various strands of inquiry is required. For Bell and Gray these extended well beyond the theoretical practice of pure science and included a variety of technologies; but the range of these is not so wide as to constitute a random field. Bell or Gray were working in a well-established, coherent tradition.

Robert Hooke, the English experimental physicist, wrote in 1665:

I can assure the reader that I have, by the help of a distended wire, propagated the sound a very considerable distance in an instant, or with seemingly as quick a motion as that of light, at least incomparably quicker that that which at the same time was propagated through air; and this was not only in a straight line or direct, but in one bended in many angles.[2]

Hooke was secretary of the Royal Society, the first person to posit that the motion of heavenly bodies must be purely mechanical and among the earliest to glimpse the undulatory nature of light and the nature of gravitation. Most importantly he also designed the earliest balance spring clock mechanism.[3] The string telephone which he was describing enjoys, in the toybox, a popularity which has persisted to this day. In the decade before the first demonstration of electrical telephone, these string toys were in vogue as an adult diversion, **Lovers' telegraphs**. They worked up to 170 yards, the size and nature of the cord having some effect on their efficiency, silk being better than hemp. Such toys belong here because they all depended upon the attachment of a thread or a wire to a stretched membrane.[4] They arrived, as the latest fad, in Chicago in December of 1875 and Elisha Gray was well aware of them.[5] (It is curious, also, that following the demonstration of the electric telephone in 1876 a number of reputable scientists, including Heaviside (of the layer), devoted energies to improving the string telephones, on getting a device to operate over nearly half a mile of wire.[6])

Another line of inquiry dating from the seventeenth century is also of importance since it gives rise to the word 'telephone'. A contemporary of Hooke's, Sir Samuel Morland (who had shown Pepys the not very useful 'invention for casting up of sums') developed what he called a:

Tuba Stentoro-Phonica an Instrument of Excellent Use as well at sea as at land; Invented and variously experimented with in the year 1670 and Humbly Presented to the King's most Excellent Majesty Charles II in the year 1671 By S. Morland.

Pepys tried an early version of the new invention and noted that it 'was only a great glass bottle broke at the bottom'; but nevertheless he put the neck to his ear 'and there did plainly hear the dancing of the oars of the boats in the Thames at Arundel Gallery window, which, without it, I could not in the least do.'[7]

Morland's trumpets were demonstrated as transmitters as well as receivers. One, 5 feet 6 inches long.

was tried at Deal Castle by the Governor thereof, the voice being plainly heard off at Sea as far as the King's ships usually ride, which is between two and three miles, at a time when the wind blew from the shore.[8]

Kircher, a German contemporary of Morland and the man who refined and popularised the magic lantern, also claims to have made an ear trumpet, even earlier in 1649.

At the end of the eighteenth century, mindful of the limitations inherent in visual signalling systems such as the then popular **Semaphore**, another German, Huth, proposed a system of megaphones for long-distance work. He wrote that the difference between visual and acoustic methods of signalling

might deserve a different name and it might become necessary to give a different name to telegraphic communication by means of speaking tubes. What would be more appropriate here than a word derived [like SEMAPHORE] also from the Greek: Telephon or *Fernsprecher*. [our brackets][9]

During the next century the word *Telephon* came to denote in German all speaking-tube devices whether electrical or not. Similar usages can be found in French, a communication code using the tonic sol-fa musical scale being described as *téléphonie* in 1828. Its inventor, Sudre, apparently refined the system of notation so that by 1854 his 'very curious system of telephony' relied on only three notes placed at set intervals. Sound-from-afar, in that classically educated age, was then more likely than not to be named telephony if any acoustic element was involved in the communication process, irrespective of the hardware used. In 1858 a telegraph device was designated *téléphonie-électrique* in a Paris publication.[10] The English, then being every bit as well educated as the continental Europeans in the classical languages, adopted the same nomenclature.

Captain John Taylor RN perfected in 1845 an instrument 'for conveying signals during foggy weather by sounds produced by means of compressed air forced through trumpets'. He called these foghorns **telephones**. The four notes he produced could be heard at distances of up to 6 miles. Francis Wilsher in 1849 offered speaking tubes which used gutta percha as a substitute for bell-ringing systems in private houses. He exhibited the device at the Great Exhibition of 1851 (and whereas it was referred to as a telephone in 1849, for the Crystal Palace it became the far more learned Telekouphonon.[11]) By the middle years of the century in German and to a lesser extent in French and English all devices which relayed sound through whatever medium were known in both scientific and popular literature as telephones. In the meantime, in other developments, the electric telegraph had been introduced.

Using magnetic force to move a piece of metal or needle was a trick known in antiquity. St Augustine mentions the phenomenon in *De Civitate Dei* and the use of such a system to create false oracles – by marking letters around a bowl of water, say, in which floated a cork-born needle manipulated by a hidden magnet – was considered an 'abuse' by the sixteenth century.[12] The first glimmer of idea of the telegraph appears in print at this time. The Neapolitan Baptista Porta, who made this complaint about charlatans with magnets, also wrote:

> Lastly, owing to the convenience afforded by the magnet, persons can converse together through long distances. . . . I do not fear that with a long absent friend, even though he be confined by prison walls, we can communicate what we wish by means of two compass needles circumscribed with an alphabet.[13]

In 1635, Schwenteer, in his *Délassements physico-mathematiques*, describes a system using a magnetic needle along these lines, but the experiments which would demonstrate the viability of his idea were not conducted until 1819.[14] The telegraph was side-tracked into schemes such as one proposed by an anonymous correspondent of the *Scots' Magazine* writing from Renfrew in 1753. Signalling was to be effected by 26 wires with 26 electroscopes in the form of mounted pith balls, each to represent one letter and each to be moved by static electricity. It was the first of many such ideas, one, by Bozolus, a Jesuit, being explained in Latin verse. The device existed in experimental form by the 1780s. One of the brothers Chappe, whose semaphore gave armies a practical distant signalling device, had begun his telecommunication experiments with thoughts of such a friction telegraph, before perfecting the 'optical-mechanical' system. (Navies had a long-range signalling capacity since the Duke of York, a rather better admiral than he was to be a king, instituted a code of flags during the Anglo-Dutch wars in the seventeenth century.)

Lines of semaphore stations were established. Nicholas I connected St Petersburg to Warsaw and the German border, with a branch to Moscow, by towers 5 to 6 miles apart, 220 towers each with 6 men. In

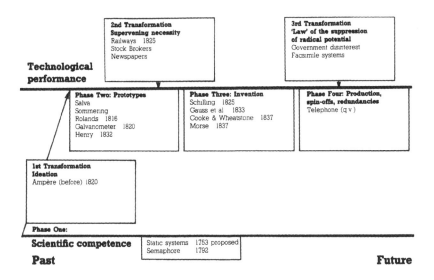

Figure 9 Telegraphy

France, by the 1840s, there were over 3000 miles of semaphore lines, all operated by the War Department. A law of 1837 established a French government monopoly in long-distance communication systems.[15] Pre-electric telegraphs, like any other technology, created a certain inertia, and research on electrical alternatives was inhibited. One of the Wedgwoods, Ralph, the creator of a useful device to write with many pens at once, planned an electric telegraph for the benefit of the Admiralty in 1814 but was turned away. Their Lordship's lackeys wrote, 'the war being at an end, and money scarce, the old system [of shutter-semaphores] was sufficient for the country' [brackets in original]. The shutter semaphore had been developed in an Admiralty competition by Lord George Murray to improve upon the French device.[16] The inventor of the most elegant of these early proposals suffered a similar fate.

In 1816, Francis Rolands demonstrated an electrical telegraph system that worked over 8 miles of wire strung up on frames in his

London garden. He mounted the clock mechanisms at either end of the wire. In place of the clock hands he had an engraved disk with letters, numbers and other instructions inscribed and in place of the glass was an opaque disk in which an aperture was cut. The clocks being exactly synchronised, the operator waited for the required letter or instruction to appear in the aperture, made the circuit and moved the electroscope, a pith ball. The receiver, seeing what letter was in the second clock's aperture as the ball moved, could note it down. Within two days of receiving notice of this apparatus, Barrow, the secretary of the Admiralty wrote:

> Mr. Barrow presents his compliments to Mr. Rolands, and acquaints him, with reference to his note of the 3rd inst., that telegraphs of any kind are now wholly unnecessary, and that no other than the one now in use will be adopted.[17]

The last static electrical telegraph was proposed, a true redundancy, in 1873, 46 years after the dynamic version was *invented*.

Systems based on dynamic electricity were proposed by the Spaniard Salva and the Austrian Sommering in the first decade of the nineteenth century, exploiting the fact that water decomposes, giving off bubbles, when electricity is introduced into it. Using a Voltaic pile and various arrangements of glass flasks, it was possible, with a series of discrete wires, to indicate letters by these bubbles.

The ideation of the modern telegraph had occurred in Schwenteer's suggestion but this was clearly forgotten; for, 175 years later, Ampère had the same sort of idea and proposed that 'one could by means of as many pairs of conducting wires and magnetic needles as there are letters' establish a signalling system. In 1819, Oersted noticed that an electric current would deflect magnetic needles and Faraday discovered that a freely moving magnetised needle when surrounded by a wire coil will move in response to the power of the electrical current in the coil. A device, the **Galvanometer**, to measure currents was built and the would-be electrical telegraphers acquired another potential signalling instrument. The bubbles were dispersed and the pith balls put aside as the prototype phase of telegraphy ended.

In 1809, Richard Trevithick took up to London the latest wonder of the country's mining areas, an iron wagon-way upon which a steam locomotive ran. At Euston Square he built a round track within a wooden fence and charged 1 shilling the ride.[18] In 1825, the first passenger train to go anywhere ran between Stockton and Darlington. The railway age began somewhat fitfully. Between 1833 and 1843 money was raised to build 2300 miles of railway in the UK, about a quarter of which was constructed during that time.[19] Early railways were single-track affairs which necessitated, for the first time, instantaneous signalling methods. One of the many who can lay claim to have *invented* the telegraph, Edward Davy, saw this clearly. In 1838 he wrote:

The numerous accidents which have occurred on railways seem to call for a remedy of some kind; and when future improvements shall have augmented the speed of railway travelling to a velocity which cannot at present be deemed safe, then every aid which science can afford must be called in to promote this object. Now, there is a contrivance (the telegraph) . . . by which, at every station along the railway line, it may be seen, by mere inspection of a dial, what is the exact situation of the engines running, either towards, or from, that station, and at what speeds they are travelling. (our brackets)[20]

Davy (who is not to be confused with Sir Humphry Davy of the miner's lamp) was eager to have the railway interests exploit his telegraph and the earliest wires did indeed run beside railway tracks. In 1840, the first telegram to excite London, that the Queen by giving birth to a child had removed the unpopular King Ernest of Hanover as heir-presumptive, was carried from Windsor on the Great Western Railway's telegraph line, developed by Cooke and Wheatstone. Four years later, 'What hath God wrought', Morse's first public message, was carried down a telegraph wire running from Washington to Baltimore along the side of railtracks. (The wire was supplied by the Philadelphia corset stays factory of Moore, the eponymous benefactor of the School of Engineering where ENIAC was to be built.) In that May, the Democratic National Convention was meeting in Baltimore and Silas Wright, its nominee for vice-president, declined by telegram from Washington.[21] A committee was dispatched by train to check the truth of this communication. The first French wire ran beside the tracks from Paris to St Germain.[22]

The same year which saw the inauguration of the supervening necessity in the form of the Stockton and Darlington railway, also witnessed the *invention* of the telegraph. Baron Pawel Schilling, a Russian diplomat in Germany, had seen Sommerring's apparatus. Using a battery-powered galvanometer, Schilling designed a device that worked in code. Right and left deflections of the needle indicated the letters – e.g. A = RL, B = RRR, C = RLL and so on. Thus the telegraph, like the television, radio, aeronautics and astronautics, turns out to be a Russian invention. However,

the Emperor Nicholas saw in it only an instrument of subversion, and by an *ukase* it was, during his reign, absolutely prohibited to give the public any information relative to electric telegraph apparatus, a prohibition which extended even to the translation of the notices respecting it, which, at this time, were appearing in the European journals.[23]

The Tsar also had a vested interest in his optical system.

The binary code which is commonly supposed to have been *invented* by Morse was not *invented* by Schilling either but again dates back to antiquity; and in the 6th book of *The Advancement and Proficience of Learning* (1604) Bacon gives an example, using the letters A and B as the binary base.[24]

In the University of Göttingen, in 1833, Gauss and some colleagues rigged up a telegraph from the physics department offices to the university observatory and the magnetic lab, a distance of 1.25 miles. Using a system along Schilling's lines, the Göttingen faculty evolved a four-bit right/left code. In 1835 their apparatus became the first to be powered by a 'magneto-electric machine', a proto-dynamo.

Galvanometers were used by Cooke and Wheatstone to construct an elegant alphabetic system, but one which needed initially 5, and later 2, wires to operate. The patent was granted on 12 June 1837 and eventually, by 1840, they had five galvanometers set in a line across the centre of a lozenge-shaped board on which were painted twenty letters By deflecting any two needles, one letter could be isolated. A Scotsman, William Alexander, on the very day of the initial patent, wrote to Lord John Russell the then Home Secretary with a proposal for a telegraph between London and Edinburgh. Three days later an acknowledgment was sent but no action was taken. In December, somewhat unwillingly, Alexander inspected the Wheatstone telegraph and admitted its superiority to his own. But he was not the only English experimenter in the field and Wheatstone was writing to his partner Cooke the following month:

> Davy has advertised an exhibition of an electric telegraph at Exeter Hall . . . I am told he employs six wires, by means of which he obtains upwards of two hundred simple and compound signals, and that he rings a bell. I scarcely think that he can effect either of these things without infringing our patent.[25]

Edward Davy, the son of a West Country doctor and inventor of 'Davy's Diamond Cement' for mending broken china and glass, had lodged a caveat against rumours of Wheatstone's work the previous March and it seems as if his was the superior scheme. Only the scientific inadequacies of the Solicitor General, who thought the devices were different when in fact they were not, allowed Wheatstone and Cooke their patent. Davy strenuously struggled to have this decision overturned and to exploit his version with the aid of supporters among the railway men. In the midst of this struggle in the summer of 1838, he wrote to his father, 'I have notice of another application for a patent by a person named *Morse*.' (Italics in original)[26] Davy succeeded in obtaining a patent but not in having his rivals denied, and, upon his emigrating to Australia where he practised his father's profession of medicine, the diffusion of his design ceased. (Thus an Australian also *invented* the telegraph.) Cooke and Wheatstone's model was adopted by many British railway companies but they were not to triumph elsewhere. Their nemesis was not to be Davy but rather the 'person named *Morse*'.

S.F.B. Morse, who was to become known as 'the American Leonardo', was the son of a New England Congregationalist minister. After Yale, where he had exhibited a talent for art, he had become a professional portraitist and eventually a professor of painting at the forerunner of New York University, the University of the City of New York. A daguerreotypist who took the first photographic portrait in the USA, he was also a child of his times, rabidly anti-immigrant, i.e. anti-Irish and anti-Catholic. His best-known paintings were *Lafayette* and *The House of Representatives* and his understanding of electricity informal. Crucial to his interest in telegraphy were the fame and proximity of Joseph Henry, subsequently secretary of the Smithsonian but then, in the late 1820s, a professor at the Albany Institute.

By substituting numerous small voltaic cells for the large one usually employed in such experiments, Henry had been able to create an electromagnetic pull sufficiently strong to move an arm with bell attached. Henry, who was called to the chair of Natural Philosophy in the College of New Jersey (Princeton) in 1832, publicly demonstrated bell ringing from afar but did not patent the device. Morse, using Henry's apparatus as a starting point, and the expertise of two of his friends who possessed a broader grounding in electrical studies, built a contrivance wherein the electrical current deflected a marker across a narrow strip of paper, a recording telegraph. The sender used notched sticks which were pulled across the electrical contact to transmit the impulses. In September 1837, some 2 years after he had first made a working model and 5 years after he had began his experiments, he filed a caveat; the Morse system, with its code, was ready. (Morse's contribution to codes was to send a backer and his assistant Vail to a printer where the relative frequency of the letters was gauged by examining type-fonts. Previous systems, such as Schilling's, seemed to rely on commonsense with the vowels represented by the shortest number of impulses but little further refinement.)

The diffusion of the telegraph was to be a most vexed affair, awaiting the full development of the railways in the mid-century 'mania' but even then not being fully exploited in that context. Other electrical signalling systems for track control came to be developed and telegraphy was relegated to carrying internal operational messages. In fact the telegraph exhibits as well as any telecommunications technology the operation of the 'law' of the suppression of radical potential. Morse came close to libelling Henry when the latter began to give expert evidence against him in various patent cases. The Morse patent was denied altogether in England; the apparatus, unenthusiastically received by the French government, was nationalised there. Even in the US, Morse, who had raised government funds to build the experimental Washington-Baltimore line, had considerable difficulty in diffusing the system further.

In 1845, the line had been operated by the US Post Office and in the first six months of the year it had cost $3284.17 and had brought in $413.44. The Postmaster General asked the crucial question:

> How far the government will allow individuals to divide with it the business of transmitting intelligence – an important duty confided to it by the Constitution, necessarily and properly exclusive. Or will it purchase the telegraph, and conduct its operations for the benefit of the public?[27]

But the public had thus far shown very little interest in benefiting from the telegraph and the Congress demurred at further involvement with such an economic white elephant. Despite the Postmaster General's talk of 'an instrument so powerful for good or evil' which could not 'with safety be left in the hands of private individuals uncontrolled by law', a crucial privatising precedent was set in American public communications policy. The telegraph was returned to Morse and his backers.

This policy was not reversed when telegraph use became commonplace a year or so later and private enterprise was creating a system racked with patent disputes, geographical dislocations and redundant duplicated wires. Nor did Congress feel moved to involve itself in either the emergence of the great national private monopoly (Western Union) or the international developments that followed. The principle established with the telegraph was to hold good through all the subsequent technologies until, as we have seen, a century and quarter later Congress allowed commercial corporations into space. What had initially been adopted to avoid wasting taxpayers' dollars became a self-denying ordinance – the government would not engage in profitable communications enterprises. In time, this would help underpin the received opinion that government enterprises in telecommunications and elsewhere cannot, by their very nature, be profitable.

From 1845 on, Morse and his partners began to build the network by licensing themselves and other entrepreneurs to construct the lines.

Henry O'Reilly, a Philadelphian newspaperman, secured the rights to the western half of the system on terms which the patentors subsequently regretted and sought to annul. In more sparsely populated country the wires were soon seen as necessities and O'Reilly did better than anticipated. By 1847, in certain sections, between Louisville and Nashville for instance, O'Reilly and Morse's licensee raced side by side to complete the line. O'Reilly turned to other telegraphic instruments of which there were a number available. Cooke and Wheatstone had a prior patent to Morse's in America and there was another machine, that of House, also on the market; but it was Bain's chemical telegraph, the facsimile prototype, which O'Reilly exploited. By 1849 there were rival Bain and Morse lines between Boston and Washington and New York and Buffalo.

The courts intervened, preventing further chaos by declaring, in 1851, the Bain patent invalid and his machine an infringement of Morse's. With this declaration, the logic of monopoly inherent in all public communications systems began to assert itself. Three large telegraph holding companies existed by the end of the decade. Finally William Orton, head of one of these, United States Telegraph, merged it into Western Union in 1866. *The Telegrapher* lamented: 'With the power which this great monopoly puts into the hands of its officers, salaries will decrease as surely and as naturally as water runs down hill.'[28] The last of the major concerns, the American Telegraph Company, followed Orton's lead into Western Union. The monster had absorbed more than 150 lines to establish a system extending 37,380 miles with 2250 telegraph offices and capital of over $41,000,000. Professor Morse wrote that the telegraph interests were now 'very much as . . . I wished them to be at the outset, not cut up in O'Reilly fashion into irresponsible parts but making one great whole like the Post Office system.'[29] (In Europe the telegraph was seen as an extension of postal, i.e. state, services.)

It had taken the better part of three decades to establish the system and slowly, always in obedience to the 'law' of the suppression of radical potential, uses were discovered which had noticeable social effects. Stock speculators were early on the scene and the beginnings of that process which today allows long-distance simultaneous participation in a multitude of international money and commodity markets was demonstrated for the first time. Newspapers became avid consumers of telegrams, which had a considerable effect on their contents. No longer, as in the eighteenth century, was it possible to scoop a rival with 'late intelligences'; now, the telegraph rendered news, like soft fruit, perishable – useless if delayed. As the century declined, the telegraph worked to stifle the analytic in favour of the reactive press. It helped the modern popular paper to emerge. And thus was Morse, although by no means the telegraph's *inventor*, responsible for the form in which it was eventually diffused. Only certain extremely conservative English railway companies stuck with alphabetic telegraphs beyond the midpoint of the century.

The telegraph affords the model of an electrical signalling system and is clearly crucial in the development of the telephone, but other strands of inquiry into various acoustic phenomena are of equal importance. Sir Charles Wheatstone, who came from a family of musical instrument makers, gave the world the concertina as well as the alphabetic telegraph and was interested in the acoustic aspects of long-range signals. In 1821 he demonstrated the **Enchanted Lyre**. If the sounding boards of two instruments were joined together Wheatstone showed that notes played on one would be reproduced on the other. For about two years the machine was exhibited in London; and

> so perfect was the illusion (wrote Sir Charles) in this instance from the intense vibratory state of the reciprocating instruments, and from the interception of the sounds of the distant exciting one, that it was universally imagined to be one of the highest efforts of ingenuity in musical mechanism.[30]

In fact it was an extension of a physical phenomenon known to anybody who had ever put ear to ground, which, as that procedure is described in Pliny, means considerable numbers of people. In a letter to his partner Cooke in 1840 referencing the line of experiments begun with the Lyre, Wheatstone says:

> When I made, in 1823, my important discovery, that sounds of all kinds might be transmitted perfectly and powerfully through solid wires, and reproduced in distant places, I thought that I had the most efficient and economical means of establishing a telegraphic (or rather telephonic) communication between two points that could be thought of.[31]

The relationship between resonance experiments and the development of the electric telegraph is twofold. As Wheatstone indicates, both are concerned with distant signalling. Secondly, the discovery of certain acoustic phenomena, held to be of the resonance type but connected with electromagnets, promised considerable practical advantage in melding sound to electricity.

The first electromagnetic device which converted electrical waves into sound is credited to a Dr C.G. Page of Massachusetts in 1837. He achieved this effect, which he called 'Galvanic Music' (but which the rest of the world named for him, 'the Page effect') by revolving the armature of an electromagnet in front of a negative and positive electrical pole. Loud sounds were emitted, which could be varied by altering the strength of the current in the poles.[32] In 1846, M. Froment of Paris showed a device which was designed not to create sounds but to analyse those made by Page's effect. This vibrating bar arrangement was a direct precursor of Helmholtz's experiment in the following decade to show that electrical impulses could be sent down a line and cause a tuning fork to resonate on the principle of sympathetic vibration. Varley in 1870 demonstrated a similar acoustic phenomenon using capacitors where different notes were produced as the charge was varied.

The investigations were seen as having important economic consequences for telegraphy. If a number of Morse senders and receivers could be variously tuned, it was theoretically possible that all these signals, sounding different notes, could be sent down the same wire simultaneously but independently of each other. Increasing the capacity of the wires without physically stringing more of them was obviously desirable. Since, although the message was still encoded in dots and dashes, an acoustic element was involved, all the devices for the improvement of telegraphy along these lines were referred to as *telephones* – **harmonic** or **musical telephones**.

To move from harmonic telephones to the speaking or articulating telephone, required, as did the development of television, an understanding of the variable resistance to electricity found in substances, from sulphuric acid to carbon, under different physical conditions. The Comte Du Moncel, telephony's first historian (1879), had experimented with variable resistance from 1856 on and had published his results.[33] In 1866, M. Clarac of the French telegraphic administration 'constructed tubes containing powdered carbon, the electrical resistance of which could be regulated by increasing the pressure upon it by means of an adjusting screw.' The purpose of the device was simply to demonstrate the variable resistance phenomenon and it flowed from work by Sir William Thomson (Lord Kelvin) which showed that resistance to current in a wire could be varied by putting the wire under tension.[34]

Finally, to move from devices where signalling was accomplished with discrete electrical pulses to systems which offered electrical analogues of sound waves required an understanding of human speech. Helmholtz is of importance to this strand of competence too, what today would be called psycholinguistics. He published, in 1862, his seminal work *Sensations of Tone* and the fixed pitch theory of vowel tones he enunciated was part of the underpinning of the experiments with harmonic telephones. This work was part of an ongoing widespread interest in the production and reception of human speech.

Alexander Graham Bell was an elocution teacher concerned with deafness. The received impression is that such a man invents the telephone as the very pattern of inspired nineteenth-century amateur;

and further his amateur status is in complete contrast to the way in which things are done these days, not least at the mighty research laboratories that bear his name. But this impression is wrong. There was a continuum between the scientific investigation of electro-magnetic phenomena and human communication, as Helmholtz's interests indicate. Bell's father and grandfather, both in his profession, had made distinguished contributions to the psycholinguistic literature of the day (at least in its practical elocutionist aspect). His father had invented a universal system of orthography called 'visible speech'. Bell from his childhood was aware of this work, the theory of speech production and the creation of machines to produce human sounds.

Work on **speech synthesisers** goes back to the eighteenth century. In 1779 the St Petersburg Academy offered a prize for a machine which could reproduce vowel sounds. In 1791 De Kempelen of Vienna built a more elaborate device which attempted consonantial sounds as well as vowels. By 1843 Reale was able to show the American Philosophical Society a machine which could enunciate a few words, but destroyed it (after sixteen years of work) 'in frenzy'.[35] Another Viennese, Faber, exhibited a machine in London in 1846 which was considered the best of these. His device in essence copied as closely as possible in mechanical form the structure of the human larynx and attendant parts. Fourteen notes arranged on a keyboard allowed the device to utter all the vowels and a selection of consonants. Du Moncel reported, 'the public will believe that the assertions of ventriloquism are unfounded when I add that I have myself made the machine speak.'[36]

Following experiments by Young at the beginning of the century, Scott's **phonautograph** of 1855, a contrivance designed to listen (as it were) rather than to talk, was to be of some consequence both for the telephone and the phonograph.

> This [phonautograph] consisted of a cone over the smaller end of which was tightly stretched a membrane, and hinged at the end of this was a long wooden lever. At the other end of the lever a short pig's bristle was attached and suspended just above the surface of a sheet of glass covered with lampblack. By speaking into the cone and moving the glass, the pig's bristle would trace the pattern of the sound waves. [our brackets][37]

Bell was well enough immersed in this tradition to warrant a place in its history, albeit small, even if he had not gone on to the telephone.

His father offered the three Bell boys prizes, after the fashion of the Academy of St Petersburg, for speaking machines. Alexander appar-ently produced a crude version using the model of a human skull stuffed with India rubber and cotton. His father was acquainted with

Sir Charles Wheatstone and borrowed scientific literature from him. Wheatstone himself demonstrated the De Kempelen **Automaton Speaking Machine** to the young Bell.[38] In the 1860s after meeting Alexander Ellis, the man who was to translate Helmholtz into English, Bell also developed an interest in electricity. He read Helmholtz himself in French in 1870, five years before the Ellis translation appeared.[39]

By 1872 Bell was experimenting along Helmholtzian lines. In the summer of 1874, while immersed in the problems of the harmonic telephone, he took time to build a macabre version of Scott's phonautograph. He obtained the ear of a deceased man and rigged it up to a metal horn with the armature and stylus attached to the ossicles. He got very good tracings from the device.[40] He was also aware of Koenig's **manometic capsule**. In this machine the voice acting on a membrane caused a gas flame to flicker. The flame was reflected in a continuously revolving mirror which converted the flickerings into seamless bands of light. Both these devices, like his father's 'visible speech' system of writing, were part of a search to allow the deaf to 'see' speech. Since the patterns produced by these two contrivances differed radically, Bell gave up on them as potentially useful tools in his professional work as a teacher of the deaf. And anyway, as his harmonic telephone researches indicated, his mind was on other matters.[41]

The first transformation: ideation

Sir Charles Wheatstone invents the TELEPHONE, 1831

String phones, megaphones and ear trumpets; sound resonators, the telegraph and electromagnetic phenomena, variable resistance to electrical current in different substances, the wave theory of electro-magnetic phenomena and the problems of the reception and produc-tion of speech, together form the ground of scientific competence which leads to telephony. In effect all these elements constitute a tradition of research, both theoretical and practical, which can be transformed into the **electric speaking telephone**.

The ideation transformation works on elements within the ground of scientific competence and might or might not lead to devices in the metal during the technological performance phases. In the case of the telephone, but not the television, say, no recorded theoretical notions (including those of Bell himself) envisage a fully practical machine. In such circumstances ideation demonstrates, as it were, the possibility that against a specific scientific background a technology might be envisaged in a certain way – exercises in generation which stop short of performance.

Wheatstone in 1831 wrote:

When sound is allowed to diffuse itself in all directions as from a
centre, its intensity, according to theory, decreases as the square of
the distance increases; but if it be confined to one rectilinear
direction, no diminution of intensity ought to take place. But this is
on the supposition that the conducting body possesses perfect
homogeneity. . . . Could any conducting substance be rendered
perfectly equal in density and elasticity so as to allow the
undulations to proceed with uniform velocity without any reflections
or interferences, it would be as easy to transmit sounds through such
conductors from Aberdeen to London as it is now to establish
communication from one chamber to another. . . . The almost
hopeless difficulty of communicating sounds produced in air with
sufficient intensity to solid bodies might induce us to despair of
further success; but could articulations similar to those enounced by
the human organs of speech be produced immediately in solid
bodies, their transmission might be effected with any required
degree of intensity. Some recent investigations lead us to hope that
we are not far from effecting these desiderata.[42]

Although Wheatstone, less than a decade away from his telegraph
and in the midst of his success with the Enchanted Lyre, did not seem
to have electrical signalling in mind, the references to undulations and
the notion of creating an analogue of the human voice are essential to
the idea of the telephone. By the mid-years of the century, the
connection to electricity was being made. Du Moncel, in 1853,
envisaged using Page's effect to create with different tunings on
various metal plates 'an electric harmonica'.[43] A year later Charles
Bourseul published the following in Paris:

I have for example, asked myself whether speech itself may not be
transmitted by electricity – in a word if what is spoken in Vienna
may not be heard in Paris. The thing is practical in this way:
    We know that sounds are made by vibrations. . . . Suppose a man
speaks near a moveable disk, sufficiently flexible to lose none of the
vibrations of the voice, that this disk alternately makes and breaks
the currents from a battery; you may have at a distance another
disk, which will simultaneously execute the same vibrations. . . . It
is, at all events, impossible in the present condition of science to
prove the impossibility of transmitting sound by electricity.
Everything tends to show, on the contrary, that there is such a
possibility. . . . I have made some experiments in this direction: they
are delicate, and demand time and patience, but the approximations
obtained promise a favorable result.[44]

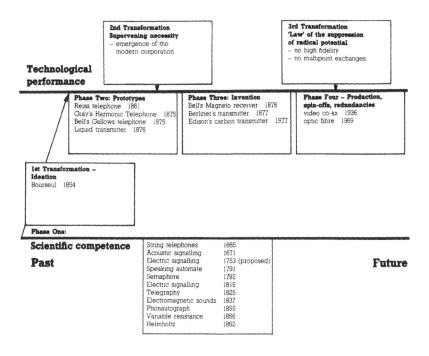

Figure 10 Telephony

It is fairly clear that Bourseul's 'approximations' did not amount to an actual machine, but the idea of electric telephony is clearly enunciated. Omitted is the need to impress an analogue of the voice on the current; Bourseul's 'makes and breaks' suggest a digital mode, as in the telegraph.

There was one non-technical impediment to this work. The notion of electrically based voice communication was, from the 1840s on, regarded as chimerical – 'a sort of term of reproach' – and an indication, in any searching for it, of a measure of mental disturbance. Thus when Cooke and Wheatstone went to arbitration to determine which of them should have the credit for their telegraph, it is Cooke's solicitor who constantly uses the word 'telephone' to emphasise, as it were, the impracticality of Wheatstone's ideas.[45]

The telephone sacrificed its respectability as an object of proper scientific inquiry not only because of a succession of, more or less, non-speaking devices between Bourseul and Bell of the harmonic telephone type but also because of the long tradition of subterfuge associated with remote-controlled speaking figures going back, perhaps, to the oracles of the ancient world. In such a tradition, 'assertions of ventriloquism' continued to find their mark.

A fear of ridicule inhibited Bell. He was working on the harmonic telephone but he explained his failure to build a particular speaking telephone prototype in the summer of 1874 in part thus:

> Fearing ridicule would be attached to the idea of transmitting vocal sounds telephonically, especially by those unacquainted with Helmholtz's experiments, I said little or nothing of this plan.[46]

The plan in question combined two elements. Bell knew that a piano note will sympathetically resonate with a human voice and hoped that a series of tuned reeds might be simultaneously vibrated to produce different electrical waves. Such a device, the **harp telephone** as it came to be called, could represent a good solution to the harmonic problem, each reed allowing a different morse message to be sent. But Bell, eager to push his conception of the speaking telephone back as far as possible, subsequently stated the harp was the 'first form of speaking telephone that occurred to my mind'. This claim was characterised by counsel in the government case against AT&T in the 1880s as 'incompetent and argumentative'.[47] Given Bell's bald statement, there is some justice to that objection, but Bell had further variations in mind.

He understood that within the human ear, in the hair-like organ of Corti, were hundreds of fibres which Helmholtz believed were mathematically related to sound reception: '4200 [fibres] for the seven octaves of musical instruments, that is 600 for each octave, 50 for every semi-tone.' [our brackets][48]

Bell came to think of a sort of piano-sized musical box comb with between 3000 and 5000 tines – the number Helmholtz had led him to believe was needed to receive human speech. Such a device went well beyond the needs of the harmonic telephone and clearly indicated his interest in voice transmission, but he claimed he was inhibited both by expense and fear of ridicule from producing this object in the metal. In the winter of 1874, having secured some backing, he continued to pursue the commercially attractive harmonic telephone and began to have his prototypes built for him at a professional telegraphic equipment manufacturers in Boston, where he was assigned the services of a young electrician, Thomas Watson. Watson attests to Bell's hesitancy about telephony and its dubious status in 1875.

> We discussed the possibility of constructing [a telephone], but nothing was ever done about it for Thomas Sanders and Gardiner

G. Hubbard, the two men who were financially and otherwise supporting Bell in his experiments, were urging him to perfect his telegraph, assuring him that then he would have money and time enough to play with his speech-by-telegraph vagary all he pleased. So we pegged away at the telegraph and dreamed about the other vastly more wonderful thing. [our brackets][49]

Clearly Bell felt confident enough about Watson to share his thinking with him in some detail.

Bell had a remarkable power for clear and terse explanation. The words he used in giving me the essence of his great idea have remained with me ever since. 'Watson', he said, 'if I can get a mechanism which will make a current of electricity vary in its intensity, as air varies in intensity when a sound is passing through it, I can telegraph any sound, even the sound of speech.' He went on to describe a machine which he thought might do this. It was an apparatus with a multitude of tuned strings, reeds and other vibrating things, all of steel or iron combined with many magnets. It was big, perhaps, as an upright piano.[50]

This account is less lucid than Watson suggests and its inconsistencies find expression first in the patent Bell was awarded a year later and subsequently in the more than 600 legal actions he and his partners fought to defend it.

The 'multitude of tuned strings' has nothing to do with the eventual solution and does not imply a need for a 'mechanism which will make an electric current vary in its intensity'. In fact the rationale Bell gave Watson is somewhat at odds with the device he described, crucially in this matter of variable resistance. Variable resistance was to be at the heart of the telephone transmitter in its *invented* form, but Bell and his supporters had some considerable difficulty laying claim to the concept prior to February 1876. The record yields only one conclusive mention of variable resistance before this date, in a letter by Bell written in May 1875.

Another experiment has occurred to me which, if successful, will pave the way for still greater results than any yet obtained. The strings of a musical instrument in vibrating undergo great changes of *molecular tension*. In fact, the vibration represents the struggle between the tension of the string and the moving force impressed upon it. I have read somewhere that the resistance offered by a wire to the passage of an electric current is affected by the *tension of the wire*. If this is so, a *continuous current of electricity* passed through a vibrating wire should meet with varying resistance, and hence a pulsatory action should be induced in the current. If this turns out to be the case, the oscillations of the current should correspond in *amplitude*, as well as in the rate of movement, to the vibrations of the string. One consequence would be that the *timbre* of a sound could be transmitted. (italics in original)[51]

It is curious that Bell should refer to an experiment by Kelvin, later to be one of Bell's most influential supporters and already a most famous scientist, in such vague terms: 'I have read somewhere. . . .'; but that he did not conduct any experiments with wires under tension for four years and, most importantly, makes only glancing reference to variable resistance in the first patent, is even stranger and of extreme significance.

In 1875 it is clear that Elisha Gray was conceptually closer to the telephone than Bell was. Eight years earlier, Gray had made his first experiments in the area of Page's effect and claimed to conceive then of

the idea of a telephonic system based on the differences of resistance effected in a circuit completed by a liquid where the layer of liquid interposed between the electrodes varies in thickness under the influence of the telegraphic plate which is in connection with one of the electrodes.[52]

It would seem reasonable to suggest that Gray too is being tendentious. The concept is most unlikely to have been so well formed at the early date claimed, especially since Gray, by his own account, did not pursue the idea for some years. Gray went on to achieve a great deal with his harmonic telephone researches, but he also continued with less focussed work. In 1874 he was conducting some experiments with a battery and the zinc lining of a bathtub which demonstrated that he could amplify the galvanic noise by friction and pressure.[53] Then the following December, in Milwaukee, he saw the lovers' telegraph which he stated:

proved to my mind that the movements of a single point on the diaphragm corresponded accurately with the movements of the air produced by any spoken word or sound. I saw that if I could reproduce electronically the same motions that were made mechanically at the centre of the diaphragm by speaking upon it, such electrical vibrations would be reproduced upon a common receiver in the same manner that musical tones were. . . . The fact that the longitudinal movement (in water or other fluid of poor conducting quality) of a wire or some good conductor of electricity, with reference to another wire or metal conductor, produced variations in the resistance of an electric circuit proportional to the amplitude of movement, was old in the art of that time; so that the last link of knowledge necessary to solve the problem in my mind was furnished in the capabilities of the longitudinal vibrations of the string in the before mentioned so called lovers' telegraph.[54]

Without any experimentation in the metal, Gray simply, in February 1876, designed a system along these lines and filed it as a caveat – an incomplete *invention* – a few hours (apparently) after Bell filed for his patent. The caveat was quite clear that the object of the device was 'to transmit the tones of the human voice through a telegraphic circuit and reproduce them at the receiving end of the line, so that actual conversations can be carried on by persons at long distances apart.'[55]

The transmitter consisted of a tube, in the shape of a flat-bottomed flask, the bottom of which was covered with a membrane. A platinum wire ran from the centre of the membrane into a small container of sulphuric acid which was connected to a battery. The other battery terminal was a button set into the bottom of the container. His planned receiver was an electromagnetic device of less sophistication than that which Bell had already built. Because of this, although it was clear that the transmitter was better than Bell's (as was proved by Bell's building one virtually identical to it on 10 March, 1876) it was equally true that Gray, like Bell, was some way from a system.

Apart from the simultaneous arrival of the plans at the patent office, there is a distinction of style between the two men. Bell patented a telegraphic device, which was finished as such and had been built but was not a fully practical speaking telephone system. He described it as an improvement in telegraphy not only because that what it was but also presumably because he needed to satisfy his backers that he was not wasting their money on a 'speech-by-telegraph vagary'. Nevertheless there was a certain boldness about his approach.

I was so satisfied in my own mind that I had solved the problem of the transmission of articulate speech, that I ventured to describe and claim my method and apparatus in a United States patent, without waiting for better results; in full confidence that the problem had been solved, and that my instruments *would turn out to be* operative speaking telephones. I was more concerned about taking out a caveat or patent than about further experiments. I believed I had experimented sufficiently to entitle me to a patent. (our italics)[56]

In the event, Bell's confidence was somewhat misplaced. His instruments turned out to be operative *listening* telephones, i.e. receivers. They were never to work very well as transmitters.

Gray, on the other hand, built nothing but was on conceptually sounder ground, at least as far as the transmitter was concerned, and was clear about his telephonic intentions. There was no talk of harmonic telegraphy in the caveat. It can be strongly argued that as far as telephony was concerned Bell too should have petitioned for a caveat. These confusions were confounded when a year later Gray wrote, in words he was increasingly to regret, to Bell:

> I do not, however, claim even the credit of inventing it [the telephone], as I do not believe a mere description of an idea that has never been reduced to practice – in the strict sense of that phrase – should be dignified with the name of invention.[57]

It is true that originally patents were granted to people who, in terms of the oldest statute on this matter, that of the republic of Venice, 1474, '*build* any new and ingenious device in this City.' (our italics)[58] The statute was designed to give to practical innovators exactly the protection given by copyright to authors, or rather publishers. It arose simultaneously with the introduction of printing and the earliest surviving patent was granted to John of Speyer exactly for the introduction of moveable type into Venice. A century later the concept reached the Common Law, was conformed to the pre-existing English system of grants from the crown, in the form of an open (or patent) letter for a monopoly, and was first codified in 1623 as the Statute of Monopolies. As the law developed Patent Offices were established, that of the United States in 1790, and *built* was glossed to include processes. The patent system under which Gray and Bell were operating required either partial or full specifications of the contrivance, not the contrivance itself nor proof that it really worked. Gray's scrupulous disclaimer was therefore quite unnecessary in law. His was not 'a mere description', although such would have sufficed, but a rather more fully described system than Bell's, yet he chose to register it as a partial innovation while Bell claimed his partial innovation was complete. Boldness (and, as is explained below, some bureaucratic help) was to carry the day.

Both Bell and Gray met one essential patent requirement, that the contrivance be useful. The issue is which of them met the other essential, that it be novel? It is obvious from this account that the ideation of telephony was within the state of art at this time and not dependent on advances in any technological or scientific area. It should therefore come as no surprise that the indefatigable Du Moncel, in his exhaustive catalogue of early workers, mentions that neither Bell nor Gray was first with the liquid version of variable resistance.

> It is a curious fact worth recording here, that Mr. Yates of Dublin, in 1865 when trying to improve Reiss's telephone, realized to a

certain extent Mr. Gray's concept of the liquid transmitter for he introduced into the platinum contacts of Mr. Reiss's instruments a drop of water, which adapted it for the reproduction of articulate sounds. However no notice was then taken of this result.[59]

If built, Yates's machine would have been a prototype.

Phase two: technological performance – prototypes

Herr Professor Philip Reiss invents the TELEPHONE, 1861

Mr Yates was tinkering with the device that was to cause Bell even more trouble that Gray's caveat, a widely known apparatus called a telephone, the creation of Philip Reiss. 'Probably the physics laboratory of every well-equipped college in the world had one in 1870', according to Thomas Watson.[60]

During the prototype phase, apparatus will often already be to hand, serving other purposes. Just as Rozing adapted an oscilloscope for the earliest electrically scanned television receiver or Marconi took Hertz's coherer to create a radio, so here in telephony can be found a parallel prototype of the same kind. In 1860 by putting together the noise-emitting electromagnets of Page with the vibrating diaphragm of Scott's phonautograph, Reiss created, for the advanced study of electromagnetic phenomena, a telephone. He was a teacher in Frankfurt who seemed to have been aware of both Helmholtz's early formulation of a sound wave theory, knowing perhaps of a popular lecture given by Helmholtz three years previously, and Bourseul's outline of an electric telephone on 'makes and breaks' principles.[61]

Reiss built a sounding box with a speaking tube attached to one side. In the top of the box was a large circular opening, across which was stretched a membrane. At the centre of this was fitted a thin disk of platinum. A metallic point rested above the platinum – a hinged 'moveable lever touching the membrane'. At its other end the lever acted upon a morse key whose impulses were sent via an electro-magnet to a receiver.[62]

From the point of view of another technology, Reiss's is in fact an *invention* in a line of devices, technological performances, all designed to demonstrate one or another aspect of wave theory. The culmination of this line is the Hertz *invention* of 1888, a wave coherer. (This machine is also a radio but, because it was not seen as such, must be classed as a prototype performance of that technology.) From this standpoint, the Reiss telephone occupies a similar position. It was, in the history of telephony, a parallel prototype – a device designed for other purposes, whose supervening necessity was Helmholtz's theoretical discussion of waves.

Reiss knew about waves but did not know how to impress sound waves on an electrical current. He did not in effect understand the need to vary the carrying current by means of a variable resistance mechanism. Clearly the device attempts, despite Reiss's understanding of the waveform that sound takes, a digitalisation of the sound wave – nearly a century in advance of the successful resolution of this possibility.[63] Nevertheless, the Reiss transmitter had a number of sophisticated features, the most important of which was that it relied on a single diaphragm. It also used electromagnets to create a constant current upon which the endlessly making-and-breaking signals of the morse sender could react. Despite these severe limitations, it was close to being a working telephone and a rich source of inspiration. Reiss himself made three other variations and Mr Yates was but one of many copyists.

In 1848 a refined Reiss telephone was made by Vander Weider and demonstrated at the Cooper Union, New York. 'The vocal airs [presumably without words] were faithfully reproduced but rather weak and nasal.' [our brackets][64] Vander Weider persevered, increasing the reverberations in the sending box by means of partitions and additional coils in the receiver. Other scientists throughout the 1860s conducted similar improvements, normally involving doubling the coils in the receiving box.

Elisha Gray, whose contributions to the ideation of telephony are discussed above, had a distinguished career as an innovator, with major advances in facsimile telegraphy to his credit. In 1869, he had founded Gray and Barton, a telegraphic equipment manufacturing firm in Cleveland, which was to become Western Electric of Chicago. Obviously as a supplier to the telegraphy industry, Gray would gain tremendous commercial advantage with the marketing of a multiple harmonic telephone and Reiss's apparatus seemed an attractive jumping-off point. Backed by a Dr S.S. White who had made his fortune in dental supplies, Gray began serious work on this project in 1874. In April he constructed a transmitter on a diatonic scale, which he began demonstrating in May. By December 1874, Gray had sold the device to William Orton of Western Union, with whom he anyway had a close working relationship, and had filed for a patent on this harmonic or musical telegraph. The patent was granted in February 1875.[65] Apart from its telegraphic importance, it was complex enough, when fitted with a range of vibrating disks arranged in a scale, to be 'a new kind of instrument' upon which a piece of music, at least from the accordion or harmonium repertoire, could be played. Kelvin, on a visit to America subsequent to the *invention* of the speaking telephone, saw Gray's machine at work.

In the department of telegraphs in the United States I saw and heard

Mr. Elisha Gray's electric telephone, of wonderful construction, which can repeat four dispatches at the same time in Morse code, and, with some improvements in detail, this instrument is evidently capable of a fourfold delivery.[66]

So important to Gray was this harmonic telephone that its exploitation distracted him in his patent battle with Bell. He had been told in February 1876 of the existence of a rival telephone scheme, as the law required, and had learned in March its details and that Bell was the patentee. In May 1876, he took his harmonic telephone to the Philadelphia Centennial Exhibition where he met Bell. The following month he constructed a model of his speaking telephone and in November finally made the one exactly specified in his caveat.[67] Gray's inactivity is difficult to explain. Perhaps the fact that Orton and therefore Western Union were uninterested in telephony at this point contributed and perhaps he, like Bell in 1874, was inhibited because of telephony's poor reputation. For instance his Western Electric partner, Enos Barton, went on record as saying: 'I well remember my disgust when someone told me it was possible to send conversation along a wire.'[68] But the commonly held view of Gray, that he is a typical figure who missed out on a great fortune by being a few hours too late with a duplicate device, is not sustainable. The great fortune did not pass him by and although he was late, he had not duplicated Bell's work.

In August of 1870, following the death of his two brothers in the UK from tuberculosis, Bell's parents emigrated to Canada and his father established himself in Ontario. Bell first went to Boston in April of 1871 and finally settled in that city the following October. He opened a school for deaf-mutes and he also trained teachers of the deaf and hearing-impaired adults, in addition to being appointed to a professorship at Boston University. Despite all this activity Bell commenced the series of experiments which was to occupy him for most of the rest of the decade.

He began, in the winter of 1872, by replicating one of Helmholtz's experiments in which tuning forks of different pitches were placed between the arms of an electromagnet and thereby tied into an electric circuit. Throughout these years he continued to maintain himself by his work with the deaf. In October 1873 he moved into the home of the grandmother of one of his students, George Sanders, whose father Thomas, a leather merchant, had been much impressed with the child's progress under Bell's tutoring. A month later he had substituted metal plates, reeds, for the tuning forks he had been using for a year. These, when magnetised, could activate an exactly similarly tuned reed at the other end of the circuit. He was hoping that these tuned reeds would yield an harmonic telephone which could accommodate six or eight simultaneous transmissions; but he was disappointed:

Without going into details, I shall merely say that the great defect of this plan of multiple telegraphy was found to consist, firstly in the fact that the receiving operators were required to possess a good musical ear in order to discriminate the signals; and secondly, the signals could only pass in one direction along the line, so that two wires would be necessary in order to complete communications in both directions.[69]

By the summer of 1874, when he already knew that Gray was demonstrating a machine to telegraph industry leaders such as Orton, Bell produced another harmonic telegraph designed to overcome the two-way difficulty he had encountered. Behind the electromagnet to which he had attached the reeds, he now placed a permanent magnet. But this improvement he also felt was inadequate.

It seemed extremely doubtful whether an electromagnet current generated by the vibration of a magnetised reed in front of an electromagnet would be sufficiently powerful to produce at the receiving end of the circuit a vibration sufficiently intense to be utilised practically, on real lines, for the purposes of multiple telegraphy.[70]

He was also working this summer on his phonautograph with the dead man's ear and thinking about the harp telephone (see above page 307). Bell's devices worked well enough for him to convince the parents of two of his deaf pupils that he might have a marketable harmonic telephone in sight. Sanders, his landlady's son, joined the lawyer Gardiner Hubbard in an informal arrangement backing Bell. Hubbard's daughter, Mabel, then a teenager and also deaf, was to become Bell's wife. Money in support of the project was forthcoming following Hubbard's search of the patents in October 1874, because Gray, despite growing publicity, had not yet filed on the musical telegraph. Bell was well aware of the threat that Gray posed to his work and he wrote to Hubbard and Sanders in November 1874 that, 'It is a neck and neck race between Mr. Gray and myself who shall complete the apparatus first.'[71] This is a little disingenuous since Gray was many months further advanced.

In the midst of this contest, Bell conducted an experiment in December which put him on the road to his first patent. He revolved a set of magnets mounted on a cylinder in front of an electromagnet. He ran wires from the electromagnet to his ears which he had filled with water and heard, with difficulty over the noise of the revolving cylinder, 'a soft musical note.'[72] In February 1875 Bell learned that Gray, having at last filed a patent, was publicly exhibiting his musical telegraph. He and Watson were still having real difficulties tuning the reeds but this event forced matters. A more formal legal arrangement was struck with Hubbard and Sanders. Bell gave up all his teaching and went to Washington to file patents on his version of the harmonic telephone despite the problems he was having with it.[73] On this trip he too met with Western Union but was not encouraged; Gray was clearly winning the race. Also on March 1, in Washington he met the aged Henry and told him of the December experiment. Henry was sufficiently enthused to encourage Bell in his search for the speaking telephone. Unfortunately this was exactly what Hubbard and Sanders did not want and for Bell, who had by now given up his work with the deaf entirely, the months of tuning the telegraph went on. In a letter to his parents he wrote, 'Since I gave up professional work and devoted myself exclusively to telegraphy, I have steadily been gaining in health and strength, and am now in a fit state to encounter Mr. Gray or anybody else.'[74]

Then on June 2 the accident occurred that was to lead directly to the telephone of the first Bell patent. Despite the references in the May 1875 letter to a continuous current and variable resistance, nothing of the sort had been built into the devices Bell and Watson were tinkering with that spring. In the summer Watson accidentally created such a current when one of the tuning reeds malfunctioned and he flicked it manually – and in the other work room Bell heard the echo of the twang. Watson's account will serve to indicate the importance attached to this event by the Bell interests:

> The twang of that reed I plucked on June 2nd, 1875, marked the birth of one of the greatest of modern inventions, for when the electrically carried ghost of that twang reached Bell's ear his teeming brain shaped the first electric speaking phone the world has ever known. The sound of that twang has certainly been heard round the world and its vibration will never cease long as man exists.[75]

Bell thought again of the single membrane of the phonautograph he had been working on in 1874 and overnight Watson built to Bell's specification the Gallows telephone, so called because of the shape of its wooden frame. In it an armature was mounted so that one end was attached to an electromagnet while the other rested in the middle of a stretched membrane. In the transmitter a cone directed sounds towards the membrane; in the receiver the cone was reversed to collect sounds coming from the membrane. Using this apparatus Watson 'could unmistakably hear the tones of his [Bell's] voice and *almost catch a word now and then*. Bell was disappointed with this "meagre result".' (our italics)[76] By July, according to Bell, 'Articulated sounds were . . . unmistakably transmitted, but it was difficult to make out what was said.'[77]

Speech sounds were unmistakably produced from the receiver, and were almost intelligible. . . . While I am free to say that the character of the articulation produced was rather disappointing, the fact that any sound at all was audible . . . convinced me that the supposed difficulty which had been in my mind since the summer of 1874 – namely that magneto-electric impulses generated by the action of the voice would be too feeble to produce distinctly audible effects – was a mistake, and the results encouraged me to believe that the apparatus, if carefully constructed and tried in a quiet place, would transmit speech intelligibly, and prove to be a practically operative speaking telephone.[78]

In short, 'no really convincing demonstration of the transmission of audible and recognisable speech was made before the deposit of the specification.'[79]

But Bell was under considerable pressure. Gray was marketing the device Bell was supposed to be *inventing* and his backer, Hubbard, was denying him Mabel, with whom he had fallen in love, until he came up with something. The tuned reeds, patented in early 1875, were obviously not the answer and while the magnetos might or might not fulfil his hopes of a voice system, they were still a promising alternative to Gray. After the day of the twang, Bell and Watson abandoned tuned reeds and switched to magnets, thereby producing the contrivance which was patented as an 'Improvement in Telegraphy' the following year. Patent #174465 issued on 7 March 1876 proclaimed:

My present invention consists in the employment of a vibratory or undulatory current of electricity in contradistinction to a merely intermittent or pulsatory current and the method of, and apparatus for, producing electrical undulations upon the line wire. . . . The advantages I claim to derive from the use of an undulatory current in place of a merely intermittent one are, first, that a very much larger number of signals can be transmitted simultaneously on the same circuit; second, that a close circuit and single main battery may be used; third, that communication in both directions is established without special induction coils; fourth, that cable dispatches may be transmitted more rapidly than by means of an intermittent current or by the methods at present in use; for, as it is unnecessary to discharge the cable before a new signal can be made, the lagging of cable-signals is prevented; fifth, and that as the circuit is never broken a spark arrester becomes unnecessary.[80]

No mention of speech. The fact that Bell claimed this as the first speaking telephone is contained in the following, some pages further on in the patent:

I desire here to remark that there are many other uses to which these instruments may be put, such as the simultaneous transmission of musical notes, differing in loudness as well as in pitch, and the telegraphic transmission of noises or sounds of any kind.[81]

Again, no specific reference to speech. As it was lodged, the body of Bell's text contained no reference to variable resistance, either. Instead Bell thought to produce 'undulations in a continuous voltaic battery circuit by gradually increasing and diminishing the power of the battery.'[82] It is only in a 'For instance', added to the margin, that the clearest reference to variable resistance and a liquid is found:

For instance, let mercury or some other liquid form part of the voltaic circuit, then the more deeply the conducting wire is immersed in the mercury or other liquid, the less resistance does the liquid offer to the passage of the current. . . .[83]

The most serious attack mounted against Bell questioned the validity of these crucially important marginalia some years later when the official concerned, Zenas Wilbur, confessed before a congressional inquiry that he had illegally informed Bell's Washington attorneys of the contents of Gray's caveat, not just, as was required, its existence. But the supposition that Bell had added the references to variable resistance after receiving this information remained, in a case involving Gray that went to the Supreme Court, unsubstantiated.

Giving Bell the benefit of the doubt, whereby the sudden appearance of his interest in liquid transmitters subsequent to February 1876 is attributed entirely to his creative genius, the question then is, how close was Bell's liquid transmitter to Gray's? Even Bell's most strenuous supporters admit Gray's specification and Bell's apparatus were 'necessarily very similar in appearance' although Bell's relying on the varying depth of the rod's immersion is slightly different from Gray's varying the distance between the rod and the conductor at the base of the vessel.[84] But Bell never really satisfactorily explained why he had not even toyed with liquid transmitters earlier nor included a fuller description of variable resistance in the body of the patent. As to using liquids, Bell pointed out that he had employed water as a high resistance substance in a **spark arrester** to prevent his multiple telegraphy experiments from sparking, but this was a long way from the concept of the liquid transmitter and the mention of mercury in the patent. Bell told the court:

Almost at the last moment before sending this specification to Washington to be engrossed, I discovered that I had neglected to include in it the variable resistance mode of producing electrical undulations, upon which I had been at work in the spring of 1875.[85]

But the May 1875 letter was concerned with stretched wires, not water nor mercury. Bell explained that he intended his arrester to be combined with the variable resistance concept yet

> when I came to describe the proposed mode of affecting the external resistance by the vibration of the conducting wire in water, it occurred to me that water was not a good illustration substance to be specified in this connection, on account of its decomposibility by the action of the current. I therefore preferred to use as a typical example a liquid that could not be decomposed by the current, and specified mercury.[86]

Against this explanation is a considerable weight of evidence. A major clue is that Watson quite clearly understood the first sentence uttered on the liquid transmitter of 10 March 1876, while the Bell magneto transmitter, the **Gallows telephone**, fully described in the first patent, did not yield intelligible speech. The combination of liquid transmitter and magneto receiver dramatically improved the meagre results achieved with magnets alone. The upset acid was quite forgotten by Bell 'in his joy over the success of the new transmitter when I told him how plainly I had heard his words, and,' wrote Watson, 'his joy was increased when he went to the end of the wire and heard how distinctly my voice came through.'[87]

Compare this with the resumed work on the magneto transmitter. On March 13th, Bell demonstrated the Gallows to Hubbard and it simply did not work at all. A few weeks later, with the liquid transmitter still abandoned, Watson reported 'the telephone was talking so fluently *you did not have to repeat what you said more than half a dozen times* before almost anyone at the other end of the line could understand you perfectly.' (our italics)[88]

A new receiver was built of a cylindrical iron box wherein a coil of wire was wound round a central core, making an electromagnet. An iron lid fixed to the top acted as the diaphragm. Work was also done seeking to improve the transmitter. The developed contrivance looked in section somewhat like a squat revolver of enormous bore, its barrel being the speaking cone.[89] A non-magnetic membrane made from goldbeater's skin was mounted behind this cone with a clock spring armature glued to it. The armature connected to a coil to induce current analogous to the membrane's movements.[90] On May 10th Bell read a paper before the American Academy of Arts and Sciences and demonstrated the **iron box receiver** and **membrane transmitter**. Two weeks later he was at MIT. The decision was made to demonstrate the apparatus at the Philadelphia Centennial Exhibition. Watson made a number of instruments, a membrane transmitter as described above (which became known thereafter as the **Centennial single pole telephone**), a variant the **Centennial double pole telephone**, an iron box receiver and the liquid transmitter which was also taken.[91] On 25 June 1876 (the day of Custer's last stand), the speaking telephone was demonstrated for the first time to the public which fortuitously included Dom Pedro, the Emperor of Brazil, and Kelvin. It worked well enough for messages to be understood, but the distance was short and the utterances needed to be clichés like 'Mary had a little lamb'. As Norbert Wiener has pointed out, clichés contain less information than great poems and being highly redundant (in Information Theory terms) can be effectively guessed at.[92] Bell understood this too:

> familiar quotations – like 'To be or not to be,' etc., – counting – like 1, 2, 3, 4, 5, etc., – were very readily understood. But expectancy had a good deal to do with this, as language read from a book was not, I think, as a rule, understood, although a few words could be made out here and there.[93]

Kelvin, speaking in Glasgow upon his return, attested that he heard 'To be or not to be? There's the rub' and 'some extracts from the New York papers'. 'This discovery,' he went on, 'the wonder of wonders in electric telegraphy, is due to a young fellow-countryman of our own, Mr. Graham Bell, a native of Edinburgh, and now naturalized in New York.'[94]

Despite this testimony, the mystery of why Bell abandoned the superior liquid transmitter remains. He had been told of Gray, as had Gray been of him but without any details, on February 19th, five days after the documents had been filed. The Gray caveat was deemed to interfere with Bell's application, which was therefore being held up for ninety days. But a week later the Patent Office reversed this decision – after all Bell's was for an improvement in telegraphy not the *invention* of telephony – and determined that the application could go forward. Now in Washington, Bell visited the Patent Office on February 26th because the Examiner had discovered a further concern. Elements of the previous harmonic telegraphic patent, granted in 1875, appeared to be repeated in the current application. Bell explained to the Examiner the exact points of difference. After this visit at the latest, Bell knew, and admitted as much in a letter to Gray dated 2 March, 1876, that the latter's caveat 'had something to do with the vibration of wire in water'.[95] Bell's application was once more allowed, which resulted in the grant of March 7th. Thus the evidence is that Bell re-examined the document he had signed the previous January after he knew of Gray's vibrating wire.

Bell could have avoided the innuendo of fraudulent practice if he had another copy of the document with the variable resistance clause inscribed, which antedated the disclosure of the Gray caveat. And there was indeed another such document, the copy prepared for deposit in Britain. Bell's application had been returned to him by his lawyers for signature on 18 January, 1876 and he handed the copy over to a friend, George Brown, in New York a week later. Brown was to take it to London yet Bell, 'by some accident' forgot to add the variable resistance clause to this version as well. (The copy was never deposited because of Brown's 'fear of ridicule' which caused him not only to fail to lodge the document but to lose it altogether.)

In short, Bell had patented mercury but had used acidulated water. He had then abandoned the entire liquid transmitter technology, although it worked while his magnetos did not. He had never conducted a liquid transmitter experiment prior to March 1876, when he had learned of Gray's idea. He had never mentioned the possibility of such a transmitter anywhere except in the margin of Patent #174465, supposedly written before January 18th. He had forgotten to include variable resistance twice, in both the original and the English copy. He was in Washington in the Patent Office late in February and had the opportunity to amend his submission.

In the first years of Bell's publicising the telephone, Gray abided by his letter and never claimed the technology. Eventually he made his personal and corporate peace (with $100,000 for him and a contract for Western Electric) with the National Bell Company and only patent official Wilbur's revelations more than a decade later stirred him to contest the issue. But by then it was too late, and anyway, the matter was already moot in 1876. In fact neither he nor Bell had *invented* a telephone system. The transmitter, which still lay in the future, was to come from other minds and hands and would involve neither magnetos nor liquids.

Most of the cases fought over the patent were worthless, the protagonists naked adventurers who came forward claiming, typically, to have produced a telephone out of tin cans and broken cups at one point or another in the 1850s.[96] Given its shaky technological base and its value, it is scarcely surprising that the patent should have been so hard pressed; it was to the Bell empire, the world's richest enterprise, what the Decretals of Constantine were to the Roman Church. The entire history of the telephone, and much else, is conditioned by the patent war following Bell's announcement of 1876. This war, the first element of the operation of the 'law' of the suppression of radical potential as it applies to telephony, was occasioned not by technology *per se* but more because the patent had been awarded to a partial prototype system. It was a premature licence and that alone encouraged litigation. Bell interests made much of the flimsiness of most of the prior claims, deliberately confounding them with other more substantial cases. Well after the patent had run out Bell writers continued to pour scorn on all rivals, serious and frivolous alike; indeed the more serious, the greater the scorn. Since 'lack of novelty is fatal to a patent', Philip Reiss's apparatus was, after Gray's caveat, the greatest danger.

Reiss's telephone is described by Kingsbury, writing in 1915, as a 'knitting needle contrivance' and it is in these terms that it is often dismissed as being incapable of reproducing anything but clicks, but Kingsbury was Western Electric's man in the UK and a founder of the Standard Telephone Company Limited, Western Electric's UK incarnation. Even in 1915 when the original patent had lapsed, axes were still being ground against Reiss who had died in 1873. His knitting needle was in fact the electrical heart of a wooden sounding box and performed a function not unlike the metallic hearts of all our current speakers. The wires from his sender activated, by the principle of sympathetic resonance, another morse key whose movements sent electrical pulses down the needle, causing it to vibrate with a tin toy-trumpet-like effect. It could within those limitations nevertheless produce a variety of sounds. But could it speak?

According to Reiss, in the 1861 public lecture which popularised his device:

> Hitherto it has not been possible to reproduce human speech with sufficient distinctness. The consonants are for the most part reproduced pretty distinctly but not the vowels as yet in equal degree.[97]

In 1881 the Bell Company sued Amos Dolbear, Professor of Physics at Tufts College, for infringement of Bell patents. Dolbear had assigned his own patents for a telephone to Bell's great rival, the American Speaking Telephone Company, a subsidiary of Western Union.[98] Leaving aside the value of Dolbear's contribution for the moment, this case afforded the most protracted examination of Reiss's machine for, in essence, Dolbear's defence to the suite was that Bell's patent of February 1876 was improperly awarded since the device described would not work and that Reiss's unpatented device of 1860 would work just as well. The Reiss machine was tried in open court and was generally reported as an utter failure.

But this must be placed in context. In court, of the hundred things uttered into the Reiss telephone, only about 15 could be guessed at and only about half of those were right. Poor though this was, it was not an 'utterly unintelligible rattle' as Watson (and most others) would have it.[99] Equally, the other Dolbear contention, that the Gallows telephone did not work, has, as is reported above, some considerable merit.

All this is not to claim that the Reiss device worked and the first Bell telephone did not. It is just to call attention to the somewhat overstated claims as to the difference in the performance of the two machines which was no doubt engendered by lawyers representing interests which were already by the 1880s worth millions. One leading scientist, Kelvin, born in Belfast but a resident of Glasgow since the age of 8, claimed Bell for Scotland and pronounced his machine a 'daring invention'.[100] One judge, convinced of the superiority of the Caledonian/American over the German device, ruled that 'a century of Reiss would never have produced a speaking telephone by mere improvement in construction'. It seems that such a strong opinion is simply untrue and modern experimentation with the Reiss telephone reveals how close he came.[101]

The essential question is, how close did he want to come? That he called his device a telephone means very little given the common German usage of the term in the nineteenth century. It does seem likely, as Kingsbury claims, 'that Reiss was not looking for a still small voice'. Rather he was trying to demonstrate the reproduction of an undulatory sound wave by electricity, possibly with a view to amplifying it.

The various harmonic telephones, Reiss's apparatus, the Gallows telephone, Bell's and Gray's liquid transmitters, the Centennial single pole telephone, the Double pole, the iron box receiver, all these were prototypes for the speaking telephone but in 1876, contrary to received history, a complete practical telephone system had yet to be *invented*. And it was needed.

The second transformation: supervening necessity

Charles Williams Jr connects home and factory by TELEPHONE, April 1877

The model suggests that the availability of technology in itself is insufficient to occasion *invention* and diffusion, and the telephone is a case in point. From the introduction of telegraphy at the earliest or Reiss's telephone at the latest, there was no technological bar to a working system of telephony; but neither was there a well-defined need.

The single major factor impacting on a whole range of technological developments in the middle years of the nineteenth century was the emergence of the modern corporation. The limited liability company first came into its own in the years after the Civil War in America or, in Britain, after the Companies Act of 1862. These refined commercial operations necessitated the modern office and all the essential machinery of that office, the building to house it, the elevator to access it,[102] the typewriter, the mechanical desk calculator and the telephone to make it efficient.

Up to the 1870s, even in the USA, 5-storey streetscapes were the norm. The tallest building in the world in 1873 was the Tribune office in New York. It had 11 floors.[103] The 17-storey Mondanock building was erected in 1881.

This age found its form, as early as the eighteen-eighties in America, in a new type of office building: symbolically a sort of vertical human filing case, with uniform windows, a uniform facade, uniform accommodations, rising floor by floor in competition for light and air and above all financial prestige with other skyscrapers. The abstractions of high finance produced their exact material embodiment in these buildings.[104]

Synchronous with these developments was the emergence of the legal entity called the limited company.

In the common law chartered joint stock companies first appeared in the 1390s as the legal creations of the crown. Such charters were difficult to obtain, costly and, from the point of view of the partners, difficult to control. Against the hostility of the common law, which was grounded in the fact that fraud was encouraged by their existence, great unchartered partnerships with transferable stocks became nevertheless not unusual. Markets in their shares were well established by the late eighteenth century. The earliest phases of the industrial revolution took place under the aegis of such arrangements, where, for instance, actions against the partnership involved actions against all of them individually, if they could be known. By the eighteenth century such companies were seen as a 'common nuisance' tending to 'the common grievance prejudice and inconvenience of His Majesty's subjects or great numbers of them in trade, commerce or other lawful affairs' as the preamble to the somewhat opaque 'Bubble Act' of 1719 has it.

This legislation failed in its object, which was the outright repression of such companies, and was repealed in 1825, the year of the railway. The capital needed by the railway companies necessitated a revision of the law, which slowly took shape. The crown was empowered, in 1834, to grant by letters patent unincorporated (that is unchartered) companies the right to sue and the responsibility of being sued by a public officer. In 1837 a report on the law of partnership proposed that all companies which so wished could be incorporated by publishing their objects and constitution in a register established by the state for that purpose.[105] This was implemented in 1844 at the onset of the first major railway boom. The promotions the boom entailed further pressured reform of company law. The concept of limited liability was introduced in 1856 and the entire law of companies was consolidated in an act of 1862 which remains the basis of modern English practice.

> The railways greatly advanced joint-stock company development both by their example and by the demand they induced in potential investors in other enterprises for railway-type security of investment.[106]

The demand was emphasised by the fact that from the 1850s the railways were drying up as an object of capital investment. The proportion of total capital resources they absorbed began to fall sharply by the late 1860s, but investors wanted the same limited liability advantages they had become used to in all new enterprises.[107]

Much the same pattern can be noticed in America, although the legal aspects of the developments were somewhat more rapid. A charter of incorporation was regarded as a privilege (or *franchise* in the Norman-French) in the first century of the republic and franchises were awarded to run railways and ferries or have sole rights to trade in particular areas. Incorporation for 'any lawful business' was allowed first in Connecticut in 1837, the pressure for such an opportunity coming from the New England textile industry. However, before the Civil War, companies were still held by relatively few shareholders. The consolidation of small railways, which began in 1853 with the creation of the New York Central, led to more diffused ownerhsip and, indeed, stock market battles for control in the 1860s.[108]

> Since the Civil War, the quasi-public corporation has come to dominate the railroad field almost completely. . . . Following the lead of the railroads, in the last part of the Nineteenth Century and in the early years of the Twentieth, one aspect of economic life after another has come under corporate sway.[109]

Originally, allowing a company itself to be sued as an entity was a convenience and a safeguard against fraud; but it logically implied that the legal existence of that entity was something other than the legal existence of the individuals who from time to time might make it up. This was held to mean that no individual could be liable for the actions of the company beyond the exposure created by his or her shares. Limited liability was therefore a consequence of convenience.

From the 1850s, the joint stock company was also to be found in Europe as well as 'a new kind of finance house' which used the savings of investors too small to engage in the stock market directly, to buy shares in these new enterprises.[110]

The telephone, and all the other office devices, were born into a

'Great Depression', an economic downturn that was to last until the early 1890s. Nevertheless, these technologies were also born into a world where industrialisation was continuing apace. These were the years when cities in both the US and UK got running tap water and lighting for the first time. In Birmingham the mayor, Joseph Chamberlain, began the processes of municipal socialism – slum clearance, sewage, water, gas lighting, free libraries and art galleries. In America between 1871 and 1881 the number of federal employees doubled to more than 100,000.[111] By 1876 the American countryside no longer dominated the economy. And, everywhere, the great age of imperial expansion dawned, funded by the capital the limited companies facilitated. Between 1875 and 1900 the British Empire, the leader of this process, increased by 5 million square miles and 90 millions of people.[112]

The first telephone wire was built in April 1877 by Charles Williams Jr to connect his home in Somerville to his factory in Boston.[113] Although obviously the social uses of the telephone were appreciated from the very beginning, the relative utility of the technology for business as opposed to pleasure was reflected in the lease terms. In the advertisement announcing the availability of the telephone, the Bell Telephone Association stated:

> The terms for leasing two telephones for social purposes connecting a dwelling house with any other building will be $20 a year; for business purposes $40 a year, payable semi-annually in advance.[114]

The first use of a telephone for news reporting was on 3 April, 1877 when word of one of Bell's lectures was transmitted from Salem to the *Boston Globe*. The proprietor of a burglar alarm system installed telephones so that his customers could summon messengers and express service. Another early telephone exchange (central office) connected a Boston drug store with twenty-one local doctors. Bell himself in seeking investors, downplayed the private home and pushed the telephone

> as a means of communication between bankers, merchants, manufacturers, wholesale and retail dealers, dock companies, water companies, police offices, fire stations, newspaper offices, hospitals and public buildings, and for use in railway offices, in mines and [diving] operations. [Brackets in original][115]

Prominent among the first 778 telephone users in the spring of 1877 were those New York stockbrokers who had been early customers of the telegraph companies and whose support had created Western Union's highly profitable Gold and Stock Telegraph Company subsidiary. In Manhattan fifty years later, in the late 1920s, there were twice as many private switchboard attendants in businesses as there were Bell central exchange operators.[116]

> Although we accept the telephone as a basic component of U.S.
> households, it was primarily a business tool during the first 50 years
> of growth. It was not until after World War II that most households
> leased a telephone.[117]

And the pace is even slower in other industrialised nations. For all its obvious usefulness in the home, the telephone is a child of commerce.

Phase three: technological performance – invention

Thomas Edison invents the TELEPHONE, spring 1877

The synchronicity of the various claims to the *invention* of the **telephone** is usually held as the source of the protracted battle the Bell interests waged to protect the patent. It is here suggested the law was too quick to assign the *invention* and that this, rather, is the font of litigation in the matter. It is not that numerous researchers, led by Gray and Bell, all reached the same solution to the problem at more or less the same time; it is that the design Bell filed was for a partial prototype, an imperfectly working system with at best only an effective receiver, which therefore rendered the patent vulnerable. Had Bell really achieved a complete long-distance voice communication system, he would more easily have deterred his rivals. The publicity surrounding the device, the accident of the interest of Dom Pedro of Brazil and Kelvin's advocacy, as well as Bell's not inconsiderable effectiveness as a lecturer inscribed the **speaking telephone** on public consciousness as the latest technical marvel, but in terms of science and practicality it was, throughout 1876, barely more than a toy.

The result was that a number of researchers, Bell himself with Watson, Edison, Berliner and Blake, contrary to the received view of how nineteenth-century technologists were supposed to operate (Goodyear, say), perfected the design in a highly structured way.

> Bell seems to have started out with knowledge and devised his
> instruments with an accurate mental conception of their operation.
> Knowledge grew with experience and one step led to another, but
> each was thought out beforehand.[118]

Bell might have been as deficient in theory as any other scientist of the day, but his work, and that of his rivals, was of a piece with, say, Shockley's day-to-day proceedings. In the aftermath of the Philadelphia Exhibition he set about *inventing* a more practical, less toy-like system, one that would operate efficiently but eschew any liquid variable resistance mechanism. By July 1st, he had substituted a thin metal plate for the goldbeater's skin membrane and a day later he announced to Hubbard that he had discovered a better permanent magnet which removed the need for an electromagnet and with it the battery. The design of the armature was also changed until it became a disk nearly as large as the metal diaphragm itself. By November it was being field tested on actual telegraph lines.[119] But Watson reported that although it worked 'moderately well over a short line . . . the apparatus was delicate and complicated and it didn't talk distinctly enough for practical use.'[120] All it needed was a better magnet the secret of which Watson discovered by reading a description of a new recording telegraph in a trade magazine. Binding four steel horseshoe plates together with a soft iron core did the trick. 'To my great joy the thing talked so much better than any other telephone we had tried up to that time.'[121] The **magneto telephone**, which acted as both receiver and transmitter combined was patented, still as an 'Improvement in Electric Telegraphy' in January 1877 and offered for sale in the spring.[122]

The second patent provided a securer foundation to the Bell empire than the first, clearly stating that, after multiple telegraphic transmission, the invention's object was 'the electrical transmission . . . of articulate speech and sound of every kind'.[123] Bell had produced a viable receiver, along the lines previously suggested but now rendered fully practical. It was to prove a real obstacle to his rivals and none was able to develop an alternative. Yet the magneto telephone as a transmitter remained quite another matter. Some sense of its limitations can be gleaned from the circular announcing its availability:

> The proprietors of the telephone, the invention of Alexander
> Graham Bell, for which patents have been issued by the United
> States and Great Britain, are now prepared to furnish telephones for
> the transmission of articulate speech through instruments not more
> than twenty miles apart. Conversation can easily be carried on after
> slight practice and with occasional repetition of a word or sentence.
> On first listening to the telephone, although the sound is perfectly
> audible, the articulation seems to be indistinct, but after a few trials
> the ear becomes accustomed to the peculiar sound and finds little
> difficulty in understanding the words.[124]

This, it should be emphasised, is no hostile account of the apparatus but Bell's own advertisement.

A round of demonstrations, at 10¢ a ticket, was under way. On 28 May 1877, the people of Lawrence, Massachusetts, were afforded an exhibition in which 'vocal and instrumental music and conversation' was transmitted 27 miles from Boston by 'The Miracle Wonderful Discovery Of The Age'. Watson, who was normally at the other end of the line on these occasions, developed stentorian tones. (Indeed, after he had left Bell, flirted with utopian socialist ideas and failed in some business ventures, his final career – although he was still a wealthy man – was as a minor player on the English stage.) The publicity, although crucial to the business, also alerted a number of researchers both to the possibilities of the telephone and to the fact that Bell had not quite tied up the *invention*.

Thomas Edison claimed independently to have noticed the variable resistance of substances under pressure in 1873, some years after the result was reported by the French, Clarac and Du Moncel. He did not, however, appreciate the significance of the phenomenon until he heard about the designs of Bell and Gray. Like them, in 1875, he was working on an harmonic telegraph but directly for Western Union. He devised a dual tuning fork method by September of that year only to abandon this work when Gray's harmonic telegraph received industry acceptance. William Orton of Western Union had mentioned telephony to him during the summer, though with what degree of enthusiasm is not reported, and he had first heard of Bell's work in December. At Orton's suggestion Edison took up the articulating telephone, experimenting with liquids until February 1876, the month of the Gray caveat and the Bell patent application. He then turned to vibrating diaphragms but made so little progress that he handed the work over to an assistant.

Then in January 1877 Edison remembered the 1873 semiconductor experiment. By mounting the diaphragm above a layer of one of these substances – plumbago initially, although in Edison's usual fashion many materials were tried – which formed part of an electric circuit, Edison impressed the sound waveform on to the electrical wave via the plumbago. The sensitivity of the telephone transmitter was much improved. It was not, as was the case with Bell's magnetos, voice-driven, as it were; one did not need to bellow, but it was even less distinct. Throughout 1877, Edison tried to improve the performance of the diaphragm. He thickened it until it was not a diaphragm at all but rather a solid platinum disk above a thin layer of plumbago, and later lampblack, another form of carbon. At least ten other devices, variations of this technical theme, were produced in 1877 and included, since the technology is cognate, **microphones**.[125]

The term microphone was adapted from an acoustical usage of Wheatstone's (for a species of stethoscope) by David Hughes. Hughes, an Englishman, who was to be rebuffed in his radio experiments some years later (see page 245, above), was then working at the College of Bairdstown, Kentucky. He constructed a primitive machine which consisted of nothing but two nails mounted on a wooden baseboard, attached to the poles of a battery, with a third nail laid across them which was sensitive enough to acoustic vibrations to vary a current. He also experimented with a carbon pencil mounted vertically in between conical depressions made in two carbon blocks. A paper describing these experiments was read before the Royal Society in London in May, 1878 and microphone was adopted as the term of art for any acoustically sensitive loose contact device.[126] That he was working on these lines at the same time as Edison was to be of importance in the patent battle. Edison was to claim bitterly that the credit for the microphone should be his.

Edison filed patents on the plumbago transmitter in 1877 and on the lampblack version early in 1878. Now Orton moved to add these carbon transmitter patents to the arrangements he had previously made with Gray and a professor of physics at Tufts College, Amos Dolbear, who had some revisions of the Bell magneto receiver to hand by August of 1877. Dolbear was to be the main defendant in the case where the Reiss telephone was elaborately tested. He claimed to have been working along the lines Bell patented as early as 1864, and that Bell had stolen his ideas through the medium of a friend. As Dolbear had visited Bell's workshop and heard him lecture at the Music Hall in Boston prior to filing his own patent application, clearly his potential use to Western Union was less than Edison's or Gray's; but it was all grist to Orton's mill.[127]

Upon these bases Orton created, rivalling the emerging the Bell Companies, the American Speaking Telephone Company. He was driven by the threat encapsulated in the enthusiasm of the New York stockbrokers, some of his best telegraphy customers, for the new technology. Although difficult to mass produce, the Edison transmitter allowed Western Union via the American Speaking Telephone Co. to offer a superior service and Orton made considerable inroads into Bell's nascent business. 'Almost at once these improvements of Edison's liberated the whole early art of telephony and opened up the possibilities of an instrument that many had been inclined to regard only as an interesting toy.'[128]

But the liberation involved pirating Bell's receiver. A legal attack on the telegraph company was mounted by the telephone company, grounded both on the unlicensed use of magnetos and Bell's marginal references to variable resistance in the original 1876 patent. The fledgling Bell firm took on Western Union, then one of the world's biggest enterprises, and, at the same time, it sought a device that would use the variable resistance principle in a transmitter without copying either Gray or Edison. Researchers at Brown University, led by professors Eli Blake and John Pierce, constructed a handholdable magneto version which was more convenient and produced a more distinct sound, but this was not what was needed. Watson was deputed, in a manner normally considered to be peculiar to this century, to *invent* a non-liquid transmitter.

> I soon had an opportunity [wrote Watson] to try one of the Edison Carbon Button transmitters. I found it startlingly loud and clear in its articulations. Then, my other troubles became trifles under this new cloud on our horizon. We had no transmitter equal to the Edison but that didn't worry me much for I felt sure I could devise one as good. What troubled me the most was the Western Union's people's claim that the Edison transmitter was not subject to Bell's patents for, if they were right, the future of our business looked dark to me.[129]

In April 1878 National Bell's counsel advised that although he thought the master patent could withstand Edison, using carbon would give hostages to fortune. Watson had to *invent* a non-liquid, carbon-less variable resistance transmitter.

On May 1st the *Telegraphic Journal* published the results of a test which found Edison's device worked best and was least subject to interference on the 106-mile long line between New York and Philadelphia, thus confirming what the small market in telephones was also demonstrating – carbon transmitters were superior. Before the patent interference examiner, Edison went on record more circumspectly than had Gray, swearing that:

> he never conceived the possibility of transmitting articulated speech by talking against a diaphragm in front of an electro magnet. [He went on to state that] he did most emphatically give Mr. Bell the credit of the discovery of the transmission of articulated speech by that principle. [our brackets][130]

Indeed Edison might be generous to Bell on this point, since his alternative was the superior method. But aid for the Bell interest was at hand; Watson had seen the answer to Edison the previous summer.

Nineteen-year-old Emile Berliner had arrived in the United States from Hanover on the eve of the Franco-Prussian war of 1870. Six years later, he was a clerk in a Washington DC dry-goods store. Among his acquaintances was the chief operator at the fire alarm telegraph office. One night, while practising 'sending', his friend told him, 'You must press down the key, not simply touch it.'

> Then the telegraph man explained that in long-distance transmission, where the resistance is high, the sending key must be pressed down rather forcibly if efficient reception is to be assured.

'That's why we use men exclusively for long-distance telegraphy . . . because they naturally press down hard. They have a strong touch. Women wouldn't naturally press down hard and are therefore not adaptable to long-distance work'.[131]

1876 was the year that the American Medical Association first admitted a woman to its ranks and then spent much of its annual meeting attempting to oust her.[132] The sexism of his friend being therefore in the nature of cultural background noise, Berliner hypothesised that electrically the greater the pressure, the more current passed over the contact. Early in April 1877, even as Edison was working with the plumbago apparatus, Berliner built a **soap box telephone**. He knocked the bottom out of a wooden box, 7 × 12 inches, nailing in its place a sheet-iron diaphragm. Above this was secured a cross-bar and a screw with a polished steel button on its end which passed through the bar to rest on the diaphragm. On 14 April, 1877, he filed a caveat, although the apparatus clearly worked well enough to warrantise a patent. The caveat cost $10 and the patent $60. For a $12 a week clerk there was no option. He began negotiating with the local Bell Company in New York, which was not interested, and saved enough money to file for his patent by June. That summer, with Western Union actively acquiring telephone patents and backing researchers, National Bell's attitude changed. It was possible that a speaking telephone system totally avoiding magnets was in the offing. Watson arrived at Berliner's Washington rooming house, examined the soap box, and pronounced: 'We will want that, Mr. Berliner. You will hear from us in a few days.'[133] Watson was quite clear as to why he was so enthusiastic.

I thought he might antedate Edison . . . and advised Mr. Hubbard to make some agreement with Mr. Berliner for the use of his invention. . . . We tried to make each other believe that carbon transmitter wasn't essential to our welfare but we all knew it really was.[134]

Berliner joined the National Bell Telephone Company in September. Apart from legal difficulties, there was still much to do. The soap box was unstable, tricky to adjust and it distorted high volumes. Early in 1878 Francis Blake, a scientist formerly attached to the Geodetic Service, produced a variation of it which interested the company. He joined National Bell in the summer of 1878 and melded the Berliner soap box with the Hughes microphone in a transmitter where a carbon button and a bead of platinum were held against the diaphragm by springs.[135] This too was a temperamental device but by January 1879 Berliner, having been sent to Bell's Boston Lab specifically for this purpose, had perfected it. The National Bell Company at last had a stable transmitter that did not distort and that matched the performance of Edison's patented a year earlier. Just as importantly, it was a lot simpler to make, two hundred instruments a day being produced. Bell history acknowledges that the patent is Berliner's, but the transmitter is named for Blake rather than for his more exotic colleague.[136] (Berliner was to establish himself as an independent researcher on the strength of his Bell royalties and in the five years from 1883 he set about improving Edison's tin-foil recorder, the **phonograph**. On 12 November 1887, he obtained a patent for a **gramophone** (his term), an analogue device for reproducing sound stored on disks.)

By 1879, Dolbear, using a condenser developed by Kelvin in 1863, produced a viable non-magneto receiver.[137] Western Union and National Bell were in a stand-off position. National Bell had the superior receiver, the real source of its strength, and had acquired a comparable transmitter to Edison's which might or might not be sustained in the courts. Western Union had the advantages of its size and resources, Gray's caveat and a superior transmitter. Although it was relying on Dolbear's shakier claim to the magneto, it also had his **condenser receiver**. It was not true that, 'The American Speaking Telephone Company had no controlling patents.'[138] In advance of the spate of cases all decided in Bell's favour in the 1880s, these rival patents did not look so insufficient, even though at least one of Western Union's lawyers early convinced himself that Bell was impregnable.[139] In November 1879 the two companies reached an agreement. For the life of the Bell master patents, i.e. until 1893 and 1894, the American Speaking Telephone Company agreed to drop all actions against those patents and to hand over to National Bell its own telephone patent rights. Western Union was not to enter the telephone business. In return, National Bell agreed to drop its cases and to buy out all the American Speaking Telephone's subscribers and equipment, 56,000 'stations' in 55 cities, as well as paying 20 per cent of each rental to Western Union. In the fifteen years of the contract some $7,000,000 were to be handed over.[140] In addition Elisha Gray received $100,000 and his company, Western Electric, became, by a contract of 1882, Bell's manufacturing arm. The telephone was *invented*.

The irony is that Bell had been interested in Western Union's patronage from the very beginning. In the early part of the summer of 1875, he had been conducting experiments in the company's New York office until William Orton 'learned that Mr Gardiner G. Hubbard, who was personally obnoxious to him, was pecuniarily interested in Mr. Bell's inventions, and immediately directed that permission to conduct his experiments should be withdrawn.'[141] A year later Orton refused to consider Hubbard's offer of the Bell patents.[142] It is possible that Orton contributed to Gray's inattention to telephony during that year too; but he was clearly not averse from having his own wholly owned telephony system as evidenced by his remit to Edison.

Although the Edison device was now available to it, the Bell system stuck with the Black transmitter, making 340,000 of them in the 1880s. (Only 6000 **magneto box telephones**, as described in Bell's 1877 patent, were made.) The standardised device, virtually the contemporary telephone, using carbon granules in a technique patented by Edison in 1886, was developed by both Edison and Blake and came into production in 1895.[143]

The telephone was a system *invented* to order. Bell and Watson's creation of a viable receiver in 1876 and Berliner's perfecting a transmitter in six weeks in 1879 compare very well with the work of innovators labouring in industrial laboratories this century. If a change in the processes of innovation has taken place, clearly it must antedate the telephone; and, since telegraphy was similarly developed to plug a technological gap occasioned by the advent of the railways, it can be argued that the putative change must antedate that too. Equally clear is the point that such serendipity as the nineteenth-century telecommunication technologists exhibit is no less in evidence among their successors today. In fact, the entire technical history of telecommunications, from the late eighteenth century on, must either all be deemed to be 'modern', or the argument that innovation has recently been transformed must fall.

In the 'information revolution' literature, there is a tendency to apply Whitehead's famous remark that the greatest of nineteenth century inventions was 'the invention of the method of invention' to the end of the century instead of its beginning as he intended. While the idea of 'the invention of the method of invention' remains in the light of quotidian activities of both nineteenth- and twentieth-century innovators somewhat opaque, it is quite clear that such force as it has must apply to the whole period as Whitehead originally suggested.

The third transformation: the 'law' of the suppression of radical potential

AT&T abandons TELEGRAPHY, 19 December 1913

Bell must perforce share the credit for the telephone with others, but there is no question that he was the most highly motivated of its *inventors* and the one with the most developed 'system concept'.

Thanks to the income generated by his lectures, his backer, Hubbard, was able to make the momentous decision to rent rather than sell the 'station' equipment, an idea he took from the practice of a company he represented at law which rented shoe sewing machinery to cobblers and collected on every shoe processed.[144] But overall Hubbard, like most others, saw the business in a limited way as the installation of single point-to-point pairs of telephones. Most of those Hubbard approached to invest in telephony saw little future in it. 10,754 Bell phones, largely in pairs, were placed in service by the end of the first year. Bell's vision was far more extensive than this. On his European honeymoon with Mabel Hubbard (whom he had finally been allowed to marry in July 1877) he drew up a prospectus for investors:

> At the present time we have a perfect network of gas pipes and water pipes throughout our large cities. We have main pipes laid under the streets communicating by side pipes with various dwellings, enabling the members to draw their supplies of gas and water from a common source.
>
> In a similar manner, it is conceivable that cables of telephone wires could be laid underground, or suspended overhead, communicating by branch wires with private dwellings, country houses, shops, manufactories, etc., etc., uniting them through the main cable with a central office where the wires could be connected as desired, establishing direct communication between any two places in the city. Such a plan as this, though impracticable at the present moment, will, I firmly believe, be the outcome of the introduction of the telephone to the public. Not only so, but I believe, in the future, wires will unite the head offices of the Telephone Company in different cities, and a man in one part of the country may communicate by word of mouth with another in a distant place.[145]

To facilitate what Bell called 'this grand system', the patent association between himself, Watson and his two backers, which had obtained since February of 1875 became, on the eve of his departure, the first Bell Telephone Company. Hubbard appointed agents who placed the telephones, collected a percentage of the rental and often built the lines connecting them as independent contractors. By 1878, as the advantages of the central office concept became clearer, Hubbard created the New England Telephone Company to promote the construction of exchanges by local operating licensees. The first exchange technology was telegraphic and the earliest exchanges tended to be run by Western Union's American Speaking Telephone Company which already had the knowhow.

In September 1878 Emma Nutt was engaged as Bell's first woman central office operator.[146] In distinction to telegraphy, there was no assumption that telephone work was too hard for females; or perhaps the company considered Bell's pronouncement in the prospectus that the telephone 'requires no skill to operate' made it safe to hire her. Within three years the rightness of the decision was being celebrated by telephone exchange men everywhere.

> I would like to say right here, I've been asked by Mr. Sabin what our experience has been with young ladies' help; the service is very much superior to that of boys and men. They are steadier, do not drink beer and are always on hand.[147]

The public contributed to the constraint of these new entities by not flocking to acquire the product, although there was enough interest to sustain a measure of diffusion. Western Union, the most directly threatened organisation, took steps to suppress its new rival with a superior product – better Edison telephones and more developed exchanges. Hubbard was forced to sue. He was rapidly running out of capital and finding it hard to attract more. He merged the two Bell companies into National Bell and secured the services of Theodore Vail to be its general manager. Cousin to the Vail who had been Morse's assistant, Theodore had taken a job with the postal service in Omaha while studying law at night. Seven years later, he was superintendent of the Railway Mail Service in Washington. His boss, the assistant postmaster general, remonstrated with him over his resignation.

> Listen to the prophecy of an old fool to a friend. One or two years from hence there will be more Telephone companies in existence than there are sewing machine companies today . . . I can scarcely believe that a man of your sound judgement, one who holds an honorable and far more respectable position than any man under the P.M. Genl., with honor and respect attached to the same, should throw it up for a d——d old Yankee notion . . . called a telephone.[148]

Vail took no notice and went to Bell's in the summer of 1878 where he immediately dramatically improved the legal situation. Exploiting the claimed priority of the recently acquired Berliner patent, he brought a second legal action, an 'interference' against the Edison device which allowed Bell Telephone to begin to use its Berliner/Blake transmitters without fear of a Western Union injunction.

Bell himself was clearly uneasy about these cases and on his return from Europe to give evidence in the first of them, November 1878, Watson had to go to Quebec to escort him personally to Boston, where he was promptly admitted to the Massachusetts General Hospital for an operation.[149] He was uninterested in corporate details and obviously had no reason to relish legal attacks on his work and integrity. In the event he was to spend the next three decades defending himself.

Orton, on the other side, also had difficulties and could not give his undivided attention to the destruction of National Bell. Western Union itself was under attack from one of the archetypical entrepreneurs of the period, Jay Gould. Moreover Bell did hold the master patents to at least half the system and, under Vail, was being as deft as the telegraph company in acquiring rights to the other half. It was also becoming clear that the telephone and the telegraph could live together, the one being used for short distances and the other on longer routes.

The corporate situation by the fall of 1879 therefore matched the technological situation. It was a stand-off. The deal made out of court in November of that year is a classic expression of the 'law' of suppression in that it gave the established technology a real interest in the exploitation of the new, melded the patents so that they could be exploited effectively and rewarded Gray. The patents also worked to create a 15-year breathing space.

Although potentially a splendid source of regular revenues the leasing system required capital. In fact, telephones were a peculiar business since more stations created more value to the user but at more cost to the provider and less profit per every new subscriber. Economies of scale did not quite work. The legal actions also drained funds. Hubbard was forced to involve 'the Boston Capitalists', led by one G.L. Bradley. National Bell became American Bell and Bradley's group took half its shares. Colonel William Forbes, a 'swashbuckling and aristocratic financier in the tradition of Raleigh and Lord Nelson' became the principal fund-raiser.[150] American Bell had six times the capital of its predecessor and by the end of 1880 the number of its stations or telephones had doubled to 123,380. A year later there were only nine cities, according to its annual report, with populations of over 10,000 and only one with more than 15,000 that did not have a telephone exchange.[151] Albeit far from a revolution in daily life, this was not bad given that the service was still subject to extreme interference from trolley-cars, telegraph wires, and even thunderstorms.

The American Bell Telephone Company succeeded National Bell in 1880, and the removal of Hubbard as well as the resignation of Bell himself, both with tidy but not extraordinary fortunes, soon followed. Forbes, now president, and Vail set about bringing the system under one corporate roof. American Bell began to buy up interests in its licensees, the local operating companies, encouraging them to amalgamate as it did so. By 1934 it wholly owned 18 of the 23 entities that resulted and held a majority interest in 3 more.

Bell's 'grand system' envisioned exchanges, the building and operation of which had become the function of the local companies, but it also required longlines 'to unite the head offices of the Telephone Company' in every city. In 1884, American Bell demonstrated a viable longline between New York and Boston. The following year it created, in New York, a subsidiary to build a longline network which it called, despite the Western Union agreement, the American Telephone and *Telegraph* Company. 'Telegraph' was incorporated into the name by an act of the New York Legislature to reflect the company's leased lines business.[152] Vail regarded the work of AT&T as the best defence in the coming post-patent battle. Within ten years AT&T's system reached into the Midwest, but ten years later it had progressed no farther than Omaha and Denison, Texas.[153] It was essentially an East Coast operation for which wiring the more sparsely settled West was not that pressing a project.

Vail also made Gray's old firm Western Electric into the sole supplier of equipment for both the longline requirements of AT&T and the needs of the local Bell subsidiaries, buying out the remaining Western Union interests in the firm.[154] Finally in 1888, the court decided (by 4-3) the consolidated patent case, all the claimants having been lumped together, in favour of Bell. In fact, it held that any system of electronic speech transmission was covered by the original patents, a curious decision, given that one did not mention speech at all and the other did so in a subordinate fashion.[155]

American Bell, protected by its patents, paying substantial licensee fees to its major telecommunications competitor, grew slowly. By the mid-1880s there were still only 156,000 phones nationwide, but the firm's earnings had increased, despite the ongoing depression, fivefold to over $2,000,000. And it was positioning itself for the confusions that were bound to follow the lapse of Bell's master patents in 1893 and 1894.

Until recently the issue of telephone competition, resolved seventy years ago everywhere in favour of unified services, has been a dead one. Now there is to be, at least in Britain and the USA, telephonic 'competition', a situation whose only extended precedent is to be found in the period from the end of the Bell monopoly to the outbreak of the First World War. However, American telephone history from 1894 to 1916 actually offers little evidence to sustain the current ideological desire for telephony 'markets' and can be read as a case study in the inevitability of monopoly, at least as a concomitant of 'universal service'. Current telephone 'deregulation' merely points once more to the truth of Hegel's observation 'that people and governments never have learned anything from history, or acted on principles deduced from it.'

In 1894, the assistant postmater general's vision became a reality but by the time Vail had left Bell, 87 independent systems were established. By 1902 more than 3000 were in existence, not all of them direct competitors since some were in areas Bell did not serve and many were minute rural arrangements linking a dozen or so scattered farmhouses. Five midwestern states had over 200 telephone systems each and in general the majority of the non-Bell exchanges were in this part of the world, which had been comparatively ignored by Bell. Thus the first challenge to Bell was not a direct one but rather an attempt to make up its failures of provision. But nevertheless 45 per cent of all communities with more than 4000 people had two telephone services, two exchanges and two stations in every place that needed maximum telephone coverage. A class distinction grew up with businesses and the wealthy on the Bell exchange (with its access to AT&T's longlines) and the rest of the population was left with a more limited universe, and often inferior equipment, on the cheaper independent system. The Bureau of Census reported in 1902 that the average independent, excluding the minute farmers' phone systems of the midwest, had a mere 705 subscribers.[156]

The larger Bell rivals combined into the National Association of Independent Telephone Exchanges in 1897 to build their own longlines and by 1905 the association had '*almost* a continuous system from the eastern slope of the Rocky Mountains to the Atlantic Coast', but not one which connected the major business centres.[157] (our italics) Bell's policy during the years of monopoly had been to charge highly for its services and the removal of the patent had the predictable pricing effect. Phones did get cheaper.

> The present Columbus subscriber, who finds it either convenient or necessary to keep two such services, pays $54.00 for one and $40.00 for the other, or a total of $94.00 per year, $2.00 less annually for two than one cost him before competition, and what does he get? Connection with 22,000 telephones instead of 2,000, or, in other words, competition has brought to the alleged burdened businessman who has to keep two telephones, 20,000 more telephones to talk with and has handed him a $2.00 yearly rebate in the bargain.[158]

However, the independents, who by 1900 had 28 per cent of subscribers, did not exploit their advantages as heavily as they might have done.

> The Independent Companies, for the most part were concerned not to improve service for the public good, but rather to make money by corporate manipulations . . . their plants were poorly built. They offered service at low rates, made inadequate allowance for maintenance and depreciation, and paid dividends out of capital for

a few years, and then were compelled to raise their rates or go into the hands of receivers.[159]

In Kansas City, the company was so undercapitalised it could not extend its lines to new subscribers. In Baltimore, Bell's competitor, which obtained a city franchise on the grounds that it could provide service at $36 per year for residences and $48 per year for businesses, raised its rates to $60 and $72, claiming it was using better equipment than the contract specified. The pattern was repeated elsewhere in a curious precursor of the sort of argument to be used by cable television companies against the same municipalities three-quarters of a century later. And, just as was to happen in the 1980s, the municipalities were told in 1905 that in law they could not regulate the rates. AT&T put out a pamphlet itemising the 85 independents who had not kept their promises and had raised their rates. (Again, this exactly parallels AT&T's advertising campaign in the first year of renewed 'competition', 1984, when its television commercials drew constant attention to the promises of its rivals and their failures to match its range of service.)

There is no question that competition forced Bell to improve its service, drop its haughtiness and cut its rates, but a further issue remains as to whether or not competition was the only way of achieving these ends. American Bell had ruthlessly exploited its monopoly position, in an age when such behaviour was common and expected. Its pricing policy was the result of this unregulated atmosphere. Americans disenchanted with the service it provided looked abroad to find evidence of its rapacity.

All over Europe, from 1879 on, Bell had set up subsidiaries along the lines of its local US operators and in each country local competitors, local licensees, municipal or national governments took them over. In all these cases, the removal of the Americans was an essential prerequisite to reasonable rates.[160] Only in St Petersburg, Moscow and Warsaw was Bell left in command, with the result that these cities had the highest annual rental rates ($121) in Europe. In the early 1890s, in Washington Bell charged $100; in greater New York anything up to $240; in Philadelphia $250; in Chicago $175. But in Paris, where the state owned the system, the rate was $18 per annum and in Stockholm, where it was owned collectively by its users, $20. There were five times as many telephones per hundred people in Stockholm as there were in Washington and the system boasted the latest metallic circuit underground wires which most of the US Bell operators did not.[161] AT&T adopted the same high-charge policy for its longline services. A call from Philadelphia to Washington cost $1.25 for five minutes, whereas in England the same duration and distance cost 48 cents and in France 30 cents.[162]

It is therefore true that it took the end of monopoly in America to bring Bell prices down but it is equally true that elsewhere other monopolies were charging less without the bite of competition. During this early period (to the mid-1890s) the European states, either directly or through the operation of franchised entities, were comparatively effective diffusers of the new technology, although overall penetration outside of the great cities was far less than in America. This relative success was not, of course, universal.

In Britain, Bell had personally demonstrated the telephone to Queen Victoria, who gave her approval by accepting a pair. A well-capitalised Telephone Company (Bell Patents) Limited was established, but Edison had prior rights having lodged his patent in the UK earlier than had Bell. A rival company was set up for which the young George Bernard Shaw laboured as a 'wayleave manager'. He recalled:

> Whilst the Edison Telephone Company lasted it crowded the basement of a huge pile of offices in Queen Victoria Street with American artificers. They adored Mr. Edison as the greatest man of all time in every possible department of science, art, and philosophy, and execrated Mr. Graham Bell, the inventor of the rival telephone, as his satanic adversary.[163]

The Post Office had been given control of telegraphy in a series of Acts in the 1860s but the system remained as unprofitable as it had been in the US initially. Th GPO viewed the new technology with considerable disapproval as a further threat to its faltering telegraph service.[164] In 1880 it moved against what had become the United Telephone Company, reflecting the Western Union/Bell arrangement of the previous year. The UTC went into the court believing that a straightforward demonstration of the difference between a telephone and a telegraph would suffice to convince Mr Baron Pollock and Mr Justice Stephen that the Telegraphy Acts did not apply to the new technology. It was wrong. The court found for the Crown. The GPO then imposed a 10 per cent royalty on the UTC and licensed all comers, private companies as well as municipalities, to operate telephone services. But, unlike other parts of Europe, Sweden say, where the government encouraged local cooperative systems and only built the longlines itself, in Britain the GPO decided to operate at all levels as a competitor to its own licensees. The result was total uncertainty on the part of the commercial interests and hostile lethargy from the GPO. Unable to make telegraphy pay, it was unwilling to absorb another potential loss maker; but it was equally fearful that others could make the telephone profitable and at the inevitable further expense of the telegraph. The GPO eventually took over the longlines and then granted no local operating licences to run beyond

31 December 1911, which had the same effect as planning blight. What little expansion there was virtually ceased after the turn of the century. The GPO finally took telephony over in 1912, only to find the war halting development. It was in fact not until 1919, forty years after the *invention*, that the telephone began to be seriously diffused in the UK. Only Kingston-upon-Hull survived as an independent municipal system.

This perhaps explains why R.S. Culley, GPO Engineer-in-Chief, refused, at the outset in 1877, to see Mr Bell's representative, 'stating that his department was in possession of full knowledge of the details of the invention and that the possible use of the telephone was very limited.' – in the event a self-fulfilling prophecy.[165] The result was a perfect expression of the 'law' of the suppression of radical potential. The UK, as the world's leading industrial nation, achieved less telephone penetration and higher rates than any other Western European country.

Elsewhere, with differing details, the same result – state ownership – was, by the First World War, the norm. As the more socially adaptable United States increasingly accepted the telephone in the mid-1890s, facilitated by the arrival of the independents, its lead in penetration grew. By 1900 the US had about one phone for every 60 people; the nearest European state, Sweden, had one phone for every 115 people, France one for every 1216 people and Russia one for every 6988. A major article in telephonic faith was inscribed – that private enterprise worked better. The suggestion is that the low rates imposed by the public entities, while politically popular, failed to provide sufficient capital for a first-class service.

But the conclusion is hasty. Clearly the discrepancy between Russia and America, where in both cases the service was provided either entirely or largely by Bell, was not just due to the organisation of the institution of telephony. There were cultural factors to consider. Switzerland had a widespread state system which worked very well but the Swiss made on average a mere two calls a day. The Bank of England lived without a telephone until 1900.[166] Rural Europe clustered its farmsteads together in villages; rural America spread them out over the vast plain. To this day the range of social and business intercourse done on American telephones strikes the European observer as extremely extensive by comparison with home. In 1962 the German engineer Jipp pointed out that there was a direct relationship between *per capita* GNP and telephone penetration, a law which guarantees America's place at the top of the penetration table; but it can be noted that in the late 1970s, both Switzerland and the USSR were still, albeit slightly, below average in the number of their phones according to Jipp's Law.[167]

It was not then simply a question of a better service being provided by free-enterprise American companies (for no such better service was necessarily provided) as opposed to the European states' failure to achieve systems that worked (for many states did achieve good systems.) It was rather that different countries responded differently to the lure of the telephone for a wide range of cultural and economic reasons. But in each and every case, the logic of universal service was at work. It is all very well for current proponents of deregulation to slight the historical evidence but it is common sense that deregulated and competitive American penetration was achieved at no little cost. 'Two telephone systems in a community were a great source of inconvenience and usually of expense to the subscribers.'[168] And history is on the side of this view. The logic that drove the Europeans to a single control was to put paid to the notion of meaningful competition in America as well.

With the arrival of competition, Bell's approach to the American market had to change. In all the years of protection, it had only placed one third of a million telephones and its network had just reached Chicago.

American Bell responded aggressively to the new situation, cutting its rates in towns where there was direct competition, refusing to interconnect with rivals but offering interconnection to systems where it had no prior interests. In the later part of this era, that is to say from 1907 on, Bell began to buy up the stocks of its competitors, and not just in telephony. The end of the patent marked the end of the arrangement with Western Union too and Bell made initial moves on the telegraph company, just as Western Union bought shares in telephone concerns. When, by the late 1890s, the legal restraints imposed by the Commonwealth of Massachusetts on holding companies and trusts chafed on its burgeoning take-over activities, American Bell moved its assets to the more entrepreneurially minded state of New York and AT&T became the holding company.

Bell exercised all its considerable muscle in this fight. For instance, it was claimed that the People's Telephone Company could not establish itself in New York because New York Bell held stock in the subway operations which were required by the city to duct telephone lines. The subway company said its tunnels were unable to accommodate further wires and refused to build for the People's Telephone. The president of New York Bell swore that 'no steps had ever been taken to prevent competition from entering New York.'[169] New York, however, never got the People's Telephone.

Bell also continued to exploit its other patent advantages. The Berliner patent had been delayed until November 1891 and was therefore in force to 1905. Bell had interests too in the most effective automatic telephone exchange equipment. The assignment of numbers to subscribers had been first suggested in 1879 and the earliest exchanges were simply patchboards operated by hand. Although various adaptations from telegraphic models were made, it was not until 1889 that a viable automatic exchange was developed, and with it the possibility of the telephone dial.

Almon B. Strowger is another figure in the pantheon of Victorian amateurs, a Kansas city undertaker supposedly dreaming up the mechanism that bears his name to thwart malicious and corrupt central telephone office operators who were denying him business by not connecting clients. In fact, having had the initial idea, Strowger involved a number of engineers, A.E. Keith and the brothers Erickson, who actually perfected a working system. The first Strowger mechanism was operating experimentally in La Porte, Indiana, in November 1892. For the first time subscribers could make a call without the aid of the operator. In 1902 Western Electric developed a 10,000-line system based on another set of patents, but the machinery was not installed. In 1916, Western Electric acquired the Strowger manufacturing rights.

Too much must not be made of the automatic exchange. Bell, and the telephone industry generally, remained unconvinced as to its superiority over manual operations and questioned whether or not the public would accept dialling instead of having the operator do it. It was not until 1919, the year of the telephone in Britain, that the company finally began the extensive installation of automatic exchanges.[170] (A similar Bell conservatism can be seen at work in the matter of the combined receiver/transmitter. The first such handset had been patented in the UK in 1877, the US army were issued them by the 1890s and Bell linesmen had them in 1902. Despite the widespread diffusion of the device in Europe (they were known as 'French phones'), top Bell engineering opinion was against them. J.J. Carty, chief engineer, gave orders in 1907 that the company 'avoid taking the slightest step which might precipitate a general demand' for French phones.[171] As a result Americans got the modern telephone 'station' (i.e. a one-piece transmitter/receiver cradled on a device with a numerical dial) in 1924.[172])

The most important technological advantage Bell had over its competitors was in the development of longlines. In contrast to IBM, say, whose history contained nothing like the struggle to *invent* the telephone in the years 1876-79 and whose patents were not subjected to legal attack, Bell Telephone began with a high awareness of the importance of patented technology. (Perhaps its slowness with automatic exchanges and handsets can be attributed to the fact that these were not, *ab initio*, Bell devices.) 'Out of the dispute over Bell's claims has come the most important, the most protracted litigation that has arisen under the patent system in this country.'[173] One of AT&T's central functions within the emerging Bell system was to continue a serious programme of research and development, a tradition that led to the merging of AT&T's research arm with Western Electric's in 1907, and the foundation of Bell Telephone Laboratories in 1925. The tradition had as much to do with the protection of company interests as with the advancement of human knowledge. The Labs were to give AT&T a crucial presence in every congruent technology, as is detailed in the substance of this book. All modern communications – radio, sound motion pictures, television, coaxial cables and other wire technologies, lasers, microelectronics, satellites – has been touched (and patented) by Bell. The patent system gave the telephone company a series of major bargaining chips to be used, during the first century of its existence, in its evolving relationship with the society it serves.

Despite the distractions of competition, Bell invested much effort in long-distance circuit technology, the basic mathematics of which had been outlined by Oliver Heaviside in 1887. In 1889, an MIT graduate in the Boston laboratory of AT&T, George Campbell, began working on a bimetallic wire with 'loading coils' which preserved the vibration and resonance of the original signal in the fashion of a primitive repeater. At the same time Michael Pupin, a professor at Columbia University, produced a similar device and managed to patent it first. AT&T mounted an interference action and, as insurance against its outcome, bought Pupin's patents.[174] By 1911, using Pupin coils, the network reached Denver. To carry it to the Pacific coast, the newly invented vacuum tubes were used as repeaters, Bell acquiring De Forest's patent in 1913. Two years later, carried by 130,000 poles, the telephone line between New York and San Francisco was opened.[175]

Technical factors therefore deeply affected the competitive situation. Bell was still protected by a number of significant patents so that even if the independents had adopted a better notion of public service than they evidently did, they would have been hamstrung, especially in the matter of toll lines, by inferior technology. (Again, this is not without its parallels in the current period of renewed 'competition'.)

In fact, the relative immaturity of the whole system was crucial in determining the shape of competition, in allowing it to happen at all. It was not simply that Bell dropped its high tariffs the moment another phone company opened up shop. Competition occurred when telephony in America was only partially diffused and the provision of service was by no means universal. Much of the work of the independents was more rapidly to expand the system into areas where Bell had not yet gone. Had Bell been regulated into a wide provision and low tariff mentality before the expiration of its patents, the independents would have had far less, if any, area of manoeuvre. Even given Bell's failure to make itself impregnable before 1894, the result of this free-for-all in the market place was still the certain victory of Bell.

After nearly a decade of competition in 1902 it had 57 per cent of the stations, and 70 per cent of the wire. Five years later when all systems together, including Bell's, had three times as much wire and twice as many telephones, Bell's share remained much the same.[176] Five years after that, in 1912, Bell companies operated 55 per cent of all telephones and all but 17 per cent of independent stations were also connected to its wires. In that year, the last before the state took a hand in the telephone industry, Bell sold 11,000 subscribers and bought systems with 136,000.[177] By 1916, Bell had eliminated competition in 80 per cent of its markets. The independents had long since begun to call upon the law, in the form of the Sherman Antitrust Act, for protection. As early as 1910 the independents' association was calling for regulation:

> We do not ask the Government to fight our battles, but we do ask
> for protection against outrageous methods of warfare which are
> illegal and detrimental to the public welfare. . . . We are not afraid
> of supervision; we believe in regulation.[178]

Long before America's entry into the war, it had become clear that Bell would not be allowed to enjoy its emerging victory undisturbed although, as the Supreme Court of the State of Kansas eloquently witnesses, people were simply fed up with the inanities of competition. 'Two telephone systems serving the same constituency place a useless burden upon the community, causing sorrow of heart and vexation of spirit and are altogether undesirable.'[179] However much communities wanted just one telephone company, the wave of populist politics, which had found partial expression in the Sherman Act, also more directly sought out AT&T as an object of attack.

Theodore Vail, having made a fortune outside telephony, had returned to the board of AT&T in 1900 and was persuaded to assume presidency of the company in 1907. Now, thirty years after Alexander Bell had enunciated the 'grand system', Vail began to make it a political reality. He continued, if not intensified, the competitive strategies of the company but he also had a remarkably enlightened view of the need for public service. He stated that:

> With a large population with large potentialities, the experience of
> all industrial and utility enterprises has been that it adds to the
> permanency and undisturbed enjoyment of a business, as well as to
> the profits, if the prices are put at such a point as will create a
> maximum consumption at a small percentage of profits.[180]

In 1912, having bought Western Union and with it a dominance of the republic's non-postal communication system, Vail enunciated a new route between the various perils of untrammelled monopoly, limited and wasteful competition and government take-over. He declared himself to be in favour of the strict regulation by government of the telephone industry as long as it was 'independent, intelligent, considerate, thorough and just'.[181]

As Edison said, 'Mr. Vail is a big man. Until his day the telephone was in the hands of men of little business capacity.'[182]

Vail's was a long way from the view the company had taken in the 1880s when Forbes had written in an annual report: 'No state in fairness ought to destroy what the patent system has created. . . . Sound public policy is surely against the regulation of the price of any class of commodities by law.'[183] But times had changed. The US Attorney-General in 1908 tried for one last time to break the patent and had failed. Now, pressured by the independents, he warned AT&T that further acquisitions would indeed offend the antitrust laws. In the same month, the Interstate Commerce Commission, under whose jurisdiction telephony now fell, began an investigation of what some claimed was AT&T's attempt to take over the country's entire communication infrastructure. Britain, the last Western European state to have commercial telephone companies, closed them down in 1912. The US Postmaster General, Albert Berleson, was eager to reopen the question of 'postalisation'. He told a congressional committee,

I have never been able to understand why the use of wires should be denied for the transmission of communications, any more than the use of a man on foot, or a boy on a horse or a stage coach.[184]

In reply to these multifaceted threats, Vail said, 'All monopolies should be regulated. Government ownership would be unregulated monopoly:' and he continued to hew his statesmanlike path. For instance, he had already allowed Western Electric, breaking the original monopoly arrangements, to sell to non-Bell companies. But the forces arrayed against AT&T were not to be so easily distracted. No aspect of the giant's affairs was left in peace.

The Attorney-General now had a new concern, that Bell was directing telegraphy customers to use Western Union, its subsidiary, at the expense of its competitors. This charge was to act as the final straw. After some months of negotiation, Bell officials determined that the wisest course of action would be not just to address this latest complaint piecemeal but rather to seek the elimination of all conditions and practices 'repugnant to the Federal authorities'.[185] On 19 December 1913 an AT&T vice-president, Nathan Kingsbury, wrote the Attorney-General a letter which became known as the Kingsbury

Commitment. In it AT&T agreed to disgorge Western Union, to make its toll lines available to independents, and to work with the Interstate Commerce Commission in obtaining prior approval before acquiring any more telephone systems.

The Kingsbury Commitment closes off the period begun by the Bell patent of 1876, the end of the first phase of the operation of the 'law' of the suppression of radical potential. It outlined the basis upon which the telephone was to become part of American society and it remained in force for the better part of seventy years. It marks the acceptance of the logic of universality but it also enshrines the principle of private ownership. The result was as ideologically confusing to capitalists as it was to the populists, but it worked. It allowed for the necessary creation of a unified telephone system, yet preserved some elements outside the monopoly. Even Western Union survived strong enough to own and operate satellites in the 1980s.

This compromise was so effective that it withstood the high point of populist attack on the company when, five years later during the war, telephony was nationalised. In effect all that happened was the imposition of a government board of control and the nominal management by the Post Office of a system that remained AT&T's. AT&T's profits for the one year of nationalisation were $540,000,000, $5,000,000 more than it had managed to make by itself in 1917; but this period of government control, occasioned largely by years of populist demands for cheaper telephone rates, saw a rate increase, and with it a total justification of AT&T. (To what extent the generous terms secured by the company from the Post Office contributed to the necessity for this rise was not publicly discussed.)

With the peace, the Kingsbury Commitment was re-established as the *modus operandi* and, in 1921, the Graham Act enshrined it in law. AT&T was exempted from the provisions of the antitrust legislation and no longer prohibited from taking over competing services. (With mandated interconnection and so much of the system in its hands, it conducted itself with restraint. Philadelphia, the last city with two competing telephone systems, came into Bell's sole control as late as 1945.)

In 1931, the government, responding to the depression, ordered Bell to cut its rates in Wisconsin, and thereafter across the whole system by 5 per cent. 'Sound public policy' had come to embrace the regulation of the price of the telephone commodity, but without damaging the private corporation. 'One system, one policy and universal service' meant that what was good for AT&T was also good for America, even in these circumstances. The result of these policies was that America's early lead in numbers of telephones *per capita* was maintained. By 1914 there had been 12.8 million telephones worldwide and 9.5 million of them were in the US.[186] By 1917 there were 12 million telephones in

America, one for every ten people. On the eve of the Second World War, Bell was worth $5 billion, the largest firm in history. It controlled 83 per cent of all US telephones, 98 per cent of long-distance toll wires and had a total monopoly on overseas radio telephony.[187]

Regulation was also responsible for the fact that it provided the most sophisticated and up-to-date service in the world. Its profit margins were determined by its investments; the more it spent, the more it could earn. This, rather than private enterprise *per se*, accounted for its technical superiority over rival systems.

But its success did not mean that the essential contradictions of its position remained unchallenged. Whether universal service and a regulated monopoly also meant vertical integration, that is ownership of the equipment supplier as well as the network, was vigorously questioned by the FCC, now the regulatory agency, in the 1930s. In 1949, the Justice Department argued that Bell should divest itself of Western Electric, but failed in its attack, although the telephone company was forced to agree that its role should be limited to that of a common carrier. In 1968, the FCC ordered Bell to allow non-Bell equipment to be attached to its system. In 1971, the FCC allowed a rival microwave link between St Louis and Chicago to go into operation. In 1974 Justice returned to the attack, claiming that Bell's behaviour since 1956 was stifling competition and inflating rates. By 1982, AT&T consented to divest itself of all its local operating companies. The world's biggest firm was reduced by a third, but it was relieved of its promise not to engage in other businesses – which, given its technical dominance, could mean it will become an even greater behemoth than before especially since it no longer has the wearisome provision of local service to worry about. It immediately set about becoming a major computer manufacturer, while IBM bought into the new longlines services. The net result of this competition was simply the provision of rival lines to serve the profitable business sector at the expense of the average user and, contrary to 'revolutionary' expectations, it worked to concentrate communication lines between cities. Fourteen of America's biggest cities were expected to be 'served' by at least three rival fibre-optic nets by the 1990s.[188] At the same time the GPO's telephone department was privatised in the UK and provided with one competitor.

It is likely that, since the essential logic of universal provision remains unchanged by current technology, the institution of telephony overall will be little affected by these moves. Certainly the main purpose seems to be the provision of a measure of competitive service for only the heaviest, i.e. business, users; the establishment of competitive telephonic infrastructures for the mass is neither contemplated nor likely, and the most widespread forecast of the results of these various moves is nothing but an increase in the costs of universal service. These possibilities are of limited interest here since the model suggests the radical potential of the telephone has long since been contained.

The essential potential of the telephone was contained by the arrangements made at the time of the First World War, detailed above. Telephony was a technology which potentially allowed one person to talk to many on a democratic basis. Limiting it to a one-to-one system is the mark of its repression. That broadcasting is a function of the more controllable and more limited radio is the obverse of that mark.

In the 1870s, microphones and loudspeakers were attached to telephone wires for experimental purposes. A church service was brought to the bedside of a sick person. In Switzerland an engineer relayed Donizetti's *Don Pasquale*.[189] in 1884 a London company offered, for an annual charge of £10, four pairs of headsets through which a subscriber would be connected to theatres, concerts, lectures and church services. In Paris and Budapest, all-day news services were available. The London experiment lasted until 1904, but the telephone news channels persisted into the inter-war years.[190]

These services were technologically futile, exploiting a potential that had actually already been designed out of the system. The responsiveness of the telephone, once it worked at all, was deliberately limited in the interests of economy. The less bandwidth taken, the more conversations could be accommodated on the single wire and the cheaper was the service provided. It was discovered that the human brain needs remarkably little information in order to recognise a voice, a fact of which telephones were to take maximum advantage. Coupled with the non-provision of high fidelity was the failure to expand the interconnectibility of stations via the central office; instead each subscriber could only talk to one other station at a time. Making it possible for one subscriber to talk to many others would have enhanced the telephone as a non-hierarchical means of communication. It was not to be. The telephone was a device designed to aid commercial intercourse, not to redress imbalances in information power within society. Indeed, as events in Poland in the early 1980s revealed, even the spin-off social use of the phone, in which unsupervised conversations between only two stations can take place, are sometimes too dangerous for the state to allow. The phones were shut down. The People's Telephone Company never established itself in New York; nor indeed, one might add, anywhere else.

Phase four: technological performance – production, spin-offs, redundancies

John Walson invents CABLE TELEVISION, MAHANOY CITY, PA, 1948

By 1984, there was just about one telephone for every ten people on earth, but the radical potential of the wired network was not thereby fully exploited or even remotely exhausted. The use of wires for the transmission of information in audio/visual or non-alphanumeric (i.e. telegraphic) data forms, uses which can be characterised as spin-offs from telephony, have not been anything like as widely diffused.

That the bandwidth limitations placed upon the telephone circuit were not the result of technological limitations can be proved by the immediate development of higher fidelity wires for the retransmission of radio signals. These were in place almost as soon as radio itself. By the early 1920s, in the Netherlands, signal enhancement and the importation of distant foreign stations provided supervening necessities for radio cable systems. By 1939, 50 per cent of Dutch homes, and in some urban areas as many as 80 per cent, were wired. In Britain the 'rediffusion' of domestic radio services had begun in 1925 on the South coast where shipping signals caused enough interference to render domestic reception difficult. In the Northeast too, where the BBC was slow to provide service, relay exchanges sprang up. Reception in most places was however excellent and improving and so by 1935 only some 3 per cent of radio licence holders received service through these wires.

Peter Eckersley, the man who had built the BBC's wireless net, left the corporation in 1929 to join Rediffusion, the leading relay company. Two years later he had devised a way of using the national electrical power distribution grid and mains cables to carry six radio signals. The radical potential of this development was well understood. The smallness of the relay audience had not prevented the GPO, in 1930, from forbidding any 'wireless exchange' operator originating programmes.[191] Further restrictions were imposed. No company was to have more than 100,000 subscribers drawn from any one area of 2,000,000 and none was to operate within an area as a monopoly. As the BBC Year Book noted in 1933,

> The system, however, contains within it forces which if uncontrolled might be disruptive of the spirit and intention of the B.B.C.'s charter. The persons in charge of wireless exchanges have power, by replacing selected items of the Corporation's programmes with transmissions from abroad, to alter entirely the general drift of the B.B.C.'s programme policy. They can transmit amusing items from

the British programmes and replace talks and other matter of informative or experimental value by amusing items in programmes from abroad and so debase their programmes to a level of amusement interest only. The BBC has always provided entertainment for its listeners but it has set its face resolutely against devoting its programmes entirely to amusement.[192]

In the event neither the attraction of foreign stations (with dance bands on Sundays) nor the lacuna in the domestic transmission system was sufficient in Britain or on the continent to allow public demand to sweep away the objections of the postal and broacasting authorities. Cable radio withered on the branch.

The technology of the wideband coaxial cable capable of carrying a high-definition picture of some 400 plus lines had been developed by Bell Labs in 1936 and put into operation by AT&T in 1937 (see page 74 above). In Britain, the new video technology was grafted on the old radio system and the first television cable was operating in Gloucester by 1951. Fifteen years later more than 1,000,000 receivers were on cable; yet the introduction of the UHF network, which began in 1963, inexorably wiped out the prime raison d'être of most of these systems – poor reception. Only municipal regulations forbidding roof antennae kept cable television alive, increasing its universe to a high of 2.3 million homes in 1973, one third of them Rediffusion subscribers.

The nemesis of the UK cable industry was the GPO and broadcasting industry's insistence on a form of universal service wherein the variety of UHF signals is curtailed in the interests of maximising reception. In America, eschewing UHF, the internationally agreed 13 VHF channels (reduced to 12 by a 1952 FCC decision) were used to achieve exactly the opposite result, i.e. a maximising of the number of different signals with a concomitant disregard of fringe reception problems. That objective, after all, is what ostensibly occasioned the FCC's four-year freeze in licensing television stations between 1948 and 1952.

The result was that signal enhancement remained a viable business in America, a measure of the technical shortcomings of the broadcasting system. It was an important function, exacerbated by naturally poor reception conditions both geographical (distance and mountains) and urban (skyscrapers) and the fact that the NTSC colour system was anyway less easily received than comparable European systems.

On Tuesday and Saturday nights, in the late 1940s, the bars of Summit Hill, Pennsylvania would fill with people from Panther Valley below to watch Milton Berle and the fights. Summit Hill was high enough to receive Philadelphia's television signals 70 miles to the south. The valleys were not, and because of the 'freeze' could have no local television stations either. It was here, locked between the Blue Mountains and the Poconos, that cable television in America was born, and Panther Valley's bar-owners were among the very first to complain about its effects on business.[193]

The development of cable is perhaps the most vivid example, in this book, of the operation of the 'law' of the suppression of radical potential. It was initially, as Eckersley had suggested to Reith, nothing less than a complete alternative to wireless transmission.[194] But nowhere has this occurred. Instead careful prevarication and delay has meant that everywhere cable has been allowed only to supplement the efforts of the broadcasters. It has taken decades to achieve even this limited function but it should be noted that at no time has any of this slow diffusion been occasioned by technological constraints. Unlike some other devices we have examined, with cable there are no questions as to possible earlier or alternative applications of the hardware. Cable has stood ready to supplant broadcasting from the very beginning of both radio and television. Thus the history of cable television in America between the establishment of the first system in 1948 and the Cable Telecommunications Act 36 years later can be seen as an examplar of the 'law' of suppression, the perfect expression of the limitations of technological determinism.

Four radio dealers in Lansford, George Bright, William McDonald, Robert Tarlton and Rudolph Dubosky, established the Panther Valley Television Company. They built a tower on Summit Hill 85 feet high, charged subscribers $100 installation fee and thereafter a monthly $3. At the same time, John Walson, a maintenance worker with the local power company who had a part interest in a Mahanoy City radio shop, ran a cable from the mountain top into the store and filled its windows with working televisions. In 1948 he sold seven sets but slowly the business grew. By December 1950 both Walson's and the Panther Valley systems had been franchised by their respective communities. They had found a way to side-step FCC regulations which prohibited the most obvious technical solution to the problem of non-reception, i.e. rebroadcasting the signals received. The FCC had nothing to say, however, about cabled rebroadcasting.

Press coverage of the Pennsylvanian pioneers was extensive and again the radical implication of the technology was well understood. Interviewed by the *New York Times*, George Bright pointed out, 'There's nothing that can be patented. . . . The system just uses standard equipment and anybody can do it.' The *Times* added, 'With the present "freeze" on the construction of new television stations, which may continue until the end of the national emergency [i.e. Korean War], Mr. Bright noted that the Lansford system might be the means for vastly extending the range of television service.'

George Bright was somewhat disingenuous as regards the ready availability of equipment. While it was readily available, nevertheless it had to be specially configured to the nascent cable industry's needs. Milton Jerrold Shapp started Jerrold Electronics in Pennsylvania with his army discharge money, $500, in 1948 and expanded the business by building master television antennae on apartment buildings. One of his staff told him about the Lansford system where further developments depended upon the availability of a reliable signal amplifier. Shapp promised Robert Tarlton, one of the Lansford four, such a device and was within years the leading cable hardware manufacturer, a position his company still holds. One of Shapp's most important innovations was to insist on a five-year service contract based on an initial $5 connection fee and a monthly 25¢ service charge for each subscriber, in return for which Jerrold undertook to train service personnel, replace defective units and generally keep the system running for the duration of the agreement. Such a relationship gave the early five-channel 'ma'n'pa' operations the technical wherewithal to offer a reliable service to the public and allowed Jerrold to commandeer some 76 per cent of the early cable market.

Cable programming remained a secondary order of business throughout the developmental period although another Pennsylvanian pioneer, Martin Malarkey, did start his own local talent showcase in Pottsville. But the more significant additions to cable's programming came from the importation of distant signals. For instance, San Diego received three network VHF stations but in the 1960s became the largest cabled market when all seven Los Angeles stations (including the same three networks) were imported.[195] As five-channel systems gave way to an industry standard of 12 channels, even well-served communities like San Diego were attracted to big city television via cable.

The FCC's wavelength allocation procedures had been to privilege conurbations at the expense of the countryside, obviously because such a policy created the maximum audiences for advertisers. In terms of VHF coverage, the cost of seven stations in New York and the other 'major markets' was that generally across the country one house in five could only get two services. The slow move to colour also helped cable. Poor reception of black and white in built-up urban areas meant ghosting on the picture. With colour the signal became worse and cable thereby found an opportunity to sell itself even in the privileged conurbations, such as Manhattan. Thus, in rural areas, in smaller towns and cities as well as in great cities, broadcasting transmission standards were low enough to give cable something to sell.

The received rhetoric of cable is that growth was very slow throughout the 1950s and 1960s and that its supervening necessity was either signal enhancement nor distant importation, but additional unique entertainment services. However, on the contrary, poor reception and limited rural choice were the basis of cable's steady and consistent growth throughout the period when almost no special services were offered. By comparison with the other technologies here discussed, there is no objective reason for describing cable television's early growth as slow. Within a decade 850,000 rural subscribers received service over no less than 800 systems. By 1962 there were 1325 systems with 1,200,000 subscribers, i.e. a $40 million a year industry.

Blessed with the inattention of the urbanised broadcasting industry and its regulators, by the early 1960s cable had become a sufficiently lusty infant to demand action. It was the ultimate parasite, reselling a product that was not only expensive but also protected in law by copyright for which it paid not a cent. Throughout the 1960s, the courts were to take a very lenient view of cable. In 1961, it was held that television stations could not restrict the use of their signals after transmission unless economic harm could be proved. In 1964, an Idaho network affiliate claimed it was being damaged because the rival cable company ran the same programmes imported from distant stations. The court held for the cable company.[196] The most significant of this strand of decisions was that of the Supreme Court in Fortnightly Corporation v. United Artists Television.[197] Since the stations and the networks had failed to establish their rights to the material they transmitted, in this case an original producer, United Artists, sought relief against a cable company, Fortnightly. The inadequacies of the common law of copyright, essentially a protection for eighteenth-century booksellers, were fiercely illuminated by this decision. Since the common law did not admit the concept of intellectual property, the court began by holding that copyright was a limited right. The relevant American statute, the Copyright Act of 1909, provided protection only against an 'exclusive right to perform' and the court determined that rebroadcasting a signal whether for profit or not was not a 'performance' within the meaning of the act and that United Artists had no action. The cable company was more like an ordinary viewer than a television station. The technological naiveté of this decision is perhaps matched only by the court failing, for half-a-century, to acknowledge that telephone wiretapping constituted an invasion of privacy protected by the Fourth Amendment against illegal physical searches of property.

Clearly the disruptive potential of the cable industry was not going to be contained at law. The FCC, which had refused jurisdiction in 1959, would have to act to protect its primary clients in the broadcasting industry; and it did. In 1963 it began to exercise its traditional role as the prime agent of suppression in a volatile technological environment.

The commission was petitioned by a firm which wished to establish a community antenna (CATV) to import distant signals into a remote part of Wyoming. The local broadcasting station successfully challenged the FCC's original grant on the ground that it would be put out of business. The FCC withdrew its permission and the CATV company sued on First Amendment grounds. The court held that the commission was correct in considering economic harm.

Bolstered by this, the FCC asserted its right to regulate the entire industry and in its First Report and Order of 1965, it simply told cable operators they could not import programming which duplicated that of local stations (defined as those within 60 miles of the cable system) within 30 days of a local broadcast. The FCC also announced that it would freeze the importation of distant television signals into the top 100 markets. A year later the 30-day rule was relaxed to a one-day prohibition but the other restrictions on cable were not lifted. After the decision in Fortnightly, the FCC imposed even stricter regulations on the importation of distant signals, wiping out the legal advantages gained by the cable industry. Now cable companies would require the specific permission of copyright holders, stations, networks, producers, syndicators, etc., on a programme-by-programme basis before they could import the signals. Local rebroadcasting remained the prime permitted function but, in addition in the early 1970s, the FCC required all systems to originate material on a 'public access' basis.

By the 1970s the game was changing. No longer were the cable operators 'ma'n'pa' enterprises, outgrowths of rural electrical appliance stores. For the whole of the previous decade, the usual telecommunications interests had been buying up cable systems, becoming MSOs or multiple systems operators. The cable system was being integrated into the industry behind the shield of the FCC's draconian restrictions. The cable 'freeze', the period of maximum regulation, operated exactly as the television 'freeze' had twenty years earlier. It allowed the players to begin to sort themselves out while not entirely inhibiting growth. Indeed by 1974, despite these restrictions 11.7 per cent of all American homes were cabled. By 1980 21.7 per cent were on the wire.

During the late 1970s, in part because its immediate work was done (cable was contained) and in part because of the new rhetoric of deregulation, the FCC eased up on its requirements. The Copyright Law was rewritten and cable operators were required to pay licence fees which were to become, despite the modest percentage of production costs they represent, an ongoing source of friction with the broadcasters. Importation regulations and mandatory access requirements were dropped. Finally in the 1980s the industry disentangled tself from the cities, just as the telephone companies had done at the)eginning of the century. The 1984 Cable Telecommunications Act masculated the city franchising process and left the industry free to pursue profitability as it thought best.

Of all American homes 42.9 per cent were cabled as of 1984 but cable programming services, for all that the rhetoric claims otherwise, had not so denied the audiences or advertising revenues of the networks as to cause major realignments. The American cable industry has failed to break through the programming formats which the culture, extremely conservative in this regard, demands. Cable has, perforce, limited itself to plugging the lacuna created by the broadcasting, film and record industries – sex, videos and first-run movies (where the last two are largely synonymous with the first). Cable viewers' loyalty is still, in the main, to the rebroadcast signal and the business of providing programming for cable, as opposed to running cable systems, remained chancy. Cable programming has not drained off broadcsting advertising dollars as was being strenuously argued as a possibility only five years ago. Cable's fragmented audience could not overturn the basic economic reality behind the television image, which is that only the mass, however exploited, can yield high enough revenues to allow production at the level that same mass accepts as television – which put crudely means, by 1986, dramas that cost $1 million per hour to make. All the talk of a multiplicity of programming – specialised magazines, each with its own 24-hour channel – failed to grasp this central point.

Cable potentially provided hundreds of channels but could not programme more than a fraction of them to meet even the minimum cultural expectations of its audience and the 'television of abundance' turned out to be nothing but the most primitive and boring output across far too many outlets. By 1985, most of these channels had audiences too small to register and the biggest of them could only achieve an average rating of 1 or 2. The broadcasters had public taste on their side. Independent and public non-commercial UHF broadcast stations have benefited from the improved reception cable brings as much as have the cable industry's own programming services, if not more.

Nevertheless the cable industry has a considerable degree of public acceptance in the US; it is no longer an outsider. Basic provision of service was worth $2 billion gross by 1984.[198] Cable programme services are now being acquired by broadcast networks, ABC controlling the largest basic cable service and CBS, after failing to launch its own, having bought into a group of channels. NBC, via its parent RCA, owns satellites upon which these services are distributed. CBS ran two cable systems in the early 1980s, the FCC allowing its cross-ownership regulations to be suspended for this purpose. Time Inc owns programmme services and has 4.6 million subscribers. Warner has 2.6 million. There is a great deal of vertical integration; Viacom, one of cable's big fifteen, is not untypical. It has 1.4 million cable subscribers, four broadcast television stations, seven broadcast radio stations, a basic cable programming service, a pay cable service with 5 million subscribers, a videocassette distribution company and various production and marketing entities. Viacom, like its fellow entertainment conglomerates, is well placed to cope with the 'information revolution' – should it ever occur.

The 'law' of the suppression of radical potential has worked with cable television in America just as it worked previously with telephony itself. The new technology, over a period of forty years, has been absorbed by the institutional structures of the old. This process has not only reduced cable's disruptive potential, it also ensures that in case of any upheavals those same structures will remain profitable. They are already placed to exploit significant shifts in the market place. This is not to say responding to such shifts, even if they are on a less than seismic scale, does not require considerable managerial skills and that failure is impossible. Both RCA and Warner have conducted themselves in such a manner over these past several years as to stretch seriously the system's capacity for preventing them coming to harm. But, thus far, they and their fellows are all in place.

Cable became an opportunity for the industry not because of the possibilities of abundance but rather for signal enhancement. Significantly, cable penetration is slowing markedly as the inner city areas are left the only remaining quarters to be wired. Building cable systems in such an environment is expensive, the recipients of the service are poorer and the overall need in terms of reception is less acute. The industry is not as confident as it once was about the percentage of homes it will have connected by 1990. But nevertheless cable in America remains one of the consciousness industry's real triumphs. For now Americans pay twice, through advertisements and through subscriptions, primarily to watch TV channels they used to pay for only once.

Signal enhancement is not the only supervening necessity for cable. In culturally dependent societies the importation of distant foreign signals might occasion the creation of extensive cable systems. Belgium, where 75 per cent of all homes are passed by cable and 60 per cent subscribe and Canada, where 80 per cent are passed and 75 per cent subscribe, are cases in point. In Canada, the audience spends the majority of its time watching American television. In Belgium, though, despite cable's popularity and the comparative poverty of the two domestic television services, the majority of time is still given to the national services. It is likely that cable, satellite-fed cable and even direct broadcast satellites will have lesser rather than major effects on the loyalty of national audiences unless cultural dependency is pronounced. The model of radio, where the availability of transnational signals has never dented the dominance of the national networks, is confirmed rather than denied by the Belgian cable experience.

Cable's radical potential to destroy wireless communications is being contained. But rational telecommunications policies would suggest that wires are very much needed. Spectrum pollution requires that signals be placed in cables, just as Eckersley foresaw in the late 1920s. The problem with the market development of cable television is that, quite simply, the wrong wires are being used. It is possible to make a sober argument that the entire telephonic infrastructure, after all now a century old, will need to be replaced and upgraded. Obviously, fibre optics and light-carrier waves should be the basis of this rebuilding. But, instead, considerable resources have been diverted to establishing electromagnetic carrier systems, in essence fifty years old. The control of cable's radical potential as a destroyer of broadcasting is matched by the inhibiting effect exerted on the development of a truly forward looking broadband communications infrastructure. The first fibre optic system to be installed was operating by November 1984 in Birmingham, Alabama, fifteen years after the technology was introduced and while switching technologies were still undeveloped and expensive. Given the investment in the telephone infrastructure and the slow rate at which it is amortised, these 'problems' should come as no surprise. The 'law' of suppression is at work. By 1985, only Denmark had made a decision to rebuild its entire telephone system with fibres.

Cable television, delivered through conventional wires, is not a symbol of the 'information revolution', much less one of its cutting edges; rather it is a clear indicator of the extent to which the 'revolution' is thwarted and delayed.

IN CONCLUSION

On babble

The information revolution is an illusion, a rhetorical gambit, an expression of profound ignorance, a movement dedicated to purveying misunderstanding and disseminating disinformation. At the popular level this is particularly pronounced:

> Plugging the new technologies into this new energy base will raise to a wholly new level our entire civilisation. . . . We feel increasingly helpless, and our dependence on tv is surely a reflection of that helplessness. . . . The acceleration of change in our time is itself an elemental force. . . . The tempo of human evolution during recorded history is at least 100000 times as rapid as that of prehuman evolution. . . . Now the means are at hand to take the programming of the citizen much further.[1]

The popular literature resounds with visions of techno-glory or apocalypse, the same set of phenomena being the source for both styles of pontification. As Norman Cohn pointed out in another connection:

> It is characteristic of this kind of movement that its aims and premises are boundless. Social struggle is seen not as a struggle for specific limited objectives, but as an event of unique importance, different in kind from all other struggles known to history, a cataclysm from which the world is to emerge totally transformed and redeemed. This is the essence of the recurrent phenomenon, or, if one will, the persistent tradition – that we have called 'revolutionary millenarianism'.[2]

And what form should such a tradition take in a faithless age, if not one technologically determined?

That ranting texts such as those quoted above should be produced in such circumstances is no surprise. It is only worthy of note that supposedly scholarly work exhibits the same traits – fervid but purblind imagination, unbalanced judgments and unidimensional insights. The study of the future is as much as it ever was the domain of seers, visionaries and other extremely unreliable types. This reality, however, has been masked by the exigencies of corporate and military forward planning which has occasioned the growth of a spurious scientism, 'futurology'. In this the military, where self-fulfilling prophecies are perhaps more easily engineered than in the civilian world, has taken the lead, funding the contemporary colleges of augurs. Military necessity led to the systematising of the future in the 1960s. The auger's new *templum* ('a rectangular place in which auspices were to be sought') is the graph whose seductive curves we have already referenced; for mathematical forecasting is nothing but computerised augury. 'The difficulty with these curves . . . is either that they presume a fixed environment, or that in an open environment they become erratic.'[3] For instance, envelop curve extrapolation, a favourite of the Hudson Institute, can be used to plot, it is claimed, the rates at which speeds will increase towards the end of the century. By taking the speed of light as a limit and plotting all speeds up to those achieved by today's chemical rockets as a series of curves against time, it is possible to predict interstellar space craft travelling at near the speed of light by 2050.[4] Or not, as the case may be. The deployment of resources to achieve the historical speeds marked on the graph were all the results of complex interactions of which the procedure takes no cognisance. Those forces are still at work. The environment is indeed erratic.

Of other techniques, the Rand Corporation's so called 'Delphic Method' is the one which most closely acknowledges its antique origins:

First a carefully selected panel of appropriate experts is assembled. Each is queried independently. The responses of others are then fed back to the participant so that he has an opportunity to revise his opinions on the basis of his knowledge of their opinions. What emerges is a consensus or 'convergence' of expert opinion. This is presented in quantified form. The sophistication of the procedures can be increased by weighting the responses of individual experts according to the relevance of their expertise to the specific question, according to their previous success at prediction, according to their own sense of certainty, or some other principle.[5]

Outside of the limited variables of the military (and maybe even within it) it is hard to see how any of these activities have anything other than rhetorical use.

The eminent futurologist Herman Kahn collaborated with Philco Ford some years ago to produce a film of the shape of things to come. In this vision we saw a wife ordering supper by the 'World Box', bills being paid automatically in a house that could grow as its family grew. One sequence showed homemaker Mum and wage-earner Dad entertaining. On a large screen TV Dad punched up a singer with the remark that the shots – synchronous interiors filmed to a professional standard – were taken while on holiday in Puerto Rico. The screen was filled with a Sinatra-style crooner. The environment and all its hardware is transformed, but there is still a home and the woman is still the homemaker, Puerto Rico is still a vacation paradise, a Sinatra is the unchanging norm for popular music. The film starred a 1950s nuclear WASP family doing all the things a nuclear WASP family then did and the next half-century of technological upheaval made no apparent difference to them, nor presumably to Philco Ford.

Technology by itself is comparatively easy to hypothesise about, but such activity is of limited value. The futurologists are simply not good at variables and in consequence all they do is draw attention to the difficulties of the prediction game and the narrow limits of their own imaginations.

The technological determinist vision, in both its promise and curse variation, proposes revolutionary change but in a very curious way. The vision does considerable violence to every aspect of our culture. It supposes the end of productive work, the institutions of representational democracy, the school and the shop, to name but a few. But those who hold this view have no difficulty in preserving through all technological storms one essential and dominant institution which remains at the centre and indeed acts as the engine of upheaval – the modern corporation. Because the vision is entirely technologically determined it assumes these changes can occur without any concomitant alterations in social relations.

Technological determinists generally have a very poor grasp of the world and the people in it. For instance, the influential James Martin, a lucid and prolific IBM propagandist, believes that media images of American affluence could incite a rabid jealousy among the poor of the earth.[6] He also articulates the commonly held claim that falling intercontinental telecommunications charges 'together with the worldwide spread of identical-looking hotel chains, bars, multinational advertising, and multinational corporations, makes the planet seem much smaller. . . . We can make one firm prediction. The change in communications will have a major effect on the life of every individual.'[7]

Martin is the *pontifex maximus* of such matters.

Most people still feel stronger emotional ties to their country than they do to their corporation [he has written]. Volkswagen employees in Germany feel more strongly that they are Germans than that they are Volkswagen people. But patriotism is declining and may decline more with decades of advanced global communications. Some people will feel more loyalty to their global cultural thread than to their country. An English computer specialist with IBM feels more in common with a Japanese IMBer (sic) than with an English poet. American and Russian astronauts together exhibit understanding and loyalty that transcends national differences. . . . If the technology described in this book had come forty years earlier, the British commonwealth might never have disintegrated.[8]

Such babble offers no insight into the real impact of technological change in communications – the use of the transistor in the Iranian revolution to create an anti-Western, anti-technological backlash would be but one example of an 'information revolution'. The political repercussions in the UK of the South Atlantic War of 1982, the persistence of Christian fundamentalism in the USA would be other examples; above all, the essentially unchanging nature of the media themselves gives these widely retailed sophisms the lie.

Beyond the pervasive fallacious interpretation of specific phenomena, three central assertions are made about the nature of innovation in telecommunications (and indeed in general). They are:

(a) that there is an exponential increase in the amount of information in the world;

(b) that the pace of change is faster now than it was in the past and is looking to get faster yet;

(c) that the nature and locus of innovation is now very different from what it was even in the immediate past of the last century.

To take these points in order:

More information

Certainly there is more information in the world – there are, after all, more people in the world too. The significance of the rising tide of information was widely noticed first in the 1960s. Daniel Bell, in *The Coming Post-Industrial Society* – a *locus classicus* for this argument – itemises every aspect of the growth of the knowledge industries. But the measures he employs, the most thorough and sophisticated available, are nevertheless impossibly crude; i.e. that university research libraries are doubling in holdings every 13 years, that scientific periodicals increase by a factor of ten every 50 years, that professional and technical workers doubled between 1947 and 1963 in the US, that the number of students as a percentage of the peer age group went from 22 to 44 per cent in the same period, and that in the century from 1870 the total number of degrees awarded by American universities went from 9372 to 1,025,400 and so on.[9]

If the American academy was first nationalised so that it was less tuition-fee dependent and then, second, became convinced that those areas of vocational training it had subsumed – the teaching of business, trades and the industrialised arts – were, after all, inappropriate to its mission, what then would happen to its inflated student rolls? If that one academy alone decided that teaching skills (or athletic prowess, physical appearance, articulateness or whatever) rather than publications were to be made the basis of tenurial decisions, what then the fate of these burgeoning journals and books whose outpourings of information are supposedly creating new polders of human knowledge freshly claimed from the sea of ignorance? In short, what about quality?

> I recall his [Sholokhov's] pungent joke about the Tula Section of the [Soviet] Writer's Union, with its 2000 per cent increase! 'Superb progress; today there are in Tula *twenty-three* writers; in 1910 there was *only one*! He was called Lev Tolstoy.' (italics in original)[10]

There is not a scrap of evidence that our capacity to absorb information has markedly changed because of this monstrous regiment of data, print and images, nor that the regular provision of 'facts' beyond our limited capaicty to absorb them materially alters our ability to run our lives and societies. And the peculiarities of the younger generation's attention span, if peculiarities they are, will not do service for a case in this regard. The supposed aliteracy of the young is bound up not only with the 'information revolution' but also a curious post-McLuhanesque argument about orality and literacy generally in cultures. From the reading aloud of the medieval monk, we have passed, it is suggested, through the age of silent reading into a new period of no reading at all, of visual communication systems. As to the first of these changes, the 'proof' can be put in the following terms:

> some cultures have the technology of speech – these are oral
> cultures; others have the more advanced technology of writing –
> these are writing cultures. Their superior technology fundamentally
> alters the mental and social organisation of these written cultures
> just as the application of iron or steam revolutionizes human
> ideology and social structure.[11]

And the rebuttal can be put in thus:

This appealing argument cannot survive exposure to common sense.

All cultures are by definition oral cultures. When men learn to write they do not then forget how to speak. Even with writing, much information – probably most information necessary for fundamental human activity – continues to be passed along solely by speech and held in mind without written records.[12]

Robert Pattison's rebuttal can be extended to the next supposed shift from literacy to visual communication. Written information is no more tossed aside with the new modes than it itself swept away orality – and that assumes that there is such a shift to visual communication. Those who believe television is a visual medium should attempt to decode its messages with the sound removed. (To make the exercise more reasonable, let them avoid news broadcasts. There the lack of sound is more than compensated for, in most 'writing-cultures', by the writing on the screen.) In short, 'the kids are all right!' As they have always done, querulous historians and other social critics claim, on the basis of what are largely journalistic anecdotes, that the next generation is of a different species; but the same figures Bell quotes on the increasing scope of academic achievement contradict that view. These 'post-literate' beings are the very ones responsible for the information flood.

If anecdotage is permitted, then there is just as much to suggest the flood means very little – the ignorance of politicians (and indeed electorates) seems unchanged. Economic, social and urban planning is not transformed. When one ceases to compare the entertainment and literary practices of the nineteenth-century Western bourgeoisie with the twentieth-century proletariat (a most fruitful source of all moral panic about the decline of our civilisation), the uses of entertainment and literacy look stable. To take one example as a microcosm: the computerisation of audience survey figures has allowed the American television industry to have on a daily basis statistics which were once, a few decades ago, available only monthly or quarterly. This information has increased the industry's capacity to create publicly acceptable product not one iota; if anything, each year the twenty new shows in network prime-time yield ever fewer hits.

The fact of more information, like the fact of miniaturisation, is meaningless of itself; and there is little discussion of the quality or understanding of the effects of increased information beyond the absurd assertion that it throws us into a state of 'shock'. The concept of 'increase' itself rests on the quantifications that Information Theory made popular which exactly purvey an idea of information drained of semantic and social content. Mechanemorphic conceits have contributed to this picture the image of the human being as a computer about to burn out because of input overload – something computers, never mind people, do not do. People, at least the vast majority of those born into this culture, and machines just switch off, mostly without too much effort.

As with a belief in the flatness of the earth, it is easy to cleave to the 'revolution' position but there is really no objective reason for so doing. The very people who claim that the world is out of control in this regard are themselves sufficiently in command to write long logically organised texts making satisfying patterns out of the welter of data that is apparently drowning everybody else.

This of course does not address the argument that, aside from Bell's measures, economic analysis reveals a shift into a sector which can be designated, in distinction to Service, Industry and Agriculture, Information. There is no question that this has happened, but the issue is when and how massively? The information revolutionaries would have us believe that we are in the midst of an explosive growth in this employment sector, yet a case can be made for the explosion, if indeed it was one, to have taken place far earlier.

Between 1870 and 1930 the percentage of office and sales clerks in Philadelphia's workforce jumped from 3.4 to 13.9 per cent.[13] US Census figures suggest that between 1900 and 1940 the percentage of 'white collar' workers increased nationally from 17.6 to 31.1 per cent. In the next 40 years to 1980 growth was from 31.1 to 51.2 per cent, the major rise taking place during the 1950s. Current US Bureau of Labor Statistics thinking is that the 'white-collar' sector – now described as (i) Managerial and professional, speciality, (ii) Technical, sales and administrative support, (iii) Service – is levelling off.[14] Thus the curve is steady throughout this century, if not before, and there is no American evidence of a violent upsurge in this type of employment consequent upon the arrival of 'info tech' in the last 15 years.

But this is not the essential point. What about the basic classification of 'white collar' used in these statistics? The 50 per cent figure includes all who work in the management offices of factories, mines, etc. Even in the least 'white collar' of industries (construction or manufacturing for example), office workers account for up to a third of the total numbers employed. They have always been there, to one degree or another, but they have not always been counted as such by the statisticians; historically they have been grouped with their blue-collar colleagues. If more modern classifications are extrapolated into the past, then, for instance, 12.4 per cent of the workforce could be classed as information workers in 1890.[15] The 'white-collar' growth curve flattens.

Panko would flatten it further, reducing the contemporary official American 'white-collar' figure by excluding from it such groups as foresters and conservationists, nurses, dieticians and therapists, school teachers, mail carriers, newspaper vendors, cashiers, hucksters and peddlers. However, he leaves in, for example, physicians, dentists and college teachers – thus illuminating the essential class nature of the classification process. 'White collar' means clean hands and which hands are 'clean' becomes pretty subjective. However, by his proceeding the number in the 'white-collar' category falls from 50 to 40 per cent. Clearly, though, a further considerable percentage could be removed without doing violence to what commonsense would suggest as appropriate to a category designated 'information worker'. Lancing the inflated contemporary roll and counting in the forgotten factory clerks of the past shows something more like a doubling in 'information work' over the century to a current figure much nearer a third of the workforce than a half.

This denial of explosive growth is important only because the substitution of information production for other sorts of production is central to the 'information revolution'. The 'revolution' claims to have provided 'infojobs' – the first point in what Panko calls 'the Basic Litany'[16] – but when one discounts the hordes of people serving hamburgers and the like, it becomes clear that it has not delivered them.

Another view of the supposed flood of information is possible. It could be that, as with too much vitamin C, we simply discard this excess contentless data. Bell quotes Borges on the image of the 'universe (which others call the Library)' to the effect that '*The Library is unlimited and cyclical*' without seemingly appreciating the force of '*cyclical*' and the point of Borges's insight.

> If an eternal traveller were to cross it [the Library] in any direction, after centuries he would see the same volumes were repeated in the same disorder (which, thus repeated, would be an order: the Order). (italics in original)[17]

The Borgesian library is, eventually, tautological; likewise the information revolution's data flow is, more quickly, *redundant*.

More change

The refutation of the second assertion, that the pace of change has become faster, has been one of the main purposes of this book. Positivist research, on behalf of the US government, produced very definite evidence of this contention, that the lapse from 'initial discovery' to the 'recognition of its commercial potential' has dropped from thirty years in the period from 1880 to end of the First World War, to sixteen years between the wars, to nine years in the period from 1945. During this three-quarters of a century, the time taken to move from 'basic technical discovery' to 'commercial exploitation' has dropped from seven to five years.[18] Thus Bell claims:

What is true is that the rate at which technology is diffused through our economy has accelerated somewhat in the past seventy five years, and this is one measure of the popular conception of the increase in the rate of change.[19]

The business of such computations has been largely in the hands of economists whose literature on innovation is rather more sombre than that of the futurologists; however, beyond the marketing level, it is no better at predictions. Nowhere in the economic literature are there definitions of terms like 'initial discovery', 'recognition of commercial potential' and, most important of all, 'diffusion' remains vague.[20] The argument that R&D has reduced the 'incubation time' for new products substantially while the marketing time has remained fairly constant could be the result of these vaguenesses. The economic literature is at its best when dealing with the diffusion of innovation within a given market – the object or process to be studied is established, concrete; and the market provides a detailed measure of its success. 'Incubation time' is very much less easily grasped. Thus, in calculating the 'incubation time' of, say, the transistor, does one start with the establishment of the Shockley semiconductor group after the war, or Shockley's pre-war briefing, or the German researches of the 1920s and 1930s? Alternatively, can one date the telephone from the twang of June 1875 and commercial exploitation from the establishment of the first Bell company a mere 24 months later? It would seem that all pronouncements about rates of change need to be examined very carefully.

In general, it is possible to take a quite opposing view and argue that increasing complexity is slowing the date of innovation and diffusion. Galbraith, for instance, has suggested that, 'An increasing time-span separates the beginning from the completion of any task,' because technology costs ever more money, requires more specialists and specialists in the organisation of specialists, and results in inflexibility.[21] All of this is subsumed by the concept 'planning'. Planning is the process whereby one ensures 'that what is ultimately foreseen eventuates in fact' and conversely nothing unforeseen (which in the nature of the case includes *invention* especially in its eureka guise) should occur. The effect of all this is to slow down the rate of change.

In the early days of Ford, the future was very near at hand. Only days elapsed between the commitment of machinery and materials to production and their appearance as a car. . . . The earliest cars, as they came on the market, did not meet with complete customer approval: there were complaints. . . . These defects were promptly remedied. They did the reputation of the car no lasting harm. Such shortcomings in the Mustang would have been unpleasant. And they would have been subject to no such quick simple inexpensive remedy.[22]

This is not the place to explore the ramifications of this argument; suffice it to say, the evidence of speed-up is by no means as compelling as it seems at first sight. As far as the development of telecommunications technologies is concerned, it is the case that when an attempt is made to compare the historical like with like, as has been done here, there is simply no evidence of speed-up at all. In chapter 2 we showed how electronic television was described in 1908, partially demonstrated in 1911, patented in 1923, perfected (or *invented*) a decade later and how it took a further quarter of a century for it to be diffused. Videorecording was proposed in 1927 and demonstrated in 1951. Its potential use as a home recording device was noted in 1953. The first cassette appeared in 1969. Diffusion was to take a further decade and a half. As is detailed in chapter 3, the idea of the computer as a stored programme device was first made in 1944, although glimmers of it as a possibility can be traced back over the previous century. Five years later the first giant machines worked: 10 years later there were a couple of hundred computers in the world: 20 years later, some 30,000. In 1976 the first microcomputer appears, although throughout the 40-year period of the giants, various 'baby' machines had been stillborn. Five more years pass before the PC achieves a significant measure of diffusion. In chapter 4 a similar time frame is described for the microprocessor heart of the minicomputer. A suggestion for a solid-state device is made in 1925 and various partial prototypes are demonstrated from the 1930s to the 1950s. The idea of a solid state circuit is articulated in 1952. Contrary to popular opinion, the move from primitive transistor to CPU takes place virtually independently of the computer industry; and it takes two decades. Then a further six years elapses before anybody puts a CPU into a small computer. In chapter 5 is evidence that the Nazis could have orbited a satellite, if they had so desired, in 1944. It takes twenty years from the clear statement of the idea of a geostationary communications satellite to the start of service and the availability of small receiving dishes. Twenty years after that there is no exploitation of the technology's full potential, that is DBS in commercial service.

All this compares very well with parallel nineteenth-century developments described in chapter 6. The electric telegraph: proposed first in the seventeenth century, but re-proposed in 1820, demonstrated in 1825, perfected in 1837 and diffused over the following three decades. The electric telephone: proposed in 1854, demonstrated in 1876, *invented* in 1877 and widely available by 1914. Radio: suggested in 1872, demonstrated in 1879, *invented* in 1895 and diffused over the next 40 years.

And the introduction of the laser, in the midst of which we currently find ourselves, suggests that the pattern still holds good: ideation – 1947, *invention* 1960, and slow diffusion from then on.

These regularities not only refute, for telecommunications at least, the idea of a general technological speed-up; they are at the heart of the model of technological change suggested in this book.

More inventing

We have also had occasion in the body of the book to question the third assertion, that the nature and locale of innovation and the type of inventor have significantly changed. This can be stated classically in the following terms:

> Compare Watt's utilization of the theory of latent heat in his invention of the separate condensing chamber for steam engines, or Marconi's exploitation of developments in electromagnetics with Carother's work which led to nylon. Shockley's work which led to the transistor, or recent technological advances in drugs and military aircraft. In the earlier cases the scientific research that created the breakthrough was completely autonomous to the inventive effort. In the later cases, much of the underlying scientific knowledge was won in the course of efforts specifically aimed at providing the basic understanding and data needed to achieve further technological advances. Carother's basic research at DuPont which led to nylon was financed by management in the hope that improvements in the understanding of long polymers would lead to important or new improved chemical products. Shockley's Bell Laboratories project was undertaken in the belief that improved knowledge of semi-conductors would lead to better electrical devices.[23]
>
> The pith of the modern view is, therefore, that in the nineteenth century most invention came from the individual inventor who had little or no scientific training, and who worked largely with simple equipment and by empirical methods and unsystematic hunches. The link between science and technology was slight. Manufacturing businesses did not concern themselves with research. In the twentieth century the characteristic features of the nineteenth are passing away. The individual inventor is becoming rare. . . . Useful invention is to an ever-increasing degree issuing from the research laboratories of large firms which alone can afford to operate on an appropriate scale.[24]

The pith of the modern view is wrong. First, as we have shown with the telecommunications pioneers and as can be demonstrated in other fields, the nineteenth-century technologists were not isolated from contemporary science and, second, their twentieth-century successors were very often a good deal less than 'scientific' in their procedures. Jewkes, Sawers and Stillerman, in an important survey of innovation, describe numerous contemporary technologists who were as intuitive, as individual and as scientifically well (or ill) informed as their predecessors. 'The histories of these inventors, therefore, do not give the impression that a fundamental change in the methods and character-istics of inventors has taken place in the present century . . .'[25]

Goodyear, for all that vulcanisation was 'accidentally' discovered, worked with leading scientists of the day and the Commissioner of Patents opined that he was 'so much the master of the subject of rubber that nothing could miss his attention'. Edison, after building his lab, continued to use Princeton faculty as research staff to scour the learned journals for him.[26] 'Accidental' nineteenth-century discoveries, such as aniline dye or vulcanised rubber, are matched by twentieth century accidents such as polyethylene, Freon refrigiants and Duco lacquers. Modern innovation is not necessarily better focussed and less wasteful – the A bomb cost $1500 million, a sizeable percentage of which was thrown away. The individual has not disappeared from world of inventing. In the twentieth century solo inventors, inside and outside the industrial lab, have been responsible for, among other things, air conditioning, ballpoint pens, the electron microscope, the helicopter, insulin, the jet engine, penicillin, streptomycin, titanium, xerography and zip fasteners. From the communications area can be added Cinerama, Kodachrome (and all colour-film systems), magnetic recording, the personal computer, the hologram and Polaroid.

The innovator, whatever his or her characteristics actually are, is unchanged; but what of the method of innovation? 'It is a commonplace of modern technology that there is a high measure of certainty that problems have solutions before there is knowledge of how they are to be solved.'[27] Remove the word 'modern' and the commonplace stands. Leave the word be and the statement is historically unjustifiable. The men who made the telegraph and the telephone were every bit as aware that solutions existed to the problems they had determined to solve as their descendants who grappled with television, microelectronics, lasers and the rest. Take the history of the filament electric bulb, the cartoonists' icon of the bright idea, between 1880 and 1920.

Edison, in his private lab at Melno Park, reputedly tried over 6000 substances in his two-year search for a filament, the prime example of the Edisonian try-and-try-again method of invention. After human hairs and blades of grass, he eventually used charred sewing thread and for 45 hours in mid-October 1879 it burned – longer than any filament ever had before. By 1900, Edison had sold 24,000,000 sewing thread lamps.[28]

The search for a better incandescent lamp was continued by employees of the giant corporation founded on Edison's patents, the General Electric Company. GE had established the first industrial research laboratory in America in 1900, following the lead of Bayer (aspirin) in Germany a decade before.[29] Willis Whitney, a professor at MIT (whence he had his first degree, his doctorate being from Leipzig), undertook the part-time direction of the new lab.

One of the first things Whitney had done when he arrived at Schenectady was to build a small, high temperature furnace. . . . Into it he had put a good many different materials and substances just to see what would happen to them under the influence of extreme heat. One of the things he put into the furnace was a carbonized cellulose filament. On removing it, he discovered that the filament had changed its character and had become, in effect, metallized.[30]

That the furnace was in an industrial lab made no more difference to the procedure than Whitney's formal qualifications in physics. The new lamp he produced was 17 per cent more efficient than Edison's original filaments on which the patent had anyway expired. GE built a million-dollar plant to exploit the competitive advantage Whitney had given them.

Whitney invited Coolidge, like himself a graduate of MIT and Leipzig, to continue experiments on filaments even before the new plant was open. After 4 years of effort with 20 assistants and at a cost in materials alone of $150,000 Coolidge created the first tungsten filament. He proceeded much as had Whitney, mixing the difficult-to-work metal with various substances and eventually stumbling upon a way to make it ductile. By 1914 85 per cent of all bulbs had tungsten filaments.[31]

Now Langmuir (Columbia and Göttingen) took up the challenge. Coolidge's lamp, like its forerunners, tended to blacken and Langmuir's investigation of this phenomenon on behalf of GE, begun in the year Coolidge cracked the ductility of tungsten, proved that the blackening was caused, not as had been thought by imperfect vacua, but rather by the vaporising of the filament. By a process, again, of trial and error, he then discovered that if argon gas were used instead of a vacuum it would prevent blackening and that, further, coiling the filament would also improve performance.

When he understood this much, he turned his findings over to twenty-five other men who took a year and a half to move the new lamp through the development stage into production. He himself continued those studies of gases, electrical discharges . . . and led in time to his being awarded the Nobel prize.[32]

The dazzle of the Nobel prize however must not obscure the fact that Langmuir, for all his theoretical understanding of the problem, produced the improved design much as Edison had produced the original and much as Shockley and Co. were to produce the transistor, that cynosure of the supposed modern method of innovation. It is nothing but assertion to state that 'research aimed at opening up new possibilities has substituted both for chance development in the relevant sciences, and for the classical major inventive effort aimed at cracking open a problem through direct attack.'[33] Development in science is not a matter of 'chance' and there is no evidence that the relationship of science and technology is changed markedly so that innovation is now science-led, except in a most obvious and banal way. It is true that in the dance technology weaves around science, today science is quicker to offer explanations; but it leads no more and no less than it did in the past. Intuition and accident, individual insight, still have their place and they remain paramount. The gap between Goodyear and Marconi on the one hand and Carothers and Shockley on the other is merely rhetorical. We have now dealt with three of this quartet and have established at length that what Shockley calls 'the will to think' is the mark of the unchanging intuitive nature of creation in any field (see chapter 4, pp. 191 et seq. above). But what of Carothers and **nylon**?

Carothers was a brilliant chemist specifically hired by DuPont to undertake a programme of pure research. While still at Harvard he had been concerned with the synthesis of long-chain molecules.

> At first this information was only of academic value, since Carothers and his associates were primarily interested in the phenomenon of polymerisation. But something occurred in 1930 during an experiment with a polyester made from ethylene glycol and sebacic acid which was destined to be of great practical value. In an attempt to remove a sample of the heated polymer from the vessel in which it was prepared, Dr. Julian W. Hill, working under the direction of Carothers, noted that the molten polymer could be drawn out in the form of a long fibre. More important, he found that even after the fibre was cold it could be further drawn to several times its original length, and that such drawing greatly increased the fibre's strength and elasticity.[34]

The only problem was that the polymer softened in hot water, which for a textile was a considerable disadvantage. Trying to find a superpolymer that had the characteristics accidentally discovered by Hill so discouraged Carothers that he simply gave up the search while DuPont considered closing the operation down. Persuaded to review the work, Carothers turned from alcohol and polyesters to amins (ammonia compounds) and polyamides. In 1935, using Hill's cold-drawing process, Carothers produced a polyamide, named '66' polymer, which was eventually to become nylon. The first stockings were spun experimentally in April of 1937 and the pilot plant constructed in 1938. Mass production began in December 1939. The whole development process had taken 11 years. Carothers had committed suicide in 1937.[35]

Pretending that all is now organisation and logical method might be necessary to assure stockholders that their money is being well spent (nylon cost DuPont, at the height of the Depression, $6,000,000 to research) but it is no basis for understanding the processes of innovation. Nylon does illustrate the two distinct functions of the industrial lab in its guise as the site for 'structured' innovation – the creation of new products and the protection of the parent company's base. Nylon, the new DuPont monopoly, was a brilliant product but it also (slowly) destroyed the market in rayon, once a DuPont monopoly.

Of the two tasks, protecting the business, as the history of telecommunications reveals, is done more efficiently than inventing new products. The latter are still produced by processes ('something happened') too serendipitous to be industrialised, yet the incorporation of research and development into business is nevertheless crucial for protection. It is not that R&D primarily provides a cutting edge of innovation, whereby profits from new products can be crudely maximised; rather it is that research protects the firm from being overtaken by others and that within any high-tech market, all research efforts (from different firms) have the combined (and conservative) effect of ensuring stability, and with it profitability over the long haul. This is exactly the clue to the history of television development, largely a struggle by everybody's technologists to contain the disruption promised by the breakthroughs of RCA's technologists. The Bell Labs function as much as a brake as an accelerator (to use Braudel's terminology) across the whole field we have been examining.

Although delays and failures of exploitation can occur at any stage of the process of innovation (as when the scientific agenda – and with it the ground of scientific competence – is distorted by particular needs of the moment, e.g. war), we locate the operation of the 'law' of the suppression of radical potential in our model at the point of diffusion, for it is then that the market is most in need of protection and the 'law' can be most clearly in operation.

It is further erroneous to suggest that the industrial lab is an institution wholly new and particular to this century. Such labs are rather nothing but the latest development of the technologists' workplace, the proceedings of which are well established by three centuries of practice. The modern lab, like any other modern institution, is the prisoner of its history. W.H.G. Armytage suggests that the learned academies and the royal monopolies – and their immediate predecessors the botanical gardens – constituted the original site of modern technological labour. There were by 1790 some 220 societies whose members were actively 'involved in improving guns, powders or mines, commerce and medicine, or in making palaces, fountains, dyes and medicaments'. They were international; the Berlin Academy was only one quarter German, and Benjamin Franklin belonged to more than twenty of them. They cooperated in celestial observations, published in the common languages of French or Latin and in general the very fact of their existence and activities denies the received idea of the lonely amateur eighteenth- and nineteenth-century researcher.[36]

The savants were not isolated from each other nor from the industrial implications of their researches. For instance, soda was traditionally obtained from burning seaweed or other plants. In the 1770s a number of workers sought a method of making it from salt heated with sulphuric acid. One of them, Nicholas Leblanc, perfected a system by 1787 which was in commercial production by 1790.

In the French Revolution, the Ecole Polytechnique (itself modelled on the mid-eighteenth century School of Bridges and Roads) was training students in pure science and such academies began to supplement the learned societies. In these years the German tradition of the ersatz began; Goethe was the state official who equipped a laboratory at Jena for the use of scientists seeking to negate the consequences of the British blockade by producing sugar from alcohol and carbon dioxide as well as creating a new method for making sulphuric acid. New universities along Ecole Polytechnique lines were founded at Berlin and Bonn.

At the end of the Napoleonic wars, the learned societies and the newer polytechniques/technical universities were joined by the first Association for the Promotion of Industrial Knowledge in Germany. The Association for the Advancement of Science was founded in England, directly at the instigation of Babbage. He had campaigned for Royal Society appointment to be by merit but was voted down; so he and others founded the Association in York in 1831. By 1848 half its monies were devoted to the provision of research funds and equipment.

The professionalisation of engineering was necessitated by the railroads, a charter for an Institute of Civil Engineers being secured as early as 1828. When George Stephenson, eighteen years later, could not show on paper his qualifications to be a member, the Institution of Mechanical Engineers was founded to accommodate him. Other professional societies followed.

By the midyears of the century, even the intellectually moribund

English universities were awakening to the need for science. Maxwell, the first Cavendish Professor, complained 'I have no place to erect my chair but move about like a cuckoo'; yet soon the status of science in tertiary education was to be a public scandal leading to the creation of no less than eleven new colleges in the last quarter of the century – which became the redbrick universities.[37] In this Britain was only catching up with the US and Germany.

The industrial lab in one form or another therefore emerges, in Armytage's convincing account, with a history as long as the scientific revolution itself.

> To sum up: A significant proportion of twentieth-century inventions have not come from institutions where research will tend to be guided towards defined ends. There are many similarities between the present and the past century in the type of men who invent and the conditions under which they do so. Many of the twentieth-century stories could be transplanted to the nineteenth without appearing incongruous to the time or the circumstances; far too many, indeed, to render the idea of a sharp and complete break between periods.[38]

It is perfectly possible to stress the continuity of the processes of innovation in our culture throughout the entire modern period: the words of Sir William Petty three centuries ago still stand as a model for how vexed technological change was then and still is:

> So as when a new invention is first propounded, in the beginning every man objects, and the poor inventor runs the gauntlet of all petulant wits; every mind finding his several flaw, no man approving it, unless mended according to his own advice: now not one in a hundred outlives this torture, and those that do are at length so changed by the various contrivances of others, that not any one man can pretend to the invention of the whole, nor will agree about their respective shares in the parts.[39]

The major underlying assertions of the 'information revolution' – increased information, increased pace of change, structured industrialised innovation – are no more sustainable than its detailed surface arguments. The flood of information is less significant than is claimed, the pace of change has not increased, the nature of innovation is unchanged. The rhetoric of 'revolution' is, as Nisbet points out, nothing but a hangover of a style of thought which reached its zenith in the last century – a quasi-religious belief in the idea of progress.

One of the more interesting legacies of the idea of progress of old is the great vogue for what is called futurism or futurology. It too is inseparable from a foundation of imagined progress from past to present to future. Much is made of our computer technology, allowing us, it is solemnly said, to foresee scientifically what the Tocquevilles and Marxes of the past could only guess at. . . . Basically, futurology has not changed. All invocations of computerology, systematic institute-based research, and econometric models to the contrary notwithstanding, what today's forecasters work from, essentially, is the same kind of identification of a trend or of trends in the present and then the extension of these into the future with whatever modifications insight and intuition may suggest.[40]

The positive information revolutionists are thus identified. The negative popular literature is, by Nisbet analysis, simply a little more *au courant*, dealing with the same topics but expressing the current decline of belief in those five values he isolates as being constant to the idea of progress from the ancient Greeks to the present; belief in the value of the past, in the nobility or superiority of the West, in the benignity of scientific and technological growth, in reason and in the value of individual life.

This book has been concerned with neither of these attitudes to the implications of telecommunications developments. Rather the substance of those developments themselves has been our subject and we have been at pains to understand the forces at work producing particular technological solutions at particular times. At the outset we suggested that a better historical understanding would serve more accurately to influence insight and intuition in identifying trends in telecommunications. Rather than a revolution, we suggested a range of more modest possibilities which can now, by way of conclusion, be revisited in the light of what has come between.

Misunderstanding media

Let us return to the provocations with which we began.

It is now possible to see the limits of entertainment-led cable television in numerous actual situations. The cable tide in North America has peaked and elsewhere current events reveal cable will have to compete not only with other new technologies but most significantly with traditional broadcasting. The revolutionists as late as the beginning of this decade were still suggesting that cable would transform broadcasting. They were wrong.

The videocassette recorder is now fulfilling the rhetorical function that cable can no longer sustain. But for all that the revolutionists have belatedly discovered it, it is a device of considerable significance to the moving image industry and one that is potentially, because of the degree of control it offers its owners, more disruptive than cable. All the other new delivery systems including DBS will occupy, at best, profitable but minor positions.

The essential conservatism of the culture in its audiovisual tastes is being well demonstrated by these developments, not one of which has produced a single image outside of the mainstream of bourgeois illusionism or the fad for the sixty-year-old repertoire of visual surrealism. The new technologies of television are far more productive of hardware than of software and a sequence of marketing strategies to sell first component television (with stereo sound) and camcorder 'homemovie' apparatus, then flat screens, followed by 1200 to 2000 line high-definition television, will keep the industry and consumers busy until well into the first quarter of the next century. Holography, by this timetable, can be expected at about the same time as the Hudson Institute expected stellar space-craft. Such improvements in hardware will sell because the consumers' addiction to high fidelity and high definition is well established and very deeply rooted in the culture. (This addiction will allow the misuse of laser video in CD audio recordings.)

The computer, in its most diffused form, having yielded no domestic uses beyond game-playing will simply replace the typewriter. Computing will continue to encroach on and enhance business and academic life but not in ways that threaten the structure of those activities. Its primary purpose will remain what it has been from the beginning, war; and its fifth generation will simply understand us better but be otherwise no smarter.

There will be an end to the redundant business of putting writing on the screen – except in connection with computers where specialised (i.e. stock market, legal references) videotext services will have a place. There will be an end to the culturally unacceptable and unnecessary provision of two-way video-services, to the electronic pollution of space through communications satellites, to silver-nitrate film, to banking via one's home computer, to non-specialist high-cost computer information services.

The only really interesting question, and it is scarcely revolutionary, is whether or not any society on earth (apart from the Danish) will approach rationally the rebuilding of its telephonic infrastructure so that the maximum use of digitalised, optically carried signals will occur. On this one can only say for certain that the current vogue for 'deregulation' looks like a singularly irrational way to set about this task.

These conclusions are the result of insight and intuition. Such privilege as can be claimed for them is grounded in the historical account we have offered. This sought to describe the pattern discernible in the events of telecommunications' technological development. It is a complex pattern, balancing the demands of social necessity against the constraints of the 'law' of the suppression of radical potential across all the central technologies of electronic information. It is this explanation that remains paramount; for whatever the future holds, we cannot hope to come to terms with it on the basis of our current widespread ignorances of our immediate technological past. The conclusions drawn are thus, we would argue, more firmly based than the hectic visions of the information revolutionaries which have nothing to sustain them but the outpourings of industrial public relations officers or the jeremiads of discombobulated social observers. We must simply stop misunderstanding the media, otherwise we are doomed to behave like Walter Benjamin's angel:

> A Klee painting named 'Angelus Novus' shows an angel looking as though he is about to move away from something he is fixedly contemplating. His eyes are staring, his mouth is open, his wings are spread. This is how one pictures the angel of history. His face is turned towards the past. Where we perceive a chain of events, he sees one single catastrophe which keeps piling wreckage upon wreckage and hurls it in front of his feet. The angel would like to stay, awaken the dead, and make whole what has been smashed. But a storm is blowing from Paradise; it has got caught in his wings with such violence that the angel can no longer close them. This storm irresistibly propels him into the future to which his back is turned, while the pile of debris before him grows skyward. This storm is what we call progress.[41]

NOTES

Abbreviations

Bell Journal	*Bell System Technical Journal*, NY, AT&T.
Forester	T. Forester (ed.), *The Microelectronics Revolution*, Oxford, Blackwell, 1980.
IEEE	*Institute of Electrical and Electronic Engineers, Proceedings of*: New York.
IEE	*Institution of Electrical Engineers, Proceedings of*: London.
IRE	*Institute of Radio Engineers, Proceedings of*: New York.
Metropolis et al.	N. Metropolis, J. Howlett, G.-C. Rota (eds), *A History of Computing in the Twentieth Century*, New York, Academic Press, 1980.
Randell	B. Randell (ed.), *The Origins of Digital Computers, Selected Papers*, Berlin, Springer-Verlag, 1973.
Shiers	G. Shiers (ed.), *Technical Development of Television*, New York, Arno Press, 1977.
SMP(T)E	*Society of Motion Picture (and Television) Engineers, Journal of*: New York.

Introduction

1 *The Times*, 24 April 1984, p. 16.
2 UPI, *The Daily Freeman* (Kingston, New York), 16 March 1984, p. 4.
3 R.T. Tripp (comp.), *The International Thesaurus of Quotations*, Harmondsworth, Middlesex, UK, Penguin Books, 1976, p. 564.
4 J.B. Priestley, *Thoughts in the Wilderness*, 1957, quoted in Tripp, op. cit.
5 R. Porter, *English Society in the Eighteenth Century*, Harmondsworth, Middlesex, UK, Penguin Books, 1982, p. 124.
6 My understanding of the earliest period of American television (from whence comes this ambition of Weaver's) is largely conditioned by William Boddy's NYU dissertation, 'From Golden Age to Vast Waste Land: The Struggles Over Market Power and Dramatic Formats in 1950s Television', an NYU Doctoral Dissertation, 1984 to be published by the Illinois University Press.
7 R.W. Hubbell, *4000 Years of Television*, NY, Putnam, 1942, p. 55.
8 S. Handel, *The Electronic Revolution*, Harmondsworth, Middlesex, UK, Penguin Books, 1967, p. 128.
9 G.R.M. Garratt and A.H. Mumford, 'The History of Television', *IEE*, vol. 99, Part IIIA, London, 1952, p. 26 in Shiers.

10 P.K. Gorokhov, 'History of Modern Television', *Radiotekhnika* translated as *Radio Engineering*, New York, 1961, p. 75 in Shiers.
11 B. Nichols and S.J. Lederman, 'Flicker and Motion in Film' in T. de Lauretis and S. Heath (eds), *The Cinematic Apparatus*, New York, St Martin's Press, 1980, pp. 97ff.
12 Garratt and Mumford, op. cit., p. 38.
13 Ibid., p. 41.
14 C.H. Owen, 'Television Broadcasting', *IRE*, vol. 50, New York, May 1962, p. 820 in Shiers.

Chapter 1   Breakages Limited

1 Raymond Williams, *Keywords*, New York, Oxford University Press, 1976, pp. 229 et seq.
2 C.G. Hempel, 'The Function of General Laws in History', *Journal of Philosophy*, vol. 34, no. 1, January 1942, p. 35.
3 G.B. Shaw, 'Lysistrata' in *The Applecart, The Complete Plays*, London, Odhams Press, 1937, p. 1027.
4 G. Gutting, *Paradigms and Revolutions*, Notre Dame, University of Notre Dame Press, c.1980, p. 117.
5 Thomas Kuhn, *The Structure of Scientific Revlutions*, Chicago, University of Chicago Press, 1962, p. 23.
6 Ibid., p. 10.
7 Ibid., p. 17.
8 Ibid., p. 24.
9 Karl Popper, *The Logic of Scientific Discovery*, London, Hutchinson, 1959, p. 431.
10 Kuhn, op. cit., p. 90 et seq.
11 James Clerk Maxwell, 'A dynamic theory of electromagnetic field', *Philosophical Transactions*, vol. 155, London, 1865, p. 460.
12 Kuhn op. cit., p. 74.
13 Gutting, op. cit., p. 10.
14 Ibid., pp. 43, 227, 141.
15 Ibid., p. 107.
16 Ibid., p. 117.
17 Ibid., p. 303.
18 B. Barnes, *T.S. Kuhn and Social Science*, New York, Columbia University Press, 1982, p. 115.
19 Gutting, op. cit., p. 274.
20 D. Gingerich, 'A Copernican Perspective' in O. Gingerich (ed.), *Nature of Scientific Discovery*, Washington, Smithsonian Institution Press, 1975, p. 21.

Chapter 2   Fugitive pictures

1 J.C. Wilson, 'Twenty Five Years' Change in Television', *Journal of the Television Society*, London, 1935, vol. 2, p. 86, in Shiers.
2 Garratt and Mumford, op. cit., p. 26.
3 Ibid., p. 25.
4 Quoted in Handel, op. cit., p. 39.
5 Garratt and Mumford, op. cit., p. 26.

6 G.G. Blake, *History of Radio Telegraphy and Telephony*, London, Chapman & Hall, 1928, pp. 125ff.
7 Garratt and Mumford, op. cit., p. 27.
8 Hubbell, op. cit., p. 65.
9 Garratt and Mumford, op. cit., p. 31.
10 *Journal of the Rontgen Society*, p. 15, London, January 1912.
11 Garratt and Mumford, op. cit.
12 A.G. Jensen, 'The Evolution of Modern Television', *SMPTE*, New York, 1954, vol. 63, p. 175 in Shiers.
13 Gorokhov, op. cit., pp. 71 and passim.
14 Garratt and Mumford, op. cit., pp. 27 et seq.
15 Jensen, op. cit., p. 174.
16 R.F. Tiltman, *Baird of Television*, London, Seeley Service, 1933, p. 62.
17 Ibid., after p. 72.
18 J.D. Percy, 'John L. Baird: The Founder of British Television', *Journal of the Television Society*, London, 1952, passim in Shiers.
19 Asa Briggs, *The History of Broadcasting in the United Kingdom*, vol. II, 'The Golden Age of Wireless', Oxford University Press, 1961, p. 530, n3.
20 B. Norman, *Here's Looking at You*, London, BBC/Royal Television Society, 1984, p. 61.
21 C.F. Jenkins, 'Radio Vision', *IRE*, New York, 1927, vol. 15, p. 960 in Shiers.
22 R.C. Biting, Jr, 'Creating an Industry', *SMPTE*, New York, 1965, vol. 74, p. 1017 in Shiers.
23 (i) M.D. Fagen, ed., *History of Engineering and Science in the Bell System: The Early Years 1875-1925*, Bell Laboratories, 1975, p. 790.
   (ii) H.E. Ives, 'Television', *Bell Journal*, Oct. 1927, p. 551 in Shiers.
24 H.E. Ives, F. Gray and M.W. Baldwin, 'Two Way Television', *IE(E)E*, vol. 49, New York, 1930, passim, in Shiers.
25 A. Dinsdale, 'Television in America Today', *Journal of the Television Society*, London, 1932, vol. 1, p. 139 in Shiers.
26 D.G. Fink, 'Television Broadcast Practice in America – 1927 to 1944', *IEE*, London, 1945, vol. 92, part III, p. 146 in Shiers.
27 Percy, op. cit., p. 14.
28 H. Gibas, 'Television in Germany', *IRE*, New York, 1936, vol. 24, p. 741 in Shiers.
29 Briggs, op. cit., p. 553.
30 Ibid., p. 583.
31 Norman, op. cit., p. 129.
32 Boddy, op. cit., p. 55.
33 Hubbell, op. cit., p. xi.
34 Biting, op. cit.
35 E. Barnouw, *Tube of Plenty*, New York, Oxford University Press, 1975, p. 27 et seq.
36 E.g. R. Scholes and R. Kellogg, *The Nature of Narrative*, New York, Oxford University Press, 1968, p. 85 et seq.
37 Norman, op. cit., pp. 210 et seq.
38 Briggs, op. cit., pp. 611 et seq.
39 J.H. Udelson, *The Great Television Race*, University of Alabama Press, 1982, p. 96. I am grateful to William Boddy for this reference.
40 Norman, op. cit.
41 For instance, in the UK a change from VHF to UHF (concomitant with a move from 405 lines to 625 lines) and then a further change from B&W to

colour were all accommodated on an evolving series of sets – 405, 405/625, 625 colour. These changes conditioned the British public into a belief in the instability of television which led to the rise of the TV-set rental industry.

42 F.C. Waldrop and J. Borkin, *Television: A Struggle for Power*, New York, William Morrow, 1938, p. 213.

43 V.K. Zworykin, 'Television with Cathode-Ray Tube for Receiver', *Radio Engineering*, New York, Dec. 1929 vol. 9, p. 38 in Shiers.

44 V.K. Zworykin, 'Television', *Journal of the Franklin Institute*, Philadelphia, Jan. 1934, vol. 217, pp. 19 et seq. in Shiers.

45 Schlafly, 'Some Comparative Factors of Picture Resolution in Television and Film Industries', *SMPTE*, New York, Jan. 1951, p. 50.

46 S.J. Preston, 'The Birth of a High Definition Television Service', *Journal of the Television Society*, London, 1953, vol. 7, p. 115 in Shiers.

47 Norman, op. cit., p. 107.

48 Ibid., p. 105.

49 Ibid.

50 Ibid.

51 Preston, op. cit., p. 123.

52 J.D. McGee, 'Distant Electrical Vision', *IRE*, London, June 1950, vol. 38, p. 598 in Shiers.

53 Ibid., pp. 599 et seq.

54 Preston, op. cit., p. 119.

55 Udelson, op. cit., pp. 110ff.

56 Jensen, op. cit., p. 176.

57 Norman, op. cit., p. 30.

58 (i) Waldrop and Borkin, op. cit., p. 211.
    (ii) G. Everson, *The Story of Television: The Life of Philo T. Farnsworth*, New York, W.W. Norton, 1949, p. 15 et seq.

59 Everson, op. cit., p. 200.

60 Waldrop and Borkin, op. cit., p. 219.

61 Udelson, op. cit., p. 107.

62 Ibid., p. 105.

63 Ibid., p. 112.

64 H. Iams, G.A. Morton and V.K. Zworykin, 'The Image Iconoscope', *IRE*, New York, Sep. 1939, p. 541. (In the same issue, Zworykin's debt to Farnsworth is even more overtly acknowledged in: V.K. Zworykin and J.A. Rajchman 'The Electrostatic Electron Multiplier', ibid., p. 558. J.A. Rajchman was to make a contribution to computing with the first custom-built electrostatic memory device.)

65 Everson, op. cit., p. 246.

66 Barnouw, op. cit., pp. 40 et seq.

67 Ibid., pp. 78 et seq.

68 D.G. Fink, ed., *Television Standards and Practice*, New York, 1943, p. 11 in Shiers.

69 Ibid., p. 12.

70 Ibid., p. 3.

71 Waldrop and Borkin, op. cit., p. 172.

72 I. de Sola Pool, *Technologies of Freedom*, Cambridge, Massachusetts, Belknap Press, 1983, pp. 121 et seq.

73 Waldrop and Borkin, op. cit., p. 134.

74 Ibid., pp. 127 et seq.

75 Ibid., p. 128.

76 Ibid.
77 W.I. Greenwald, 'The Impact of Sound Upon the Film Industry: A Case Study in Innovation', *Explorations in Entrepreneurial History*, vol. 4, no. 4, 15 May 1952, p. 185.
78 (i) Udelson, op. cit.. p. 92.
   (ii) Fink (1945), op. cit., p. 147.
79 Report of the Television Committee, Lord Selsdon (ch), Television Advisory Committee, London, HMSO, 1935, Cmd 4793 (The Selsdon Report), para. 53 p. 16.
80 Waldrop and Borkin, op. cit., p. 126.
81 Anon, 'Color Television Demonstrated by CBS Engineers', *Electronics*, Oct. 1940, pp. 32 et seq. in Shiers.
82 Norman, op. cit., p. 50.
83 H.E. Ives, 'Television in colours', *Bell Laboratories Record*, New York, 1929, vol. 7, passim in Shiers.
84 Fink (1943), op. cit., p. 41.
85 B. Schwartz, 'Antitrust and the FCC: The Problem of Network Dominance', *University of Pennsylvania Law Review*, vol. 1097, no. 6, p. 789.
86 US Senate Report, 81st Congress, 2nd Session, Document no. 197, passim.
87 L. Brown, *New York Times Encyclopaedia of Television*, New York, Times Books, 1977, p. 91.
88 Schwartz, op. cit., p. 788.
89 Boddy, op. cit., pp. 68, 70 et seq.
90 Ibid., pp. 76ff.
91 G.W.A. Dummer, *Electronic Inventions and Discoveries*, Oxford, Pergamon, 1983, p. 119.
92 M. Hilmes, 'Subscription Television and Theater-TV: The Film Industry Struggles for Control', unpublished paper, 1983, passim.
93 Ibid., pp. 6 et seq.
94 *Variety*, 13 June 1984, p. 37.
95 H.M. Beville Jr, 'VCR Penetration: Will it surpass cable by 1990?', *TV Radio Age*, 9 July 1984, vol. 31, no. 25, p. 28.
96 *Variety*, 23 August 1984.
97 *Encyclopaedia Britannica*, Cambridge, 1911, vol. 21, p. 467.
98 A. Abramson, 'A Short History of Television Recording', *SMPTE*, New York, 1955, vol. 64, passim, in Shiers.
99 (i) Ibid., p. 72.
   (ii) Norman, op. cit., pp. 46 et seq.
100 Lee De Forest, *Television Today and Tomorrow*, New York, Dial Press, 1942, p. 71.
101 A. Abramson, 'A Short History of Television Recording: Part II', *SMPTE*, New York, 1973, vol. 82, passim, in Shiers.
102 (i) J. Jewes, D. Sawers, R. Stillerman, *The Sources of Invention*, London, Macmillan, 1958, pp. 324ff.
    (ii) *New York Times*, 12 February 1984, p. 25.
103 (i) *TV Radio Age*, op. cit., pp. 30 et seq.
    (ii) E.G. Addeo, 'The Story of Ampex', unpublished internal document, Ampex Corporation, n.d. pp. 11-40. I am again indebted to Aaron Nmugwen for this reference.
104 Abramson (1973), op. cit., pp. 189 et seq.
105 Aaron Nmugwen found this statement in David Sarnoff, *Collected*

Speeches, RCA, n.d. p. 891.
106 *Encyclopaedia Britannica Yearbook of the Year 1968*, Chicago, Encyclopaedia Britannica, 1968, p. 322.
107 Abramson (1973), op. cit., pp. 195ff.
108 *UPI*, 14 August 1984.
109 *Variety*, 15 August 1984.
110 (i) C. Sandbank and I. Childs, 'The Evolution Towards High-Definition Television', *IEEE*, April 1985, p. 639.
    (ii) *Broadcasting*, Washington, 17 Jun 1985, vol. 108, no. 24, p. 62.
111 Pixel calculation from Senate Report on Color, quoted in Fink (1943), op. cit. p. 7.
112 Dummer, op. cit., p. 198.
113 Ibid., p. 157.
114 *Listener*, 24 May 1984, p. 7.
115 (i) *Sunday Times* (UK), 22 April 1984.
    (ii) *Listener*, 29 November 1984.
116 Paul Walton, *Space Light*, Doubleday, Sydney, NSW, 1982, passim.

Chapter 3   'Inventions for casting up sums, very pretty'

 1 H.S. Tropp, 'The Smithsonian Computer History Project and Some Personal Recollections', in Metropolis et al., op. cit., p. 119.
 2 N. Metropolis, 'The MANIAC', in Metropolis et al., p. 549.
 3 E.A. Burtt, *The Metaphysical Foundations of Modern Physical Science*, London, Routledge & Kegan Paul, 1967, pp. 113 et seq.
 4 P. Hazard, *The European Mind 1680-1715*, Harmondsworth, Middlesex, UK, Penguin Books, 1964, p. 157.
 5 E.L. Eisenstein, *The Printing Revolution in Early Modern Europe*, New York, Cambridge UP, 1983, p. 22.
 6 (i) A.M. Turing, 'On Computable Numbers, with an application to the Entscheidungsproblem', *Proceedings of the London Mathematical Society*, London, vol. 42, 12 November 1936, pp. 230-65.
    (ii) A.M. Turing, 'On Computable Numbers, with an application to the Entscheidungsproblem. A Correction', *Proceedings of the London Mathematical Society*, London, vol. 43, 1947, pp. 544-6.
 7 M. Kline, *Mathematics – The Loss of Certainty*, New York, Oxford University Press, 1980, p. 192.
 8 E. Nagel and J.R. Newman, *Godel's Proof*, New York, New York University Press, 1958, p. 13.
 9 Kline, op. cit., p. 261.
10 Nagel and Newman, op. cit., p. 58.
11 Ibid.
12 Turing (1936), op. cit., p. 241.
13 B. Randell, 'The COLOSSUS', in Metropolis et al., p. 78.
14 Turing (1936), op. cit., p. 250.
15 Ibid.
16 S. Toulmin, 'Fall of A Genius', *New York Review of Books*, 19 January 1984, p. 3.
17 Turing (1936), op. cit., p. 231.
18 E.I. Post, 'Finite Combinatory Processes – Formulation 1', *Journal of Symbolic Logic*, vol. 1, no. 3, September 1936, pp. 103 et seq.

19 Ibid.
20 Boole, quoted in H.H. Goldstine, *The Computer from Pascal to von Neumann*, Princeton, Princeton UP, 1972, p. 36.
21 A. Hodges, 'Alan Turin – The Enigma', New York, Simon & Schuster, 1983, p. 11.
22 Goldstine, op. cit., p. 37.
23 C. Shannon, 'A Symbolic Analysis of Relay and Switching Circuits', *Transactions of the American Institute of Electrical Engineers*, vol. 327, 1938, pp. 713ff.
24 C. Cherry, *On Human Communication*, New York, John Wiley, 1961, p. 168.
25 W. Weaver and C. Shannon, *The Mathematical Theory of Communication*, Urbana, University of Illinois Press, 1949, p. 11.
26 B. Winston, *Dangling Conversations – The Image of the Media*, London, Davis Poynter, 1973, p. 83.
27 R.W. Hamming, 'We would know what they thought when they did it', in Metropolis et al., op. cit., p. 7.
28 J. von Neumann, First Draft of a report on the EDVAC Contract no. W-760-)RD-4926, Philadelphia, Morse School of Electrical Engineering, University of Pennsylvania, 30 June 1945, in Randell, pp. 359 et seq.
29 J. von Neumann, *The Computer and the Brain*, New Haven, Yale UP, 1958, p. 51.
30 N. Wiener, *The Human Use of Human Beings*, New York, Doubleday, 1954, p. 12.
31 A.M. Turing, 'Computing Machinery and Intelligence', *Mind*, vol. 59, no. 236, October 1950, p. 436.
32 H. Wulforst, *Breakthrough to the Computer Age*, New York, Charles Scribner, 1982, p. 111.
33 J.R. Searle, 'Minds, Brains and Programs', *Mind Design*, ed. J. Haugeland, Cambridge, Massachusetts, MIT Press, 1981, p. 302.
34 M.V. Wilkes, *Automatic Digital Computers*, London, Methuen, 1956, p. 293.
35 Wulforst, op. cit., pp. 60 et seq.
36 J.W. Mauchly, 'Preparation of problems for EDVAC-type machines', *Annals of the Computation Laboratory of Harvard University*, Cambridge, Massachusetts, Harvard UP, 1948, vol. 16, p. 203 in Randell, pp. 365.
37 J. Presper Eckert Jr, 'The ENIAC' in Metropolis et al., op. cit., pp. 537 et seq.
38 N. Metropolis and J. Worlton, 'A Trilogy on (sic) Errors in the History of Computing', First USA-Japan Computer Conference, 1972, Session 21-2, p. 687.
39 A.W. Burks, 'From ENIAC to the Stored-Program Computer: Two Revolutions in Computers', in Metropolis et al., p. 339.
40 Goldstine, op. cit., pp. 191 et seq.
41 Ibid.
42 Ibid., pp. 197 et seq.
43 Neumann (1945), op. cit., p. 356.
44 Burks, op. cit., p. 341.
45 J. Mauchly, 'The ENIAC', in Metropolis et al., p. 547.
46 B. Randell, in Randell, p. 159.
47 Quoted in ibid., p. 11.
48 H.P. Babbage, *Babbage's Calculating Engines*, London, E. & F.N. Spon, 1889. This claim is one of the 'trilogy' of errors Metropolis and Worlton

seek to extirpate. In this connection they detail the extent to which Babbage references were known, op. cit., p. 683 et seq. The other two errors concern common failures to distinguish the stored programme concept, its origins and importance, and the misdesignation of the Princeton computer as MANIAC.

49 (i) Goldstine, op. cit., p. 11.

(ii) S. Augarten, *Bit by Bit*, New York, Ticknow & Fields, 1984, p. 37.

50 Ibid., p. 40.

51 Ibid., p. 42.

52 Goldstine, op. cit., p. 6.

53 C.W. Merryfield, *Report of the Committee . . . appointed to consider the advisability and to estimate the expense of constructing Mr. Babbage's Analytical Machine, and of printing tables by its means*, in Randell, p. 61.

54 P.E. Ludgate, 'On a proposed analytical machine', Royal Dublin Society, 1909, in Randell, pp. 71ff.

55 Goldstine, op. cit., p. 15.

56 Augarten, op. cit., p. 79.

57 D.F. Noble, *Forces of Production*, New York, Knopf, 1984, p. 147.

58 W. Rogers, *Think*, New York, Stein & Day, 1969, p. 69.

59 Goldstine, op. cit., p. 43.

60 Quoted in ibid., pp. 47 et seq.

61 R.J. Slutz, 'Memories of the Bureau of Standards' SEAC', in Metropolis et al., op. cit., p. 471.

62 C.C. Hurd, 'Computer Development at IBM', in Metropolis et al., p. 397.

63 Randell, in Metropolis et al., p. 81.

64 L. Torres Y Quevedo, 'Essais sur l'automatique', *Revue Générale des Sciences Pures et Appliquées*, 15 November 1915, in Randell, p. 91.

65 Ibid., p. 89.

66 G.E. Schindler (ed.), *A History of Engineering and Science in the Bell System – Switching Technology (1925-1975)*, Bell Telephone Laboratories, 1982, p. 2.

67 G.R. Stibitz, 'Early Computers', in Metropolis et al., p. 481.

68 (i) See ibid., passim.

(ii) Wulforst, op. cit., pp. 43ff.

69 K. Zuse, 'Some Remarks on the History of Computing in Germany', in Metropolis et al., op. cit., p. 611.

70 (i) Randell in Randell, p. 155.

(ii) Wulforst, op. cit., pp. 93 et seq.

71 Zuse, op. cit., p. 612.

72 Wulforst, op. cit., p. 95.

73 H. Schreyer, 'Technical Computing Machines', 1939, in Randell, p. 169.

74 Goldstine, op. cit., pp. 250 et seq. and note.

75 Randell in Randell, op. cit., p. 156.

76 (i) K. Zuse, 'The Outline of Computer Development from Mechanics to Electronics', in Randell, p. 183.

(ii) Augarten, op. cit.. p. 96.

77 H.H. Aitken, 'Proposed automatic calculating machine', *IEEE Spectrum*, August 1964, in Randell, p. 197.

78 Rodgers, op. cit., passim on Watson.

79 Goldstine, op. cit., p. 109.

80 Rodgers, op. cit., p. 136.

81 Ibid., p. 171 et seq.

82 Wulforst, op. cit., pp. 39ff.
83 G. Birkoff, 'Computer Developments 1935-1955, as Seen from Cambridge, USA', in Metropolis et al., p. 22.
84 Wulforst, op. cit., p. 60.
85 Ibid., p. 54.
86 N. Wiener, *Cybernetics*, New York, John Wiley, 1948, p. 11.
87 Birkoff, op. cit., p. 24.
88 (i) J.W. Mauchly, 'The Use of High Speed Vacuum Tube Devices for Calculating' (privately circulated memorandum), August 1942, in Randell, pp. 329ff.
   (ii) Wulforst, op. cit., p. 35.
89 Mauchly (in Metropolis et al.), p. 545.
90 Augarten, op. cit., p. 113.
91 J.V. Atanasoff, 'Computing machines for the solution of large systems of linear algebraic equations' (unpublished memorandum), August 1940, in Randell, op. cit., pp. 320 et seq.
92 (i) Goldstine, op. cit., pp. 123ff.
   (ii) Augarten, op. cit., pp. 114ff.
93 (i) Ibid.
   (ii) Wulforst, op. cit., pp. 28 et seq.
94 Randell, p. 288.
95 Mauchly (in Metropolis et al.), p. 549.
96 Wulforst, op. cit., pp. 49ff.
97 Burks, op. cit., p. 314.
98 Mauchly (in Metropolis et al.), p. 544.
99 Wulforst, op. cit., p. 54.
100 A.W. Burks quoted in Augarten, op. cit., p. 99.
101 Wulforst, op. cit., p. 72.
102 M. Rejewski, 'How Polish Mathematicians Deciphered the Enigma', *Annals of the History of Computing*, vol. 3, no. 3, July 1981, pp. 213ff.
103 Wulforst, op. cit., pp. 90 et seq.
104 Goldstine, op. cit., p. 273.
105 D.H. Lehmer, 'A History of the Sieve Process', in Metropolis et al., p. 455.
106 Randell (in Metropolis et al.), p. 57.
107 Hodges, op. cit., p. 199.
108 I.J. Good, 'Pioneering Work on Computers at Bletchley', in Metropolis et al., p. 34.
109 Winston Churchill, *The Second World War*, London, Cassell, 1951, vol 4, p. 104 et seq.
110 (i) Good (in Metropolis et al.), p. 35.
    (ii) I.J. Good, untitled response to Rejewski (q.v.), *Annals of the History of Computing*, vol. 3, no. 3, July 1981, p. 233.
111 C.A. Deavours, untitled response to Rejewski (q.v.), *Annals of the History of Computing*, vol. 3, no. 3, p. 232.
112 Randell (in Metropolis et al.), p. 77.
113 Randell gives 'about 10', in Randell, p. 328.
114 Hodges, op. cit., p. 277.
115 D. Mitchie quoted in Randell (in Metropolis et al.), p. 74,
116 Ibid.
117 Hodges, op. cit., p. 277.
118 Hodges, op. cit., p. 301.
119 Wulforst, op. cit., p. 65.

120 Goldstine, op. cit., p. 225.
121 S.M. Ulam, 'Von Neumann: The Interaction of Mathematics and Computing', in Metropolis et al., p. 96.
122 Hurd, op. cit., p. 399.
123 Ibid., pp. 403 et seq.
124 Goldstine, op. cit., p. 318.
125 Slutz, op. cit., p. 472.
126 Wilkes, op. cit., p. 160.
127 J.E. Robertson, 'The ORDVAC and the ILLIAC', in Metropolis et al., pp. 356 et seq.
128 Hodges, op. cit., p. 390.
129 S.H. Lavington, 'Computer Development at Manchester University', in Metropolis et al., pp. 43ff.
130 Ibid., p. 442.
131 F.C. Williams and T. Kilburn, 'Electronic Digital Computers', *Nature*, vol. 162, no. 487, September 1948, in Randell, p. 387. But Lavington's account offers a different vision of the Baby Mark I's capabilities (Lavington, op. cit., p. 433) and his *Early British Computers*, Manchester UP, 1982.
132 Hodges, op. cit., p. 406.
133 Ibid., p. 434.
134 M.V. Wilkes, 'Early Programming Developments In Cambridge', in Metropolis et al., p. 501.
135 W.J. Eckert, 'Electrons and Computation', *IEEE*, vol. 67, no. 5, November 1948, in Randell, p. 221.
136 Hurd, op. cit., p. 397.
137 Hodges, op. cit., p. 393.
138 Ibid., pp. 305ff.
139 Ibid., p. 341.
140 Ibid., p. 351.
141 Ibid., p. 341.
142 J.H. Wilkinson, 'Turing's Work at the National Physical Laboratory and the Construction of Pilot ACE, DEUCE and ACE', in Metropolis et al., p. 102.
143 Hodges, op. cit., pp. 366 et seq.
144 Toumlin, op. cit.
145 R.R. Everett, 'WHIRLWIND', in Metropolis et al., p. 372.
146 Slutz, op. cit., p. 475.
147 J. Bigelow, 'Computer Development at the Institute for Advanced Study', in Metropolis et al., pp. 394ff.
148 A.W. Burks, H.H. Goldstine, J. von Neumann, *Preliminary Discussion of the Logical Design of an Electronic Computing Instrument*, Part I, vol. I, Princeton, Institute for Advanced Study, 28 June 1946 (revised 2 September 1947), in Randell, p. 375.
149 J. Rajchman, 'Early Research on Computers at RCA', in Metropolis et al., p. 467 et seq.
150 Goldstine, op. cit., p. 311.
151 Bigelow, op. cit., p. 304.
152 Randell in Randell, p. 186.
153 Wilkes (1956), op. cit., p. 185.
154 Everett, op. cit., p. 373.
155 Augarten, op. cit., pp. 201 et seq.
156 A.D. Booth, 'Computers in the University of London 1945-1962', in

Metropolis et al., p. 551.
157 Metropolis, op. cit., p. 458.
158 Ibid., pp. 460 et seq.; confusing this machine with the one at Princeton is the second of Metropolis' and Worlton's 'errors'.
159 J.C. Chu, 'Computer Development at Argonne National Laboratory', in Metropolis et al., p. 346.
160 Rajchman, op. cit., p. 468.
161 (i) Robertson, op. cit., p. 349.
    (ii) Goldstine, op. cit., p. 310.
162 Everett, op. cit., p. 365.
163 Noble, op. cit., p. 112.
164 Robertson, op. cit., p. 375.
165 Noble, op. cit., p. 114n.
166 Robertson, op. cit., p. 377.
167 Said of the IAS machine's control. Wulforst, op. cit., p. 142.
168 Goldstine, op. cit., p. 251.
169 Wulforst, op. cit., pp. 82 et seq.
170 E. Tomash, 'The Start of an ERA: Engineering Research Associates, Inc., 1946-1955', in Metropolis et al., p. 486.
171 Ibid., pp. 490 et seq.
172 Wulforst, op. cit., p. 96.
173 Augarten, op. cit., p. 156.
174 Wulforst, ibid., p. 97.
175 Ibid., p. 100.
176 Randell in Randell, p. 351.
177 Wulforst, op. cit., p. 97.
178 Ibid., p. 114.
179 Ibid., pp. 119ff.
180 Augarten, op. cit., p. 169.
181 H.D. Huskey, 'The SWAC: The National Bureau of Standards Western Automatic Computer', in Metropolis et al., ibid., p. 419ff.
182 Augarten, op. cit., pp. 289ff.
183 Tomash, op. cit., pp. 491ff.
184 Wulforst, op. cit., p. 164.
185 Ibid., pp. 170 et seq.
186 Hurd, op. cit., p. 392.
187 Augarten, op. cit., p. 164.
188 Rodgers, op. cit., p. 177.
189 J.W. Sheldon and L. Tatum, 'The IBM Card-Programmed Electronic Calculator', paper given at joint AIEEE-IRE Computer Conference, 10-12 December 1951, in Randell, p. 229.
190 Ibid., p. 235.
191 Rodgers, op. cit., p. 174.
192 Ibid., p. 199.
193 Ibid.
194 Wulforst, op. cit., p. 173.
195 Bigelow, op. cit., p. 307.
196 Hurd, op. cit., p. 392.
197 Augarten, op. cit., pp. 209 et seq.
198 Ibid.
199 (i) Lavington, op. cit., pp. 440 et seq.
    (ii) Hodges, op. cit., p. 438.
200 Randell, op. cit., p. 57.

201 Handel, op. cit., p. 147.
202 J.O. Harrison, Jr, *Annals of the Computation Laboratory of Harvard University*, vol. 16, note in Randell, p. 421.
203 D.E. Knuth and L.T. Pardo, 'The Early Development of Programming Languages', in Metropolis et al., p. 228.
204 J. Backus, 'Programming in America in the 1950s – Some Personal Impressions', in Metropolis et al., p. 128.
205 Knuth and Pardo, op. cit., p. 203.
206 Zuse, op. cit., p. 621.
207 Hodges, op. cit., p. 437.
208 Knuth and Pardo, op. cit., p. 215.
209 Ibid.
210 Ibid., pp. 228-31.
211 Backus, op. cit., p. 129.
212 Knuth and Pardo, op. cit., p. 241.
213 Augarten, op. cit., p. 214.
214 Knuth and Pardo, op. cit., p. 237.
215 Backus, op. cit., p. 131.
216 Knuth and Pardo, op. cit., p. 237.
217 Ibid., p. 241.
218 In Thompson's sense:

> [T]he conventional picture of the Luddism of these years [circa 1812] as a blind opposition to machinery as such becomes less and less tenable. What was at issue was the 'freedom' of the capitalist to destroy the customs of the trade, whether by new machinery, by the factory-system, or by unrestricted competition, beating-down wages, undercutting his rivals, and undermining standards of craftsmanship. (E.P. Thompson, *The Making of the English Working Class*, Harmondsworth, Middlesex, UK, Penguin Books, 1968, p. 600)

Chapter 4 Digression – 'The most remarkable technology'

1 R.N. Noyce, 'Microelectronics', *Scientific American*, vol. 237, no. 3, September 1977, in Forester, p. 30.
2 E.E. Morison, *Men, Machines and Modern Times*, Cambridge, Massachusetts, MIT Press, 1968, pp. 12 et seq.
3 S. Millman (ed.), *A History of Engineering and Science in the Bell System – Physical Sciences (1925-1980)*, Bell Telephone Laboratories 1983, p. 97.
4 E. Braun and S. Macdonald, *Revolution in Miniature*, Cambridge, Cambridge University Press, pp. 17ff.
5 Ibid., p. 25.
6 Ibid., pp. 11 et seq.
7 Briggs, op. cit., p. 310.
8 G.G. Blake, *History of Radio Telegraphy and Telephony*, London, Chapman & Hall, 1928, p. 89.
9 Braun and Macdonald, op. cit., p. 37.
10 Ibid., p. 19.
11 Ibid., p. 37.
12 (i) J. Brooks, *Telephone – The First Hundred Years*, New York, Harper & Row, 1976, p. 203.
   (ii) Braun and Macdonald, op. cit., p. 38.
13 (i) Handel, op. cit., pp. 75ff.

(ii) Dummer, op. cit., p. 94.
14 Braun and Macdonald, p. 30.
15 W. Shockley, 'The Path to the Conception of the Junction Transistor', *IEEE Transactions on Electron Devices*, vol. ED-23, no. 7, New York, *IEEE*, July, 1976, p. 604.
16 Braun and Macdonald, op. cit., pp. 37ff.
17 Ibid., pp. 38 et seq.
18 Ibid.
19 Ibid.
20 S. Macdonald and E. Braun, 'The transistor and attitude to change', *American Journal of Physics*, vol. 45, no. 11, November 1977, p. 1061.
21 L. Mumford, *The Pentagon of Power*, New York, Harcourt Brace Jovanovich, 1970, pp. 20 et seq.
22 Shockley, op. cit., p. 606.
23 Ibid., p. 609.
24 Braun and Macdonald, op. cit., p. 40.
25 Brooks, op. cit., p. 223.
26 Handel, op. cit., p. 83.
27 Shockley, op. cit., p. 615.
28 Ibid., pp. 614 et seq.
29 Macdonald and Braun, op. cit., p. 1064.
30 Braun and Macdonald, op. cit., p. 54.
31 Ibid., p. 45.
32 Shockley, op. cit., pp. 600 et seq.
33 Braun and Macdonald, op. cit., p. 43.
34 Shockley, op. cit., p. 599.
35 Braun and Macdonald, op. cit., p. 43.
36 (i) Brooks, op. cit., p. 223.
    (ii) T.A. Watson, *Exploring Life*, New York, Appleton, 1926, p. 72.
37 Braun and Macdonald, op. cit., p. 182.
38 E. Braun, 'From Transistor to Microprocessor', in Forester, p. 73.
39 Braun and Macdonald, op. cit., p. 44.
40 Ibid., pp. 42 et seq.
41 Ibid., p. 49.
42 Macdonald and Braun, op. cit., p. 1062.
43 Ibid., p. 1063 and passim.
44 Braun and Macdonald, op. cit., p. 93.
45 Ibid., p. 48.
46 Schindler, op. cit., pp. 199ff.
47 Braun and Macdonald, op. cit., p. 196.
48 'The Transistor', *Bell Laboratories Record*, vol. 26, no. 8, August 1948, p. 322.
49 Braun and Macdonald, op. cit., p. 71.
50 Ibid., p. 60.
51 Macdonald and Braun, op. cit., p. 1063.
52 Braun, op. cit., p. 76.
53 Braun and Macdonald, op. cit., p. 80.
54 Ibid., p. 81.
55 Braun, op. cit., p. 94.
56 Dummer, op. cit., p. 141.
57 Braun and Macdonald, op. cit., p. 94.
58 M.J. Wolff, 'The genesis of the integrated circuit', *IEEE Spectrum*, August 1976, p. 53.

59 Braun and Macdonald, op. cit., p. 90.
60 Wolff, op. cit., p. 47.
61 Braun and Macdonald, op. cit., p. 31.
62 Dummer, op. cit., p. 158.
63 Augarten, op. cit., p. 242.
64 D. Hanson, *The New Alchemists*, New York, Avon, 1982, p. 91.
65 (i) Brooks, op. cit., p. 224.
   (ii) *Observer*, London, 4 March 1984.
66 (i) Noyce, op. cit., p. 32.
   (ii) Augarten, op. cit., p. 241.
67 Braun, op. cit., p. 75.
68 Braun and Macdonald, op. cit., pp. 76, 81.
69 Ibid., p. 89.
70 Hanson, op. cit., pp. 97 et seq.
71 Ibid., p. 106.
72 Ibid., p. 107.
73 Ibid., pp. 100 et seq.
74 Braun and Macdonald, p. 188.
75 Braun and Macdonald, p. 190.
76 D.S. Landes, *Revolution in Time*, Cambridge, Massachusetts, Harvard University Press, 1983, p. 319.
77 Ibid., p. 353.
78 Ibid., p. 346.
79 Ibid., p. 347.
80 Hanson, op. cit., pp. 198 et seq.
81 Ibid., p. 123.
82 Ibid., pp. 106, 120.
83 Augarten, op. cit., pp. 261ff.
84 G. Bylinsky, 'Here Comes the Second Computer Revolution', *Fortune*, November, 1975, in Forester, p. 6.
85 Hanson, op. cit., p. 118.
86 Braun and Macdonald, op. cit., p. 79.
87 Kemeny, quoted in Augarten, op. cit., p. 253.
88 Wilkinson, op. cit., p. 110.
89 Hodges, op. cit., p. 391.
90 Ibid., p. 478.
91 J.H. Felker, 'Regenerative Amplifier for Digital Computer Applications', *IRE*, vol. 40, no. 11, November 1952, p. 1596.
92 E.G. Andrews, *Bell Journal*, vol. 42, no. 2, March 1963, pp. 341ff (I would like to thank Robert Horwitz for alerting me to this episode.) See Lavington, op. cit., p. 48.
93 See, for example, Braun and Macdonald, op. cit., p. 69 or Braun, op. cit., p. 76.
94 Hurd, op. cit., p. 413.
95 J.T. Soma, *The Computer Industry*, Lexington, Massachusetts, D.C. Heath, 1976, p. 39.
96 Ibid., p. 37.
97 Ibid., p. 150.
98 E. Goldwyn, 'Now the Chips are Down', *The Listener*, 6 April 1978, in Forester, p. 297. (Programme transmitted on BBC 2, March 1978.)
99 Hanson, op. cit., Ibid., p. 104 et seq.
100 Augarten, op. cit., pp. 248ff.
101 Soma, op. cit., pp. 23, 37.

102 Augarten, op. cit., pp. 256ff.
103 Ibid., p. 250.
104 Bylinsky, op. cit., pp. 8 et seq.
105 Ibid.
106 Hanson, op. cit., p. 121.
107 Ibid., p. 128.
108 Ibid., p. 124.
109 Ibid., pp. 200ff.
110 Augarten, op. cit., pp. 266ff.
111 (i) Ibid., pp. 276ff.
    (ii) P. Frieberger and M. Swaine, 'Fire in the Valley', *A+*, Zipp Davis,
    vol. 2, no. 12, December 1984, pp. 56 et seq.
112 See, for example, *Personal Computing*, Brookline, Massachusetts, Benwill
    Publishing, vol. 1, no. 2, March 1977, passim.
113 Hanson, op. cit., p. 206.
114 Augarten, op. cit., p. 280.
115 I would like to thank Martin Elton for determining this statistic for me.
116 *NYU Report*, New York University, vol. 11, no. 1, p. 2.
117 'The Experience of Word Processing', *Data Processing*, May 1978, in
    Forester, p. 211.
118 Noyce, op. cit., p. 32.
119 Ibid., p. 37.
120 G.C. Chow, 'Technological Change and the Demand for Computers',
    American Economic Review, 1966, p. 1117.
121 *Encyclopaedia Britannica*, op. cit., vol. 18, p. 920.
122 Hanson, op. cit., p. 126.
123 J. Weizenbaum, 'Once More, the Computer Revolution', M. Dertouzos
    and J. Moses (eds), *The Computer Age: A Twenty-Year View*, MIT Press,
    1979, in Forester, op. cit., pp. 568 et seq.

Chapter 5   Little bird of union and understanding

1 Sir Bernard Lovell, *The Origins and International Economics of Space
   Exploration*, Edinburgh, Edinburgh University Press, 1973, p. 9.
2 Ibid., p. 7.
3 Ibid.
4 F.J. Kreiger, *Behind The Sputniks*, Washington, Public Affairs Press
   Rand Corporation, 1958, p. 36.
5 Tsander in A.A. Blagonravov et al. (eds), *Iz istorii raketnoi teknhniki*,
   Moscow, Academy of Sciences of the USSR, Historical Institute for
   Natural Science and Engineering, translated and edited (by H.I. Needler)
   as: *Soviet Rocketry – Some Contributions to its History*, Jerusalem, Israel
   Program for Scientific Translations, 1966, p. 146.
6 Ibid., p. 137.
7 (i) Merkulov in Blagonravov, op. cit., pp. 41ff.
   (ii) Kreyer, op. cit., pp. 54.
8 Ibid., p. 33.
9 Lovell, op. cit., p. 12.
10 F.I. Ordway III and M.R. Sharpe, *The Rocket Team*, New York, Thomas
   Y. Crowell, 197, p. 18.
11 Ibid., pp. 45ff.
12 Kreiger, op. cit., p. 170.
13 Ibid., p. 177.

14 R.C. Hall, 'Early US Satellite Proposals', *Technology and Culture*, vol. 4, 1963, p. 421.
15 Quoted in Blagonravov, op. cit., p. 26.
16 Lovell, op. cit., p. 21.
17 Kreiger, op. cit., p. 9.
18 Ibid., p. 333.
19 Hall, op. cit., p. 421.
20 Ibid., p. 411.
21 R.S. Lewis, *Appointment on the Moon*, New York, Viking, p. 34.
22 (i) A.A. Needell (ed.), *The First 25 Years in Space*, Washington, Smithsonian Institute Press, 1983, p. 136.
    (ii) Hall, op. cit., p. 415n.
23 Ibid., p. 419.
24 Ibid., p. 422.
25 Ibid., p. 427.
26 Ibid., p. 432.
27 Ibid., p. 433.
28 Lewis, op. cit., pp. 37ff.
29 Lovell, op. cit., p. 22.
30 Lewis, op. cit., Ibid., p. 57ff.
31 (i) Ibid., pp. 44ff.
    (ii) Lewis, op. cit., p. 378.
32 (i) Lovell, op. cit., p. 22.
    (ii) Lewis, op. cit., pp. 54ff.
33 N. Daniloff, *The Kremlin and the Cosmos*, New York, Knopf, 1972, p. 103.
34 Lewis, op. cit., p. 56.
35 Lovell, op. cit., pp. 24ff.
36 Needell, op. cit., p. 134.
37 Arthur C. Clarke, *Voices From The Sky*, London, Mayflower, 1969, p. 109.
38 Kreiger, op. cit., p. 254.
39 G. Paul, *Die dritte Entdeckung der Erde*, translated by A. and B. Lacy as *Satellite Spin-Off*, Washington, Robert B. Luce, 1975, p. 105.
40 Clarke, op. cit., pp. 141ff.
41 Ibid., pp. 193ff.
42 Millman, op. cit., p. 102.
43 A.C. Dickieson, 'The *Telstar* Experiment', *Bell Journal*, vol. 42, no. 4, part 1, July 1963, p. 741.
44 Ordway and Sharpe, op. cit., p. 426.
45 Ibid., pp. 378 et seq.
46 J.J. Fahie, *A History of Wireless Telegraphy*, New York, Dodd Mead, 1901, p. 215.
47 Blake, op. cit., p. 39.
48 Ibid., p. 32.
49 Ibid., pp. 65, 119.
50 Fahie, op. cit., p. 260.
51 Clarke, op. cit., p. 197.
52 Lewis, op. cit., pp. 297 et seq.
53 Paul, op. cit., p. 22.
54 Fagen, op. cit.. pp. 271 et seq.
55 Paul, op. cit., p. 95.
56 J. Oslund, 'Open Shores to Open Skies', J.N. Pelton and M. Snow (eds),

Economic and Policy Problems in Satellite Communications, New York, Praeger, 1977, p. 162.
57 Ibid., p. 158.
58 Ibid., p. 160.
59 Ibid., p. 163.
60 Ibid.
61 Dickieson, op. cit., p. 746.
62 D.J. Czitrom, *Media and the American Mind*, Chapel Hill, University of North Carolina Press, 1982, p. 11.
63 Lewis, op. cit., p. 172.
64 Fahie, op. cit., p. 39.
65 Field, *The Story of the Atlantic Cable*, 1893, p. 49.
66 Ibid.
67 L.T.C. Rolt, *Isambard Kingdom Brunel*, Harmondsworth, Middlesex, UK, Penguin Books, 1970, p. 397.
68 Field, op. cit., p. 373.
69 Oslund, op. cit., p. 155.
70 Czitrom, op. cit.
71 Lewis, op. cit., pp. 309 et seq.
72 Paul, op. cit., pp. 42 et seq.
73 Ibid., pp. 64ff.
74 Ibid., p. 40.
75 Lovell, op. cit., pp. 36 et seq.
76 Hall, op. cit., p. 425.
77 H. Schiller, *Communication and Cultural Domination*, White Plains, New York, International Arts and Science Press, 1976, p. 26.
78 Field, op. cit., p. 102.
79 Ibid., pp. 104 et seq.
80 Oslund, op. cit., p. 150.
81 Field op. cit., p. 100.
82 Oslund, op. cit., p. 146.
83 Brooks, op. cit., p. 1152 et seq.
84 Oslund, op. cit., p. 151.
85 Ibid., p. 161.
86 Galloway, 'Domestic Economic Issues', in Pelton and Snow, op. cit., p. 37.
87 Ibid., p. 20.
88 Lewis, op. cit., pp. 310ff.
89 Galloway, op. cit., p. 36.
90 Ibid., p. 30.
91 Ibid., p. 28.
92 Ibid., p. 34.
93 General Assembly Resolution no. 1721 XVI.
94 US Public Law 87-624, 76 STAT.419: sec 102c.
95 Ibid., sections 304.2 and 304.3.
96 Communications Satellite System COMSAT, Treaties and Other International Acts, Series 5646 TITAS, Washington, 20 August 1964, pp. 39 et seq.
97 The late Arnold Bulka and Tim Hewat worked on just such a scheme for *World In Action* coverage of the Tokyo Olympics in 1964.
98 Galloway, op. cit., p. 78.
99 Ibid., p. 96.
100 Ibid.

101 *Encyclopaedia Britannica Year Book*, London, 1961, p. 459.
102 Galloway, op. cit., p. 129.
103 Article X, TITAS 5646, op. cit., p. 11.
104 M.S. Snow, *International Commercial Satellite Communications*, New York, Praeger, 1976, p. 137.
105 Ibid., p. 120.
106 Ibid.
107 Schiller, op. cit., see especially chapter 3 passim.
108 *Encyclopaedia Britannica Year Book*, Chicago, 1972, p. 656.
109 Galloway, op. cit., p. 62.
111 Snow, op. cit., p. 104.
110 Ibid., pp. 113ff.
112 F. Friendly, *Due to Circumstances Beyond Our Control*, New York, Vintage, 1968, p. 310.
113 Oslund, op. cit., p. 173.
114 Paul, op. cit., p. 71.
115 J. Martin, *Future Developments in Telecommunications*, Englewood Cliffs, New Jersey, Prentice-Hall, 1977, p. 215.
116 T. Hollins, *Beyond Broadcasting: Into the Cable Age*, London, BFI Publications, 1984, p. 267.
117 J.M. Eder, *Geschichte der Photographie*, Vienna, 1905 and 1932, translated by E. Epstein as: *History of Photography*, New York, Dover, 1978, p. 57.
118 *Multichannel News*, New York, Fairchild Publications, 26 November 1984.
119 Section 5.b.i Congressional Record 11 October 1984, S 14297.
120 *Coop's Satellite Digest*, West Indies Video, Grace Bay, Turks and Caicos, October 1982, p. 54.
121 *Broadcasting*, 3 December 1984, p. 36.
122 Snow, op. cit., p. 17.
123 M. Moss, 'New York isn't just New York anymore', *Intermedia*, London, IIC, Autumn 1984, pp. 11 et seq.
124 *Broadcasting*, 20 May 1984, p. 64.
125 Clarke, op. cit., p. 128.
126 HBO: *The First Ten Years* (Public relations document), Home Box Office Inc, 1982, p. 13.
127 Hollins, op. cit., p. 145.

Chapter 6   Communicate by word of mouth

1 Caveats caused considerable problems of which the Bell/Gray incident is among the most notorious. The system was abolished in 1910.
2 (i) Le Compte du Moncel, *The Telephone, The Microphone and the Phonograph*, New York, Harper & Brothers, 1879, pp. 11 et seq.
(ii) J.E. Kingsbury, *The Telephone and Telephone Exchanges*, London, Longmans, Green and Co., 1915, p. 8.
3 Landes, op. cit., p. 126.
4 Moncel, op. cit., p. 33.
5 G.B. Prescott, *Bell's Electric Speaking Telephone*, reprinted, New York, Arno Press, 1972, p. 452.
6 Moncel, op. cit., pp. 35 et seq.
7 Samuel Pepys, *Diary*, London, J.M. Dent, 1953, vol. III, p. 205 entry dated 2 April 1668.

8 Kingsbury, op. cit., p. 2.
9 F.L. Rhodes, *Beginnings of Telephony*, New York, Harper & Brothers, 1929, p. 226.
10 Ibid., p. 231.
11 (i) Kingsbury, op. cit., p. 4.
(ii) Rhodes, op. cit., pp. 229 et seq.
12 J.J. Fahie, *A History of Electric Telegraphy to the Year 1837*, London E. & F.N. Spon, 1884, p. 5.
13 Ibid., p. 6.
14 Fernand Braudel, *Civilisation and Capitalism 15-18th Century*, vol. 1, New York, Harper & Row, 1981, p. 434.
15 G.W. Brock, *The Telecommunications Industry*, Cambridge, Massachusetts, Harvard University Press, 1981, p. 136.
16 Fahie (1884), op. cit., p. 124.
17 Ibid., p. 136.
18 Asa Briggs, *From Iron Bridge to Crystal Palace*, London, Thames & Hudson, 1979, p. 90.
19 H.J. Dyos and D.H. Aldcroft, *British Transport*, Harmondsworth, Middlesex, UK, Penguin Books, 1974, p. 129.
20 Fahie (1884), op. cit., p. 407.
21 Czitrom, op. cit., p. 6.
22 R.L. Thompson, *Wiring a Continent*, Princeton, Princeton University Press, 1947, p. 15.
23 Ibid., p. 317.
24 Ibid., p. 311.
25 Fahie (1884), op. cit., p. 381.
26 Ibid., p. 431.
27 Brock, op. cit., p. 63.
28 Thompson, op. cit., p. 426.
29 Ibid.
30 Kingsbury, op. cit.
31 Rhodes, op. cit., p. 229.
32 (i) Fagen, op. cit., p. 83.
(ii) Moncel, op. cit., p. 12.
33 Ibid., p. 144.
34 Kingsbury, op. cit., pp. 114ff.
35 Presumably of despair, Kingsbury adds: op. cit., p. 7.
36 Moncel, op. cit., p. 267.
37 Blake, op. cit., p. 13.
38 (i) Kingsbury, op. cit., Ibid., pp. 19, 17.
(ii) The Deposition of Alexander Graham Bell in the suit brought by the United States to annul the Bell Patents, Boston, American Bell Telephone Company, 1908, pp. 206 et seq.
39 Ibid., pp. 7 et seq., 207.
40 Blake, op. cit., p. 15.
41 Fagen, op. cit., pp. 2 et seq.
42 Kingsbury, op. cit., pp. 12 et seq.
43 Moncel, op. cit., p. 23.
44 Moncel, op. cit., pp. 13 et seq.
45 Kingsbury, op. cit., p. 11.
46 Ibid., p. 36.
47 Deposition, op. cit., p. 37.
48 Helmholtz quoted in Kingsbury, op. cit., p. 34.

49 Watson, op. cit., pp. 62 et seq.
50 Ibid.
51 Deposition, op. cit., p. 53.
52 Moncel, op. cit., p. 86.
53 Prescott, op. cit., p. 452.
54 Ibid. et seq.
55 Ibid.
56 Deposition, op. cit., p. 339.
57 Prescott, op. cit., p. 458.
58 Joint Economic Committee, Congress of the United States, *Invention and the Patent System*, Washington, US Government Printing Office, 1964, p. 11.
59 Moncel, op. cit., p. 59.
60 Watson, op. cit., p. 167.
61 (i) Kingsbury, op. cit., p. 130.
   (ii) Prescott, op. cit., p. 407.
62 (i) Moncel, op. cit., p. 16.
   (ii) Kingsbury, op. cit., p. 128.
63 The modern form is called *Pulse Code Modulation*, PCM. Developed by French ITT engineers in the late 1930s, this system samples a continuous wave at various discrete points of amplitude. These points are assigned a digital number. In this way the wave is converted into a series of digital numbers, more compressed than the original wave signal and more resistant to cross-talk and noise.
64 Moncel, op. cit., p. 21.
65 Deposition, op. cit., p. 446.
66 Moncel, op. cit., p. 37.
67 Prescott, op. cit., p. 457.
68 Brooks, op. cit., p. 83.
69 Kingsbury, op. cit., p. 29.
70 Ibid.
71 Brooks, op. cit., p. 43.
72 Deposition, op. cit., p. 45.
73 Rhodes, op. cit., p. 19.
74 Deposition, op. cit., p. 55.
75 Watson, op. cit., p. 68.
76 Ibid., p. 71.
77 Deposition, op. cit., pp. 338.
78 Ibid., pp. 70 et seq.
79 Kingsbury, op. cit., p. 46.
80 Patent reprinted in Deposition, op. cit.
81 Ibid., pp. 458 et seq.
82 Ibid., p. 460.
83 Ibid., p. 457.
84 Kingsbury, op. cit., p. 106.
85 Deposition, op. cit., p. 86.
86 Ibid., p. 87.
87 Watson, op. cit., p. 78.
88 Watson, op. cit., p. 80.
89 Fagen, op. cit., p. 12.
90 Prescott, op. cit., p. 429.
91 Kingsbury, op. cit., p. 49.
92 Wiener (1952), op. cit., p. 31.

93 Deposition, op. cit., p. 93.
94 Moncel, op. cit., p. 37.
95 Deposition, op. cit., p. 433.
96 See Rhodes, op. cit., pp. 207ff for a summary of these cases.
97 *Encyclopaedia Britannica*, op. cit., vol. 26, p. 548.
98 Rhodes, op. cit., p. 210.
99 Ibid.
100 Moncel, op. cit., p. 37.
101 Brooks, op. cit., p. 36.
102 Otis had produced a safe elevator by 1857 but the geared hydraulic lift specifically for passenger use was introduced in 1875.
103 M. Leapman, *Companion Guide to New York*, London, Collins and Englewood Cliffs, New Jersey, Prentice-Hall, 1983, p. 50.
104 L. Mumford, *The City In History*, Harmondsworth, Middlesex, UK, Penguin Books, 1966, p. 609.
105 A. Harding, *A Social History of English Law*, Harmondsworth, Middlesex, UK, Penguin Books, 1966, p. 376.
106 Dyos and Aldcroft, op. cit., p. 203.
107 Ibid., p. 199.
108 A.A. Berke and G.C. Meons, *The Modern Corporation and Private Property*, New York, Commerce Clearing House, 1932, pp. 129, 137.
109 Ibid., p. 13.
110 W.O. Henderson, *The Industrialisation of Europe*, London, Thames & Hudson, 1969, p. 31.
111 W.R. Brock, *Conflict and Transformation: The United States 1844-1877*, Harmondsworth, Middlesex, UK, Penguin Books, 1973, p. 418.
112 G.D.H. Cole and R. Postgate, *The Common People 1746-1946*, London, Methuen, 1938 (1961), p. 403.
113 J.W. Stehman, *The Financial History of The American Telephone and Telegraph Company*, Boston, Houghton Mifflin, 1925, p. 4.
114 Ibid., p. 7.
115 Fagen, op. cit., p. 23.
116 G.W. Brock, op. cit., p. 93.
117 J. Carey and M.L. Moss, 'The Diffusion of New Telecommunications Technologies, Telecommunications Policy', June 1985, p. 4 of manuscript.
118 Kingsbury, op. cit., p. 37.
119 Prescott, op. cit., p. 432.
120 Watson, op. cit., p. 98.
121 Ibid., p. 101.
122 Fagen, op. cit., p. 17.
123 Deposition, op. cit., p. 464.
124 Stehman, op. cit., p. 6.
125 Prescott, op. cit., pp. 119ff.
126 (i) Fagen, op. cit., pp. 66 et seq.
    (ii) Prescott, op. cit., p. 142.
127 Ibid., p. 465.
128 G.S. Bryan, *Edison: The Man and His Work*, New York, Knopf, 1926, p. 76.
129 Watson, op. cit., p. 139.
130 Prescott, op. cit., p. 460.
131 F.W. Wile, *Emile Berliner: Maker of the Microphone*, Indianapolis, Bobbs-Merrill, 1926, pp. 74 et seq.
132 G.J. Barker-Benfield, 'Sexual Surgery in Late Nineteenth Century

America', in Claudia Dreifus (ed.), *Seizing Our Bodies*, New York, Vintage, 1978, p. 16.

133 Wile, op. cit., p. 112.
134 Watson, op. cit., p. 143.
135 Fagen, op. cit., p. 70.
136 Wile, op. cit., p. 129.
137 Blake, op. cit., p. 18.
138 Stehman, op. cit., p. 17.
139 Brooks, op. cit., p. 71.
140 Stehman, op. cit., p. 17n.
141 Prescott, op. cit., p. 44.
142 Stehman, op. cit., p. 14.
143 Fagen, op. cit., pp. 82 et seq.
144 W.C. Langdon, *The Early Corporate Development of the Telephone*, New York, Bell Telephone, 1935, p. 9.
145 Fagen, op. cit., pp. 22 et seq.
146 Brooks, op. cit., p. 66.
147 Fagen, op. cit., p. 484.
149 Brooks, op. cit., p. 71.
149 Watson, op. cit., pp. 151ff.
150 D.S. Evans (ed.), *Breaking Up Bell*, New York, North Holland, 1983, p. 9.
151 Stehman, op. cit., p. 18.
152 Fagen, op. cit., p. 34.
153 Ibid., p. 345.
154 Evans, op. cit., p. 11.
155 Brock, op. cit., p. 104.
156 Stehman, op. cit., p. 53.
157 Evans, op. cit., pp. 16, 19.
158 Ibid.
159 Ibid., p. 95.
160 A.R. Benett, *The Telephone Systems of the Continent of Europe*, London, Longmans, Green and Co, 1895, passim.
161 Stehman, op. cit., Ibid., p. 10n, 45.
162 Ibid., p. 48.
163 Shaw quoted in J.H. Robertson, *The Story of the Telephone; A History of the Telecommunications Industry of Britain*, London, Pitman and Sons, 1947.
164 H.L. Webb, *The Development of the Telephone in Europe*, London, Electrical Press, nd, p. 27.
165 Robertson, op. cit., p. 11.
166 Brooks, op. cit., p. 93.
167 R.J. Chapius, *100 Years of Telephone Switching*, Part 1, Amsterdam, North Holland, 1982, p. 43.
168 Stehman, op. cit., p. 234.
169 Ibid., p. 81.
170 Fagen, op. cit., pp. 545ff.
171 Brooks, op. cit., p. 130.
172 Ibid., pp. 138ff.
173 Judge Brower quoted in Watson, op. cit., p. 165.
174 Fagen, op. cit., pp. 243 et seq.
175 Ibid., p. 262.
176 Stehman, op. cit., p. 78.

177 G.W. Brock, op. cit., p. 156.
178 Ibid., p. 159.
179 Stehman, op. cit., p. 241.
180 Ibid.
181 Brooks, op. cit., p. 132.
182 Stehman, op. cit., p. 126.
183 Brooks, op. cit., p. 82.
184 Ibid., p. 149.
185 Stehman, op. cit., p. 153.
186 Kingsbury, op. cit., p. 530.
187 Brooks, op. cit., p. 205.
188 M.L. Moss, 'Telecommunications and the Future of Cities', paper given at the Landtronics Anglo/American Conference, Westminster Centre, London, 19-21 June 1985, p. 2.
189 Moncel, op. cit., p. 172.
190 Hollins, op. cit., p. 35.
191 Ibid., pp. 36ff.
192 *The BBC Year Book 1933*, London, BBC, p. 72.
193 J.W. Roman, *Cable Mania*, Englewood Cliffs, New Jersey, Prentice-Hall, 1983, pp. 1ff.
194 Briggs, op. cit., p. 358.
195 Sloan Commission on Cable Communications, *The Cable: The Television of Abundance*, New York, McGraw-Hill, 1971, p. 25.
196 Roman, op. cit., pp. 11 et seq.
197 392 US 390 1968.
198 Carey and Moss, op. cit., p. 10 of manuscript.

In conclusion

1 From: Alvin Toffler, *Future Shock*, New York, Bantam, 1970, *The Third Wave*, New York, Morrow, 1980, M. Winn, *The Plug in Drug*, New York, Bantam, 1977, John Wicklein, *The Electronic Nightmare*, New York, Viking, 1981.
2 N. Cohn, *The Pursuit of the Millennium*, London, Paladin, 1970, p. 281.
3 Daniel Bell, *The Coming of Post-Industrial Society*, New York, Basic Books, 1973, 1976, p. 201.
4 Ibid., p. 203.
5 Alvin Toffler, 'Value Impact Forecaster – A Profession of the Future', in K. Baier and N. Rescher, eds, *Values and the Future*, Toronto, The Free Press, 1969. I am indebted to Jimmy Weaver for this reference.
6 J. Martin, op. cit., p. 6.
7 Ibid., pp. 385, 23.
8 J. Martin, *The Wired Society*, Englewood Cliffs, New Jersey, Prentice-Hall, 1978, p. 273.
9 Bell, op. cit., chapter 3, passim.
10 A. Werth, *Russia: Hopes and Fears*, Harmondsworth, Middlesex, 1969, p. 315.
11 R. Pattison, *On Literacy*, New York, Oxford University Press, 1982, p. 24.
12 Ibid.
13 W.R. Brock, op. cit., p. 425.
14 (i) R.R. Panko, 'Office Work', in *Office: Technology and People*,

Amsterdam, Elsevier Science Publishers, 2, 1984, pp. 207, 211. I would like to thank Martin Elton for drawing my attention to this.
(ii) M.U. Porat, 'The Information Economy', unpublished PhD Dissertation, Stanford University. I thank Svein Bergum for illuminating this reference for me.
15 S. Bergum, 'Telematics and Economic Aspects of Diffusion', unpublished paper, NYU, January 1985, pp. 14ff.
16 Panko, op. cit., p. 232.
17 Jorge Luis Borges, 'The Library of Babel' (translated by J.E. Irby), in *Labyrinths*, New York, New Directions, 1964, p. 58.
18 Bell, op. cit., p. 209.
19 Ibid.
20 Ibid.
21 J.K. Galbraith, *The New Industrial State*, Harmondsworth, Middlesex, UK, Penguin Books, 1969, p. 24.
22 Ibid., p. 27.
23 R.R. Nelson, M.J. Peck and E.D. Kalacheck, *Technology, Economic Growth and Public Policy*, Washington, The Brookings Institution, 1967, p. 41.
24 Jewkes et al., op. cit., p. 31.
25 Ibid., p. 39.
26 W.H.G. Armytage, *The Rise of the Technocrats*, London, Routledge & Kegan Paul, 1965, p. 242.
27 Galbraith, op. cit., p. 31.
28 E.E. Morison, *From Know-how to Nowhere*, New York, Basic Books, 1974, p. 119.
29 Henderson, op. cit., p. 53.
30 Morison (1974), op. cit., pp. 131 et seq.
31 Ibid., pp. 132ff.
32 Ibid., p. 139.
33 Nelson et al., op. cit., p. 43.
34 Jewkes et al., op. cit., p. 335.
35 W.H.A. Carr, *The Du Ponts of Delaware*, London, Frederick Muller, 1965, p. 302.
36 Armytage, op. cit., pp. 37 et seq.
37 Ibid., p. 113.
38 Jewkes et al., op. cit.
39 Ibid., p. 111.
40 R. Nisbet, *History of the Idea of Progress*, New York, Basic Books, 1980, pp. 308 et seq.
41 Walter Benjamin, 'Theses on the Philosophy of History', in *Illuminations* (translated by Harry Zohn), New York, Schocken Books, 1969, pp. 257 et seq.